Proceedings of the 29th International Geological Congress Part A

Also available from VSP

PROCEEDINGS OF THE 29TH INTERNATIONAL GEOLOGICAL CONGRESS - PART B:
Reconstruction of the Paleo-Asian Ocean
Edited by R.G. Coleman
Quaternary Environmental Changes
Edited by E.H. Juvigné

PROCEEDINGS OF THE 29TH INTERNATIONAL GEOLOGICAL CONGRESS - PART C:
Siliceous, Phosphatic and Glauconitic Sediments of the Tertiary and Mesozoic
Edited by A. Iijima, A.M. Abed and R.E. Garrison

PROCEEDINGS OF THE 29TH INTERNATIONAL GEOLOGICAL CONGRESS - PART D:
Circum-Pacific Ophiolites
Edited by A. Ishiwatari, J. Malpas and H. Ishizuka

Related titles

Facies Models in Exploration and Development of Hydrocarbon and Ore Deposits
Edited by A.H. Bouma and R.M. Carter

Regional Metamorphism of Ore Deposits and Genetic Implications
Edited by P.G. Spry and L.T. Bryndzia

Tectonics of Circum-Pacific Continental Margins
Edited by J. Aubouin and J. Bourgois

PROCEEDINGS OF THE 29TH INTERNATIONAL GEOLOGICAL CONGRESS PART A

Kyoto, Japan, 24 August - 3 September, 1992

Metamorphic Reaction: Kinetics and Mass Transfer
Editors: T. Nishiyama and G.W. Fisher

Sandstone Petrology in Relation to Tectonics
Editors: F. Kumon and K.M. Yu

Evaporite and Desert Environment
Editors: Y. Watanabe and A. Motamed

Utrecht, The Netherlands, 1994

VSP BV
P.O. Box 346
3700 AH Zeist
The Netherlands

© VSP BV 1994

First published in 1994

ISBN 90-6764-173-1

All rights reserved. No part of this publication may be reproduced, stored in a retrieval system, or transmitted in any form or by any means, electronic, mechanical, photocopying, recording or otherwise, without the prior permission of the copyright owner.

CIP-DATA KONINKLIJKE BIBLIOTHEEK, DEN HAAG

Proceedings

Proceedings of the 29th International Geological Congress.
- Utrecht : VSP
Pt. A / ed.: T. Nishiyama ... [et al.].
ISBN 90-6764-173-1 bound
NUGI 816
Subject headings: geology.

Printed in The Netherlands by A-D Druk, Zeist.

CONTENTS

METAMORPHIC REACTIONS: KINETICS AND MASS TRANSFER
Editors: T. Nishiyama and G.W. Fisher

Preface	5
CDS modeling of chemical layering due to uphill diffusion in a mineral-fluid system *T. Nishiyama*	7
An effect of a varying porosity on metasomatic replacement *K. Hoshino*	17
Ostwald ripening and crystal size distribution (CSD) of garnet in various metamorphic rocks *K. Miyazaki*	27
Marbles, calc-silicate rocks and skarns of the Moldanubian unit (Bohemian Massif, Czechoslovakia): Carbon and oxygen isotope and fluid inclusion constraints on formation conditions *K. Zák and P. Sztacho*	39
Fluid-controlled granulite formation in the East Gondwana *M. Santosh and M. Yoshida*	51

SANDSTONE PETROLOGY IN RELATION TO TECTONICS
Editors: F. Kumon and K.M. Yu

Preface	67
Effect of provenance, sorting and weathering on the geochemistry of fluvial sands from different tectonic and climatic environments *S.B. Kroonenberg*	69
A Precambrian Andean-type flysch: Petrology and geochemistry of Yangzhanling flysch in southern Anhui Province, China *B. Xia, X. Ma, H. Lu, Z. Fang, and J. Zhou*	83

Petrofacies analysis and tectonic evolution of Zagroside flysch suites
from northeastern Iraq
B. Al-Qayim   97

Permian to mid-Triassic evolution of sandstone composition in a complex
back-arc extensional to foreland basin: the Bowen Basin, eastern
Queensland, Australia
C.R. Fielding, P. de Caritat, J.C. Baker, and M.M. Wilkinson   109

Compositional changes of the sandstone in relation to the evolution of a
mobile belt - an example of the Paleozoic to Mesozoic sandstones of
North Japan
K. Okami, M. Ehiro, and S. Koshiya   119

Modal and chemical compositions of the representative sandstones from
the Japanese Islands and their tectonic implications
F. Kumon and K. Kiminami   135

Serpentinite protruded into fore-arc region: implications of detrital chromian
spinels in Cretaceous sandstones of the Kanto Mountains, Japan
K.-I. Hisada and S. Arai   153

Sedimentation and paleoenvironment in the uppermost Lower to Middle
Miocene basin on the Japan Sea coast, Southwest Japan
T. Matsumoto   165

Coexistence of provenance-reflected shallow-marine and deep-marine
turbidite sandstones - sedimentation at the eastern margin of the
Niigata Neogene backarc basin, northeast Japan
S. Tokuhashi   173

Petrography of turbidite sandstones in Niigata basin, northernmost part of
Fossa Magna, central Japan
M. Tateishi, A.A.A. El Habab, and M. Shimazu   183

EVAPORITE AND DESERT ENVIRONMENT
Editors: Y. Watanabe and A. Motamed

Preface   199

Geology and climate evolution of Central Iran
A. Motamed   201

Cluster and factor analysis of geochemical data of Najmah Formation,
Upper Jurassic, Western Desert, Iraq
A.K. Jamil and M.A. Al-Hilaly   221

Shortite formation in Turkey and its geochemical properties
F. Suner   237

Mineral assemblages and formation of the Kestelek and Sultançayiri borate deposits
C. Helvaci   245

Hydrogeochemical controls on the formation of primary dolomite in some ephemeral lakes in the Coorong region of South Australia
R. Ahmad and P.B. Hostetler   265

Biomineralization of the mirabilite deposits by the exemplification of the Barkol Lake
W. Dongyan, L. Zhenmin, D. Xiaoling, and X. Shaokang   283

Halogenic basins: facial and paleotectonic models
G.A. Belenitskaya   289

An occurrence of primary inyoite at Lagunita Playa, Northern Argentina
C. Helvaci and R.N. Alonso   299

METAMORPHIC REACTIONS: KINETICS AND MASS TRANSFER

Editors
T. Nishiyama and G.W. Fisher

CONTENTS

Preface   5

CDS modeling of chemical layering due to uphill diffusion in a mineral-fluid system
T. Nishiyama   7

An effect of a varying porosity on metasomatic replacement
K. Hoshino   17

Ostwald ripening and crystal size distribution (CSD) of garnet in various metamorphic rocks
K. Miyazaki   27

Marbles, calc-silicate rocks and skarns of the Moldanubian unit (Bohemian Massif, Czechoslovakia): Carbon and oxygen isotope and fluid inclusion constraints on formation conditions
K. Zák and P. Sztacho   39

Fluid-controlled granulite formation in the East Gondwana
M. Santosh and M. Yoshida   51

Preface

Symposium entitled "Metamorphic reactions : Kinetics and mass transfer (II - 9 - 5) " convened by G.W.Fisher and T. Nishiyama was held at 29th IGC, 1992, in Kyoto City, Japan, which comprises 10 talks and 6 poster presentations. The present volume contains the 5 papers among them. The topic of symposium was chosen from one of the most vigorously evolving fields in metamorphic petrology. One of the objectives of the symposium was to clarify the nature of reactions and mass transfer from the viewpoint of kinetics. Recent development of nonequilibrium thermodynamics together with that in the technique of computer simulation enables us to understand what is happening in the metamorphism. I have been impressed by growing richness of phenomena such as chemical pattern formation in layered metamorphic rocks, growth of garnet due to Ostwald ripening, and metasomatism with variable porosity, all of which is discussed in this volume. I believe that the symposium succeeded in revealing a wealth of new problems lying in this rapidly growing field.

I would like to thank all the enthusiastic participants : G.W. Fisher, S. N. Olsen, A. Lasaga, L. Baumgardner, B. Jamtveit, G. Dipple, W. Carlson, M.J. Bickle, K. Miyazaki, K. Hoshino, V.V. Reverdatto, M. Yoshida, K. Zak, M.O. Jedrysek, and R. Giere. The symposium was made possible by cooperation with G.W. Fisher and S. Banno.

Fukuoka
February, 1993

Tadao Nishiyama

CDS modeling of chemical layering due to uphill diffusion in a mineral-fluid system

T. NISHIYAMA
Dept. of Earth and Planetary Sciences, Kyushu University 33, Hakozaki 6-10-1, Fukuoka 812, JAPAN and Dept. of Earth & Planetray Sciences, The Johns Hopkins University, Baltimore MD 21218, USA

Abstract Chemical layering in metamorphic rocks is a good example of self-organization in geology. The coupling of diffusion flows in a ternary system can cause diffusion up the concentration gradients, depending on the relative magnitude of L_{11}/L_{12} and L_{22}/L_{12}. The condition of local equilibrium between the fluid and mineral M (M=$v_1 \alpha + v_2 \beta$: α and β are fluid species) gives this kind of coupling. The relative magnitude of the stoichiometric coefficients v_1 and v_2, together with the condition $L_{11}L_{22} > (L_{12})^2$, defines the region of uphill diffusion in a L-ratio diagram (a plot of L_{11}/L_{12} vs. L_{22}/L_{12}). A critical concentration (c_i)$_{cr}$ for uphill diffusion always exists, and the relative magnitude of (c_i)$_{cr}$ and the mean concentration (c_i) in the fluid will determine the behaviour of the system. Fluctuations in concentration will grow when $c_i <$ (c_i)$_{cr}$, resulting in a periodic concentration profile. Simulation of the system by CDS (Cell Dynamical System) approach successfully shows the development of a chemical oscillation irrespective of the periodicity in the initial perturbation.

Keywords: uphill diffusion, metamorphic layering, chemical layering, Cell Dynamical Systems, self-organization

INTRODUCTION

"METAMORPHIC LAYERING" is a descriptive term for layered structures of various origins in metamorphic rocks. This paper deals with metamorphic layering unrelated to a crenulation (Robin's [1] type L), which is called as chemical layering hereafter. Chemical layering in metamorphic rocks has following characteristics. (1) The layering is recognized by a difference in volume proportions of at least two minerals. (2) The layer width varies from several hundred microns to several centimeters. The wide layer represent the first order layering, and they contain several thin horizons, resulting in the second order layering. (3) Boundaries between two kinds of layers are usually sharp, indicating a diffusion-controlled growth rather than a reaction-controlled growth of the layering. Earlier theories [1,2] for chemical layering are based on pressure solution under anisotropic stress. Mineral segregation due to pressure solution, however, will enhance a mechanical instability of the layered structure, finally leading to a boudinage. These theories are contrary to the concept of pressure solution itself which will eliminate heterogeneities of stress states , i.e., the mechanical instability in the rock. Robin [1] proposed a hypothesis, to avoid this difficulty, that a mechanical strength (competency) of a mixture of two minerals is a nonlinear function of a volume ratio of the minerals, with a minimum at an intermediate of the ratio. This hypothesis is, however, implausible, because physical properties of a mechanical mixture of two minerals will monotonously vary with the volume ratio. Pressure solution is essentially a reaction-controlled process, which is also incompatible with the diffusion-controlled structure of chemical layering. Recurrent zonations of minerals also occur in unstressed rocks such as diffusion rings [3] in weathered rocks and alternating mineral layers [4] in skarns. These line of evidence strongly suggests that some chemical mechanism unrelated to pressure solution plays a major role in formation of periodic structures in mineral-fluid systems. This paper presents a new model of chemical layering based on uphill diffusion due to cross-term effects, which will explain well the development of the chemical layering and is also applicable to other diffusion-reaction systems. Uphill diffusion due to cross-term effects has been discussed by Oishi [5] and Cooper [6]. Grant [7] introduced uphill diffusion into a steady diffusion model for the reaction zone. Trial and Spera [8] discussed the importance of cross diffusion coefficients for chemical transfer in a ternary fluid. Recently, Falkovitz and Keller [9] discussed precipitation pattern formation in initially uniform salt solutions and explained it by uphill diffusion based on nonlinear Cahn Hilliard equation (an analogy of spinodal decomposition). Thus, uphill diffusion itself is not a new idea, however, this paper stresses on two new findings: (1) uphill diffusion can be enhanced by the effect of mineral stoichiomertries incorporated into phenomenological diffusion equations and (2) compositional dependencies of cross diffusion coefficients D_{ij} ($i \neq j$) always give a critical concentration for uphill diffusion which plays an important role in pattern formation. My theory does not require nonideality of the fluid which was *a priori* assumed by Falkovitz and Keller [9], hence has wider applicability than their model.

DIFFUSION FLOW

Suppose a ternary system consisting of H_2O (solvent) and two other components (1 and 2). Pressure and temperature are assumed constant. Figure 1 gives a model of the chemical potential diagram for this system. Each of three minerals (P, Q, and M) is in equilibrium with the fluid on the equilibrium line defined for the mineral through the reversible reactions: P = α, Q = β, and M = $v_1 \alpha + v_2 \beta$. v_i stands for the stoichiometric coefficient, and α and β represent chemical species of components 1 and 2 in the fluid. The intersection of two equilibrium lines represents an invariant point where two minerals coexists with the fluid. No compositional fluctuation in the fluid can grow at the invariant point, therefore an arbitrary point on the equilibrium line of M is taken as an initial state. Compositional fluctuations from the initial state may cause diffusion. Diffusion takes place through the intergranular fluid in metamorphic rocks. Volume diffusion in solids can be negligible in most metamorphic conditions. The diffusion flows are defined [10] so as to satisfy the Gibbs-Duhem relation for the fluid ($\Sigma c_i d\mu_i = 0$), i.e., $J_i = j_i - (c_i/c_{H_2O}) j_{H_2O}$, where j_i denotes the absolute flow of component i and J_i stands for the flow relative to the solvent. The diffusion flow of component 1 can be written as:

$$J_1 = - L_{11} (\partial \mu_1/\partial x) - L_{12} (\partial \mu_2/\partial x), \quad (1)$$

in the one-dimensional form, dropping vector notations. L_{ij} represents phenomenological coefficients, and μ_i stands for a chemical potential of component i. The space coordinate x is taken perpendicular to the layering, because of the following reason. The diffusion flow is anisotropic in metamorphic rocks, depending on the stress state. The diffusion along the maximum principal stress may be much slower than those along the minimum and intermediate principal stress axes. The anisotropic nature of the metamorphic layering may reflect this effect, because the layering develops perpendicular to the maximum principal stress axis. When the fluid is in local equilibrium with mineral M, equation (1) reduces to:

$$J_1 = - (L_{11} - [v_1/v_2] L_{12}) (\partial \mu_1/\partial x), \quad (2)$$

using the Gibbs-Duhem relation for mineral M :

$$v_1 d\mu_1 + v_2 d\mu_2 = 0. \quad (3)$$

Similarly, the diffusion flow of component 2 is written as:

$$J_2 = - (L_{22} - [v_2/v_1] L_{21}) (\partial \mu_2/\partial x). \quad (4)$$

The formal expressions of diffusion flows with concentration gradients are:

$$J_1 = - D_{11} (\partial c_1/\partial x) - D_{12} (\partial c_2/\partial x), \quad (5)$$
and
$$J_2 = - D_{21} (\partial c_1/\partial x) - D_{22} (\partial c_2/\partial x), \quad (6)$$

where D_{ij} stands for a diffusion coefficient. The assumption of the ideal solution for the intergranular fluid: $\mu_i = \mu_i^\circ + RT \ln c_i$ gives the following relations [10]:

$$L_{ij} = c_j D_{ij} / RT, \quad \text{and} \quad \partial \mu_i / \partial x = (RT/c_i) (\partial c_i / \partial x). \quad (7)$$

Onsager's reciprocal relation $L_{ij} = L_{ji}$ ($i \neq j$) and (7) give:

$$c_1/c_2 = D_{12}/D_{21} \quad (8)$$

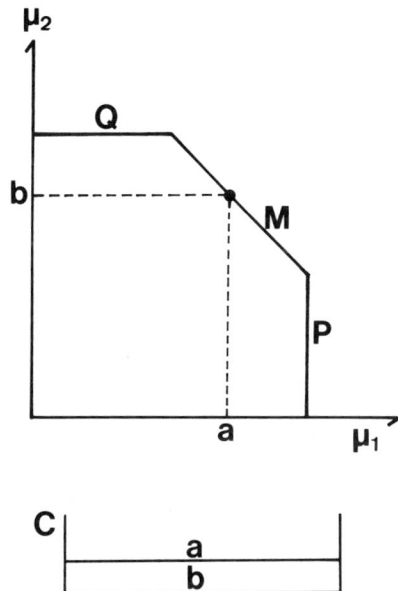

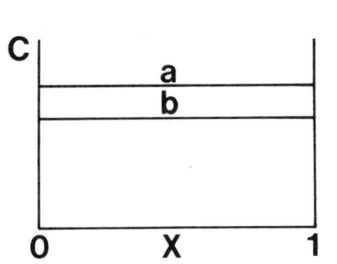

Fig. 1. (above): The chemical potential diagram for the model system as a projection from H_2O apex. Solid lines labelled as P, Q, and M indicate the saturation lines for minerals, P, Q, and M, respectively. A solid circle on the saturation line M represents a supposed initial state of the system. (below): Initial concentration profiles in the fluid versus distance (x). The fluid has uniform concentrations, a and b, for components 1 and 2, respectively. Mineral M is in equilibrium with the fluid everywhere.

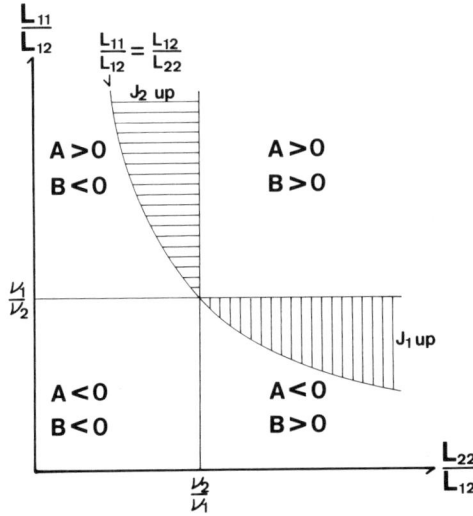

Fig. 2. The L-ratio diagram for the model system. The positive quadrant of the diagram consists of four regions divided by two lines: $A=0$ and $B=0$. The signs of A and B define whether the diffusion flows, J_1 (A) and J_2 (B), are uphill (negative) or downhill (positive). The condition of $(L_{11}/L_{12})>(L_{12}/L_{22})$ stems from positive entropy production. The domains hatched with vertical and horizontal lines represent those in which J_1 and J_2 are uphill, respectively.

Substituting (7) and (8) into (2) and (4), we get:

$$J_1 = -(D_{11} - [c_2 v_1 D_{12}]/[c_1 v_2])(\partial c_1/\partial x) = -(D_{11} - [v_1 D_{21}]/v_2)(\partial c_1/\partial x), \quad (9)$$

and
$$J_2 = -(D_{22} - [c_1 v_2 D_{21}]/[c_2 v_1])(\partial c_2/\partial x) = -(D_{22} - [v_2 D_{12}]/v_1)(\partial c_2/\partial x), \quad (10)$$

which are valid under the condition that the fluid is in local equiulibrium with M.

UPHILL DIFFUSION

Expressions of diffusive flows in the form : (2), (4), (9), and (10), make it possible to evaluate the condition for uphill diffusion. For convenience, we put

$$A = L_{11} - (v_1/v_2) L_{12} \quad \text{and} \quad B = L_{22} - (v_2/v_1) L_{21}. \quad (11)$$

The signs of A and B determine whether the diffusions down or up the chemical potential gradient. Therefore the relative magnitude of L_{ii} and L_{ij} is an important factor for the diffusion. The condition of positive entropy production gives constraints on the magnitude of L_{ii} and L_{ij} as follows [10]:

$$L_{ii} \geq 0 \quad \text{and} \quad L_{ii} L_{jj} \geq (L_{ij})^2. \quad (12)$$

If $L_{ij}(i \neq j)$ is negative, we can expect only downhill diffusion from (11). When L_{ij} is positive, (12) becomes:

$$(L_{11}/L_{12}) \geq (L_{12}/L_{22}). \quad (13)$$

Figure 2 shows an L-ratio diagram with L_{11}/L_{12} as ordinate and L_{22}/L_{12} as abscissa. The lines of $A = 0$ ($(L_{11}/L_{12}) = (v_1/v_2)$) and $B = 0$ ($(L_{22}/L_{21}) = (L_{22}/L_{12}) = (v_2/v_1)$) give four regions which represent four possible combinations of signs of A and B. (13) requires that all possible values of A and B lie above and to the right of the curve $L_{11}/L_{12} = L_{12}/L_{22}$ in Fig. 2. The intersection of two lines: $A = 0$ and $B = 0$, is always on this boundary irrespective of individual values of v_1 and v_2. J_1 and J_2 cannot both be uphill because of this constraint. J_1 becomes uphill in the domain hatched with the vertical line, and J_2 in the domain with the horizontal line. The relative area of these domains depend on the relative magnitude of v_1 and v_2. As v_1/v_2 increases, the domain of J_1 uphill becomes larger and that of J_2 uphill shrinks.

Critical Concentration

$D_{ij}(i \neq j)$ should be concentration-dependent, because the assumption of constant D_{ij} in (8) gives a linear relationship between c_1 and c_2, which is contrary to the Gibbs-Duhem relation for ideal solution:

$$v_1 d\ln c_1 + v_2 d\ln c_2 = 0. \quad (14)$$

The functional form of $D_{ij}(c_1, c_2)$ is unknown. If $D_{ij}(c_1, c_2)$ satisfies (8), however, the concentration-dependent D_{ij} always give the critical concentration (c_i)$_{cr}$ for J_i to become zero, irrespective of the functional form of $D_{ij}(c_1, c_2)$. For example, the condition for which J_1 becomes uphill is:

$$D_{11} - (c_2 v_1 D_{12})/(c_1 v_2) < 0, \quad (15)$$

from (9). c_1 is related to c_2 by (14) or its integral form:

$$c_1^{v_1} c_2^{v_2} = K, \quad (16)$$

where K denotes an equilibrium constant. Consider following three cases.
(17) $D_{12} = (D_{12}{}^o c_2{}^o)/c_2$, $D_{21} = (D_{21}{}^o c_1{}^o)/c_1$, and D_{11} and D_{22} are constants, where $D_{12}{}^o$, $D_{21}{}^o$, $c_1{}^o$, and $c_2{}^o$ are constants, satisfying $D_{12}{}^o c_2{}^o = D_{21}{}^o c_1{}^o$.
(18) $D_{12} = c_1{}^2 c_2 D_{12}{}^*$, $D_{21} = c_1 c_2{}^2 D_{21}{}^*$, and D_{11} and D_{22} are constants, where $D_{12}{}^* = (D_{12}{}^o/[c_1{}^o]^2 c_2{}^o)$ and $D_{21}{}^* = (D_{21}{}^o/c_1{}^o[c_2{}^o]^2)$ are contants satisfying $D_{12}{}^* = D_{21}{}^*$.
(19) $D_{12} = c_1{}^2 c_2 D_{12}{}^*$, $D_{21} = c_1 c_2{}^2 D_{21}{}^*$, $D_{11} = c_1 D_{11}{}^o/c_1{}^o$, and $D_{22} = c_2 D_{22}{}^o/c_2{}^o$, where $D_{12}{}^* = D_{12}{}^o/([c_1{}^o]^2 c_2{}^o)$ and $D_{21}{}^* = D_{21}{}^o/(c_1{}^o[c_2{}^o]^2)$ are constants satisfying $D_{12}{}^* = D_{21}{}^*$. $c_1{}^o$, $c_2{}^o$, $D_{11}{}^o$, and $D_{22}{}^o$ are also constants.
All of these satisfy (8). Combining (15) and (16) together with one of (17), (18), and (19), we get the condition for uphill diffusion for each case as follows.

(17') $c_1 < (v_1 D_{21}{}^o c_1{}^o)/(v_2 D_{11})$.

(18'a) $c_1 > ([v_2 D_{11}]/[v_1 D_{12}{}^* K^{2/v2}])^{v2/(v2-2v1)}$.   if $v_2 > 2v_1$.

(18'b) $c_1 < ([v_1 D_{12}{}^* K^{2/v2}]/[v_2 D_{11}])^{v2/(2v1-v2)}$.   if $v_2 < 2v_1$

(19') $c_1 < ([v_1 c_1{}^o D_{12}{}^* K^{2/v2}]/[v_2 D_{11}])^{v2/2v1}$.

Thus, the condition for J_1 uphill can be simply written as either c_1 < const. or c_1 > const., depending on the functional form of $D_{ij}(c_1,c_2)$ and the relative magnitude of v_1 and v_2. Namely, there is always a critical concentration $(c_1)_{cr}$ such as $(c_1)_{cr}$ = const., which will divide the concentration region into two parts: one where J_1 becomes uphill and the other where J_1 downhill.

BEHAVIOUR OF THE SYSTEM

We will make a qualitative discussion on the behaviour of the system in this section. Here c_1 is taken as an independent variable. Consider the simplest case: (17) and (17'). The right-hand side of (17') gives the concentration critical for uphill diffusion defined as $(c_1)_{cr}$ = $(v_1 D_{21}{}^o c_1{}^o)/(v_2 D_{11})$. The mean concentration of component 1 in the system, c_1, will be another important factor to determine the behaviour of the system. There are three cases as follows.

(a) $c_1 > (c_1)_{cr}$

Figure 3a shows a behaviour of the system when c_1 is greater than $(c_1)_{cr}$. No fluctuation can grow, because of the diffusion down the concentration gradient in the whole region.

(b) $c_1 = (c_1)_{cr}$

The fluctuation in this case will consist of two parts which will behave in a different manner. Positive fluctuations from c_1 will die, spreading towards both sides, because of diffusion down the concentration gradient. Negative fluctuations will become deeper and narrower, because of uphill diffusion. Consequently, we will expect the occurrence of sharp pulses in the system (Fig. 3b).

(c) $c_1 < (c_1)_{cr}$

Uphill diffusion will occur and make the fluctuation grow in the whole region of the system. Positive fluctuations will reach $(c_1)_{cr}$, and then spread to both sides with a flat top. Secondary fluctuations in this flat top may permit the growth of sharp pulses as in the case of $c_1 = (c_1)_{cr}$, because the local mean concentration is equal to $(c_1)_{cr}$ (Fig. 3c).

METAMORPHIC LAYERING

Above discussions showed that pattern formation due to uphill diffusion will occur in the cases when $c_1 < (c_1)_{cr}$. Here we will consider the system with dissolution-precipitation reactions, and discuss the formation of metamorphic layering. Local equilibrium assumption requires that dissolution-precipitation reactions are much faster than diffusion. In such cases, the behaviour of the system can be simply discussed by diffusion equation, because the fluid composition is constrained on the saturation line for the coexisting mineral. One may consider that uphill diffusion of component 1 should stop once downhill diffusion of component 2 flattens out the original perturbation. That is not the case if $c_1 < (c_1)_{cr}$, because M rapidly dissolutes or precipitates through the reaction: M = $v_1 \alpha + v_2 \beta$, due to the Gibbs-Duhem condition. Hence, compositional flactuation of component 2 will grow even component 2 shows downhill diffusion, as long as local equilibrium is maintained. Suppose that an initial state, c_1, on the saturation line M is between the invariant point $(c_1)_{QM}$ and $(c_1)_{cr}$ as shown in Fig. 4. Fluctuations will grow due to uphill diffusion in the whole region. Negative fluctuations will reach $(c_1)_{QM}$, where a part of M dissolves and Q precipitates, resulting in the two phase (Q+M) domain. Positive fluctuations will arrive at $(c_1)_{cr}$ and spread to both sides, resulting in a flat top accompanied by sharp pulses due to the secondary fluctuations. The sharp pulses will finally reach $(c_1)_{QM}$ and make narrow domains of two phase (Q+M) coexistence.

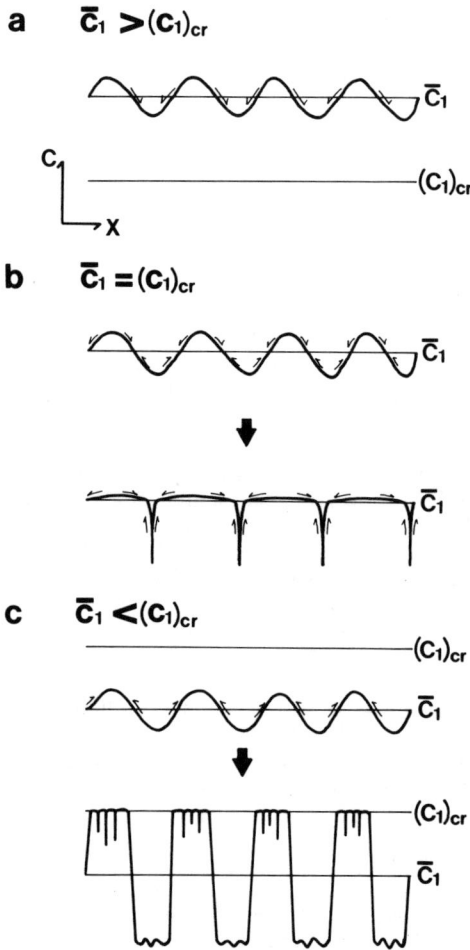

Fig. 3. Qualitative behaviour of the system with fluctuations. When the mean concentration (c_1) is larger than the concentration critical for the uphill diffusion (($c_1)_{cr}$), fluctuations will disappear by downhill diffusion (a). In the case of $c_1=(c_1)_{cr}$, sharp pulses will develop because of the death of positive fluctuations and the growth of negative fluctuations (b). If $c_1<(c_1)_{cr}$, fluctuations will grow until they will reach $(c_1)_{cr}$. Secondary fluctuations will cause sharp pulses at $(c_1)_{cr}$ (c).

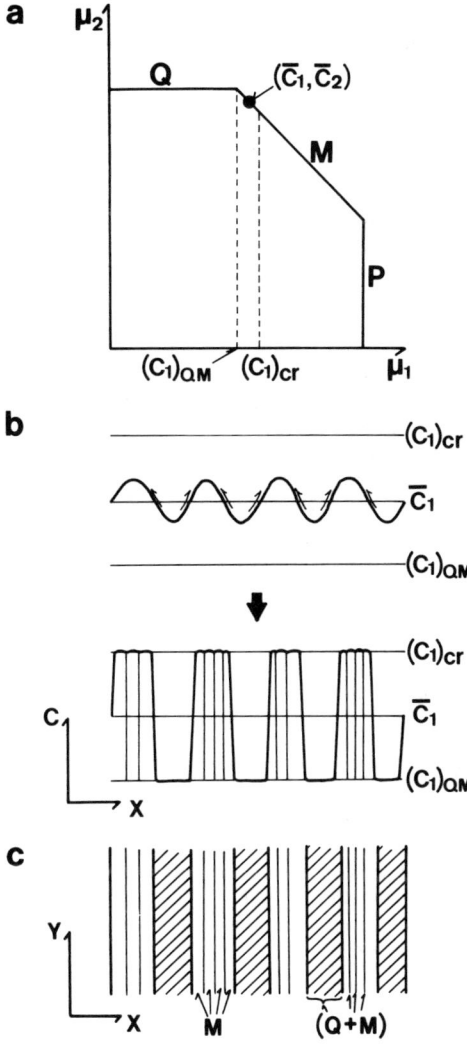

Fig. 4. Qualitative behaviour of the system with dissoluion-precipitation reactions. a. If the initial state is between the invariant point, $(c_1)_{QM}$, and the critical point, $(c_1)_{cr}$, fluctuations will grow by uphill diffusion. b. Positive fluctuations will reach $(c1)_{cr}$, resulting in a flat top with sharp pulses. Negative fluctuations and sharp pulses will finally arrive at $(c_1)_{QM}$, where a part of M dissolves and Q precipitates. c. The formation of layering consisting of M and (M+Q) layers can be expected. M also involves the second order layering indicated by the presence of thin horizons of (Q+M).

These thin two phase layers within M layers represent the secondary layering. The secondary fluctuation cannot generate at $(c_1)_{QM}$, because $(c_1)_{QM}$ is an invariant point. Thus we can expect the formation of chemical layering consisting of M and (Q+M) layers (Fig. 4c). This model can also explain why layered ($c_1 \lesssim (c_1)_{cr}$) and unlayered ($c_1 \gtrsim (c_1)_{cr}$) rocks of identical mineralogy occur in the same metamorphic region.

CELL DYNAMICAL SYSTEM MODELING

This section describes numerical modeling of development of chemical layering due to uphill diffusion. Concentration - dependent diffusion coefficients discussed above lead to nonlinear diffusion equation, which is very difficult to solve even numerically. Instead, we will adopt a Cell Dynamical System (CDS) approach [11]. CDS has been recently developed as a powerful mathematical tool to simulate the pattern formation in physics such as the segregation process of binary mixtures. CDS is much easier to handle than the nonlinear partial differential equations. CDS is a map from a set of discrete spatial patterns to itself, which can be regarded as a time evolution operator relating the configuration at time t to that at time t + 1. CDS modeling consists of three steps: modeling of the single cell dynamics, modeling of the cell interaction and modeling of the mass conservation. To delineate spatial configuration, we will divide the system into many discrete cells, in which the single value of concentration is defined. First we consider the single cell dynamics which is a map of concentration defined in a single cell at time t to that at time t + 1. Because we are interested in the concentration distribution of a component showing uphill diffusion, we adopt the following mapping function which describes the time evolution of the concentration in a single cell:

$$c(t+1) = c(t) + A\sin\pi c(t)$$

where A is an arbitrary constant. Figure 5 shows this mapping function. Behavior of the concentration due to the mapping of $c(t)$ to $c(t+1)$ depends on the initial concentration. When initial concentration is between c_m and $(c)_{cr}$, the concentration in the cell will increase with time and finally reach to $(c)_{cr}$. Similarly, sarting with the initial concentration between c_{QM} and c_m, the concentration will decrease with time and finally reach to c_{QM}. Thus the concentration region between c_{QM} and $(c)_{cr}$ represent the uphill diffusion regime. On the other hand, if the initial concentration is larger than $(c)_{cr}$ (downhill regime), the concentration will decrease until it reaches to $(c)_{cr}$. Starting with the initial concentration smaller than c_{QM} (reaction regime), the fluid composition will be buffered by mineral assemblage, resulting the equilibrium composition c_{QM}. Thus this mapping function will simulate well the single cell dynamics of the system in which uphill diffusion occur. Secondly we will take account of cell interactions, which is driven by the difference between the average concentration in the neighborhood cell and the concentration in a cell in question. We need to consider cell interaction because concentration within each layer (M layer and QM layer) should be homogeneous. Then the driving force for cell interaction is written as

$$D(<<c>> - c)$$

where $<<c>>$ denotes the local average of the concentration and D stands for a constant describing the strength of interaction between cells. There are several possibilities in choices of the local average as long as thy respect symmetry of the system. Here we take the definition:

$$<<c>> = (1/2)\Sigma(c \text{ for the nearest neighbors}) + (1/2)\Sigma(c \text{ for the next nearest neighbors})$$

in one dimension. As a third step we consider local mass balance by introducing a parameter I which describes the amount of concentration the cell obtains in the unit time step:

$$I = f(c(t)) - c(t) + D(<<c>> - c(t))$$

Local mass balance requires the following final form for the evolution of concentration:

$$c(t+1) = c(t) + I - <<I>>$$

where $<<I>>$ denotes the local average of I similarly defined as in eq. . Equation enables us to simulate the evolution of chemical layering in one dimension, in which normalized concentration is used as defined by:

$$c' = (c - c_m)/((c)_{cr} - c_m)$$

In the normalized concentration, $(c)_{cr}$ bemomes 1, c_m does 0 and c_{QM} becomes -1. Finally we will give an arbitrary initial perturbation satisfying mass balance in the system by some periodic function such as:

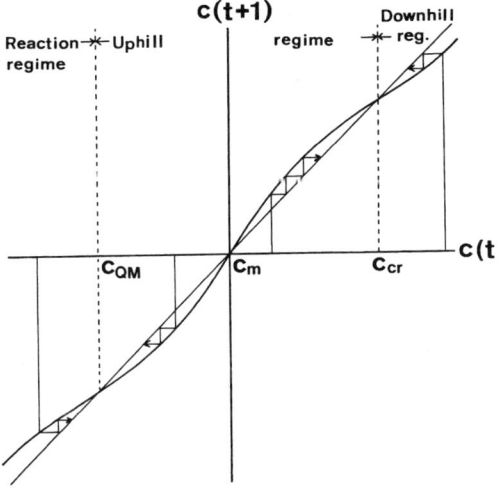

Fig. 5. Mapping function describing the evolution of the system due to uphill diffusion in one-dimension. The function maps concentratiuon at time t to that at time $t + 1$.

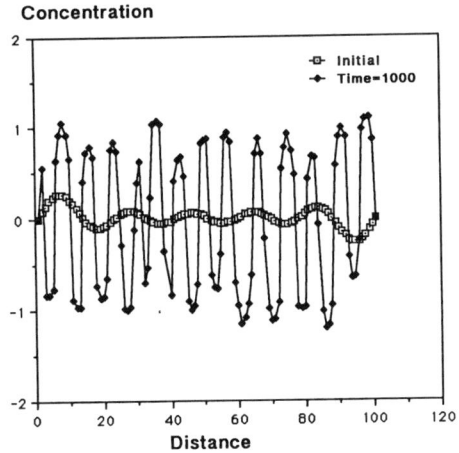

Fig. 6. Result of CDS modeling of the development of chemical layering due to uphill diffusion in one-dimension. Initial perturbations shown by open squares are arbitrarily given to satisfy mass balance. Layering occur irrespective of the wavelength of the initial perturbation.

$$c'(0, x) = \Sigma B_i \sin n_i \pi x$$

where B_i denotes arbitrary constants and n_i stands for arbitrary integers. Figure 6 shows the result at 1000 time step in which the system becomes almost stationary. Plural layers appeared irrespective of the wave length of the initial perturbation. The width of the layer depends on relative magnitude of D (the strength of cell interaction) and A (coefficient in the mapping function). Qualitative description of uphill diffusion discussed above give the layers which is exactly controlled by the initial perturbation. The difference between this qualitative view and the result of the CDS modeling is due to the growth of the secondary fluctuation at $(c)_{cr}$ in the latter case. In CDS modeling, the cell interaction plays a major role not only in homogenization of concentration within layer but also costructing layering itself by enhancement of secondary fluctuations.

Ackowledgement - Drs. G.W. Fisher and C.T. Foster kindly reviewed the earlier version of the manuscript. Suggestions by and discussions with Drs. T. Yanagi, M. Toriumi, H. Momoi, M. Obata, M. Nakada, K. Kawasaki, K. Sekimoto, K. Yoshikawa, S. Nakada, and Y. Aoki are acknowledged. I thank Drs. Sakiko Olsen and Lukas Baumgardner for their hospitality at the Hopkins. This work is an outcome of the research project on mass transfer directed by Drs. J.T. Iiyama and S. Banno, and financially supported by the Ministry of Education, Culture and Science, Japanese Government.

References

1. Robin, P.-Y. F. *Geochim. Cosmochim. Acta* **43**, 1587-1600 (1979).
2. Ortoleva P., Merino,E. & Strickholm, P. *Amer. J. Sci.* **282**, 617-643 (1982).
3. Augustithis S.S. & Vgenopoulos, A. in *Leaching and diffusion in rocks and their weathering products* (ed. Augustithis, S.S.) 151-173 (Theophrastus, Athens, 1983).
4. Kwak, T.A.P. *W-Sn skarn deposits* (Elsevier, Amsterdam, 1987).
5. Oishi, Y. *J. Chem. Phys.* **43**, 1611-1620 (1965).
6. Cooper, A.R.Jr. in *Geochemical transport and kinetics* (eds. Hofmann, A.W., Giletti, B.J., Yoder, H.S. & Yund, R.A.) 15-30 (Carnegie, Washington, 1974).
7. Grant, S.M. *Contrib. Mineral. Petrol.* 98, 49-63 (1988).
8. Trial, A.F. & Spera, F.J. *Int. J. Heat Mass Transfer* **31**, 941-955 (1988).
9. Falkovitz, M.S. & Keller, J.B. *J.Chem. Phys.* **88**, 416-421 (1988).
10. Katchalsky, A. & Curran, P.F. *Nonequilibrium thermodynamics in biophysics* (Harvard Univ. Press, Boston, 1965).
11. Oono, Y. & Puri, S. *Phys. Rev. A.* **38**, 434 (1988).

An effect of a varying porosity on metasomatic replacement

K. HOSHINO
Department of Earth and Planetary Sciences, Hiroshima University, Higashi-Hiroshima, 724, JAPAN

Abstract-- Bimetasomatism [1] has been reinvestigated considering the effect of a varying porosity on the process itself. Since the most important feature of metasomatism lies in a complete replacement of minerals, porosity may change sharply at replacement fronts. Moreover, porosity may change also in a replaced zone due to "salting out" effect. On the other hand, effective diffusion coefficients strongly depend on the porosity. Accordingly, porosity change through the process may play an important role in metasomatism.

The formation process of wollastonite skarn at a boundary between limestone and quartzite, a typical bimetasomatism, has been analyzed based on the assumptions of local equilibrium and constant volume. Porosity at each replacement front depends on initial porosity, molar densities of minerals and intrinsic diffusion coefficients of replacing and replaced components, and is independent from time and length. Consequently, internal precipitation of the reaction mineral due to the "salting out" effect causes a porosity gradient in the replaced zone. As a result, a steady state of "constant" concentrations of both replacing and replaced components ($\partial c_i / \partial$ x=0) is almost achieved near the replacement fronts. The porosity change thus acts as "braking mechanism". In other words, bimetasomatic replacement is a "self-controlled" process [2].

Keywords: bimetasomatism, skarn, porosity, self-controlled process

SYMBOLS

c_i - concentration of component i in a pore solution, (moles per unit volume)
D_i - effective diffusion coefficient of i in porous media, (length^2/time, for example, cm^2/sec)
$\overline{D}_i$ - intrinsic diffusion coefficient of i in a solution
J_i - flux of i across unit surface in unit time
K - equilibrium constant
n_m - molar density of mineral m
P - pressure
R - gas constant
s_i - concentration of i in solid phase, (moles per unit volume)
t - time
T - temperature, (°K)
V_m - volume of mineral m per total volume
x - space coordinate measured normal to the section
β - porosity
β f - porosity at replacement front
$\underline{\beta}$ - initial porosity
δ_D - constrictivity for diffusion
ρ_i - total content of i in minerals and pore solution, (moles per unit volume)
τ - tortuosity

INTRODUCTION

Since Korzhinskii [1] attempted to analyze basic features of the metasomatic columns by deriving transport equations for infiltration and diffusion metasomatism, a great deal has been written about theoretical and numerical analyses of the metasomatic processes (for example, [3-8]). In these studies, porosity and effective diffusion coefficient have been assumed to be constant through the metasomatic processes.

However, since the most important feature of metasomatism lies in a complete replacement of minerals, porosity may change sharply at replacement fronts. Moreover, porosity may change also in a zone replaced by reaction mineral(s) due to the "salting-out" effect, a term proposed by Franz and Mao [5]. Therefore, the assumption of constant porosity is not reasonable. On the other hand, the effective diffusion coefficient strongly depends on the porosity (for example, [9-11]). Accordingly, porosity change through the process, one of the fundamental features of metasomatism, does play an important role in metasomatism [2].

In contrast to analytical studies on metasomatism, those on diagenetic processes include the effect of porosity variation on material transport (for example, [9,12]). However, in those processes, the variation may be defined by initial and/or external condition(s), but does not result from material transport. On the contrary, as stated before, metasomatic replacement alters the porosity, which in turn controls the metasomatism [13]. Therefore, metasomatism is essentially a process of "self-organization" [14,15].

From this point of view, the formation process of wollastonite skarn at a boundary between limestone and quartzite, a typical bimetasomatism [1], has been analyzed to clarify the porosity change through the process and its effect on metasomatism itself.

The following assumptions are used to analyze the metasomatic process:

a) total volume is constant

b) external conditions such as f_{CO2}, temperature are also unchanged

c) total porosity is the same as effective porosity, that is, all pores are connected with each other

d) dissolution and precipitation of minerals occur instantaneously. In other words, local equilibrium [2] is achieved throughout the column.

Effects of concentrations of components on diffusion coefficients are neglected in all analyses. Other assumptions will be mentioned where needed.

TRANSPORT EQUATION

As a fundamental equation for analyses of diffusion metasomatic replacement at a boundary between two rocks which are not in chemical equilibrium, a typical diffusion "bimetasomatism" [1], the following diffusion transport equation is useful:

$$\frac{\partial \rho_i}{\partial t} = \frac{\partial}{\partial x}\left(D_i \frac{\partial c_i}{\partial x}\right) ,\qquad (II.1)$$

where,

$$\rho_i = (1-\beta)s_i + \beta c_i . \qquad (II.2)$$

It is worth noting that the above equation is equivalent to the equation for diffusion metasomatism obtained by Korzhinskii [1] when the effective diffusion coefficient and porosity are kept constant through the column.

According to [10], the effective diffusion coefficient can be described as

$$D_i = (\beta \delta_D / \tau^2) \overline{D}_i \ , \qquad (II.3)$$

the term $\beta \delta D / \tau^2$ depends only on the properties of the porous materials. Nakashima et al. [11] and Kita et al. [12] suggest that $log D_i$ is linearly correlated to $log \beta$. Therefore,

$$D_i = \beta^n \overline{D}_i \ , \qquad (II.4)$$

where n can be roughly estimated to have a value between 1 and 1.5 (from Fig. 12 in [10] and Fig. 6 in [16]). The value n will be tentatively assumed as 1 in the following numerical analyses.

By substituting (II.4) into (II.1), the final equation for analyses of porous bimetasomatic columns can be obtained as follows:

$$\frac{\partial \rho_i}{\partial t} = \overline{D}_i \frac{\partial}{\partial x} \left(\beta^n \frac{\partial c_i}{\partial x} \right) \ . \qquad (II.5)$$

ANALYZED MODEL

As a typical bimetasomatism, the formation process of wollastonite skarn at a boundary between limestone and quartzite is taken as an example to be analyzed. Physicochemical conditions of the process have been referred to those of a "vein-type" wollastonite skarn of the Fujigatani tungsten mine [2, 17]. Figs. 1a and b show stability fields of calcite, quartz and wollastonite under the conditions. The states of solvations of Ca^{++} and $SiO_{2(aq)}$ (hereafter expressed as Ca and Si, respectively) are not specified [18].

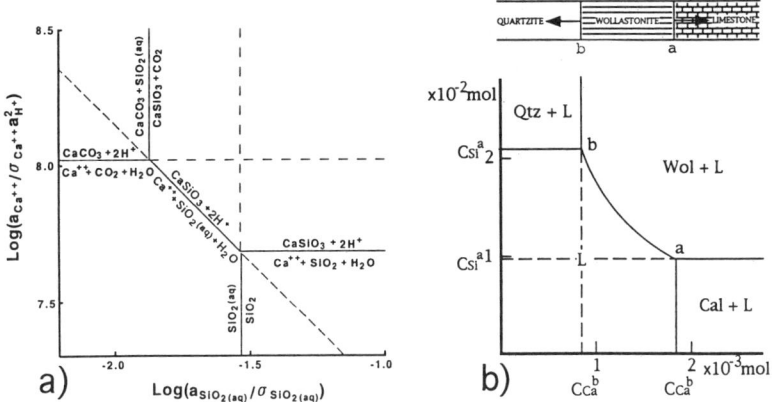

Fig. 1. a): An equilibrium activity diagram for the system H_2O-SiO_2-CaO-CO_2 [2]. For a definition of the term σi, see [18]. b): Saturation curves of calcite (Cal), quartz (Qtz) and wollastonite (Wol) [2] and an analyzed bimetasomatic column. The points a and b correspond to the initial concentrations of pore solutions in limestone and quartzite, respectively.

The pressure on the pore solution (P_{fluid}) and fco_2 are assumed constant throughout the column. Because of the assumption of constant volume, internal precipitation of a reaction mineral in pores may cause the pressure on the mineral ($P_{mineral}$) increase. Hence, the solution may become supersaturated according to the following equation [2]:

$$P_{mineral} - P_{fluid} = RTn_m ln \ (c^{ss}/c^s) \ , \qquad (III.1)$$

where c^s and c^{ss} denote saturated and supersaturated concentration, respectively. Therefore, the degree of supersaturation (c^{ss}/c^s) is a function of the excess pressure ($P_{mineral} - P_{fluid}$), which may also be a function of the porosity (β). However, since dependency of the porosity on the degree of supersaturation is not known, the following relationship is tentatively assumed in the following numerical analyses:

$$\frac{c^{ss}}{c^s} = \frac{1}{10^4 \beta} + 0.9999 \ . \qquad (III.2)$$

For simplicity, it is also assumed that initial compositions of the pore solutions in limestone and quartzite are assumed to be saturated with wollastonite (the points a and b, respectively, in Fig. 1b). Thus, the initial conditions for the analyses are

$c_{ca} = c_{ca}{}^a \ (1/10^4 \overline{\beta} + 0.9999)$,
$c_{si} = c_{si}{}^a \ (1/10^4 \overline{\beta} + 0.9999)$ at $t = 0$ in limestone,

and

$c_{ca} = c_{ca}{}^b \ (1/10^4 \overline{\beta} + 0.9999)$,
$c_{si} = c_{si}{}^b \ (1/10^4 \overline{\beta} + 0.9999)$ at $t = 0$ in quartzite.

ANALYTICAL RESULTS

Porosity at the replacement fronts
When the porosity around a replacement front is not extremely small (i.e., $c^{ss}/c^s \cong 1$), a limit value of the porosity at the front where calcite is replaced by wollastonite ($\beta_f{}^a$) can be estimated directly from the finite-difference form of (II.5) as follows [2]:

$$\beta_f{}^a = 1 - (1-\beta) \ (\alpha_a \overline{D}_{Ca} + \alpha_b \overline{D}_{Si}) / (\alpha_c \overline{D}_{Ca} + \alpha_d \overline{D}_{Si}) \ , \qquad (IV.1)$$

where
$\alpha_a = -c_{ca}{}^a c_{si}{}^a$,
$\alpha_b = c_{si}{}^a (n_{cal} - c_{ca}{}^a)$,
$\alpha_c = c_{ca}{}^a (n_{wol} - c_{si}{}^a)$,
and
$\alpha_d = c_{si}{}^a (n_{wol} - c_{ca}{}^a)$.

The porosity of the other front where quartz is replaced by wollastonite can also be estimated in the same way. As is clear in the above equation, the front porosity depends on initial values, being independent of time and distance.

A relationship between the initial porosity ($\overline{\beta}$) and the ratio of intrinsic diffusion coefficients ($\overline{D}_{Si}/\overline{D}_{Ca}$) for the case of the front porosity becoming zero is shown in Fig. 2. On the shaded areas in the figure, pore spaces in the front may be filled completely with wollastonite before perfect dissolution of calcite or quartz, with the resultant porosity

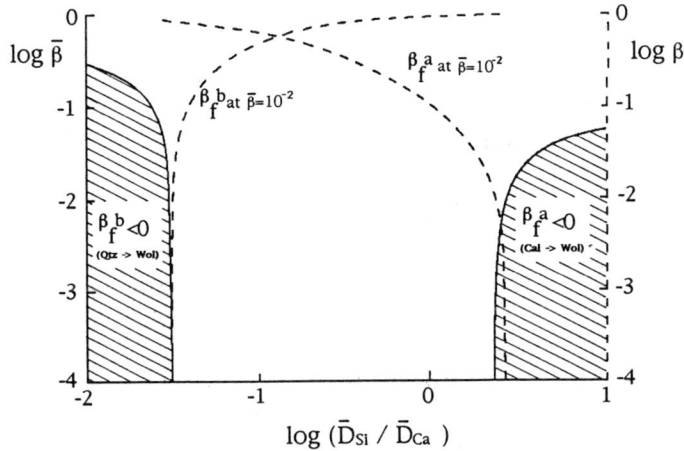

Fig. 2. Relationship between $\overline{D}_{Si} / \overline{D}_{Ca}$ and $\overline{\beta}$ at $\beta_f=0$. In the shaded fields, all cavities will be filled with wollastonite before complete dissolution of calcite (right) or quartz (left). Porosity of each replacement front in the case of $\overline{\beta} = 10^{-2}$ is also shown (broken line).

becoming zero. In this case, since the effective diffusion coefficients at the front should be zero by (II.4), no material transport can occur at the front. Accordingly, the diffusion metasomatism on those areas may be completed after dissemination of the reaction mineral in cavities in original rocks [2].

Porosity change in replaced zone

Because of the constant compositions of solutions at both ends of the replaced zone, a "steady state" may almost be achieved through the zone. That is:

$$\frac{\partial c_{Ca}}{\partial t} \cong 0 \quad , \tag{IV.2a}$$

and

$$\frac{\partial c_{Si}}{\partial t} \cong 0 \quad . \tag{IV.2b}$$

Note that a *true* steady state,

$$\beta^n \frac{\partial c_{Ca}}{\partial x} = \text{constant} \tag{IV.3a}$$

and

$$\beta^n \frac{\partial c_{Si}}{\partial x} = \text{constant} , \tag{IV.3b}$$

may not be achieved whenever the composition of pore solution varies through the zone. The reason is as follows: according to the assumption d), the following equation should hold through the zone:

$$c_{Ca} c_{Si} = K . \tag{IV.4}$$

The reaction constant K of the above equation may also vary with porosity through the zone. However, when the variation of K is negligibly small, (IV.3b) can be rewritten as

$$\beta^n \frac{dc_{Si}}{dc_{Ca}} \frac{\partial c_{Ca}}{\partial x} = \text{constant} . \tag{IV.3b'}$$

Therefore, from (IV.3a),

$$\frac{dc_{Si}}{dc_{Ca}} = \text{constant} . \tag{IV.3b''}$$

This means that the concentration gradients of both components should be zero $((\partial c_i / \partial x) = 0)$ in the state. Hoshino [2] analyzed the net mass flow of both components across surfaces at $x=\zeta$ and $x=\zeta+d\zeta$ into lamina of thickness $d\zeta$ per unit time in the wollastonite zone. According to Fick's first law, the net addition of both components to the lamina per unit area per unit time is:

$$(J_{Ca})_{x=\zeta} - (J_{Ca})_{x=\zeta+d\zeta} = -\overline{D}_{Ca}(\beta^n \partial c_{Ca} / \partial x)_{x=\zeta}$$
$$+ \overline{D}_{Ca}(\beta^n \partial c_{Ca} / \partial x)_{x=\zeta+d\zeta} , \tag{IV.5a}$$

and

$$(J_{Si})_{x=\zeta} - (J_{Si})_{x=\zeta+d\zeta} = -\overline{D}_{Si}(\beta^n \partial c_{Si} / \partial x)_{x=\zeta}$$
$$+ \overline{D}_{Si}(\beta^n \partial c_{Si} / \partial x)_{x=\zeta+d\zeta} . \tag{IV.5b}$$

Hence, according to (IV.4),

$$(J_{Si})_{x=\zeta} - (J_{Si})_{x=\zeta+d\zeta} = -\overline{D}_{Si}\{(dc_{Si} / dc_{Ca})\beta^n \partial c_{Ca} / \partial x\}_{x=\tau}$$
$$+ \overline{D}_{Si}\{(dc_{Si} / dc_{Ca})\beta^n \partial c_{Ca} / \partial x\}_{x=\zeta+d\zeta} . \tag{IV.5b'}$$

On the other hand,

$$(J_{Ca})_{x=\zeta} - (J_{Ca})_{x=\zeta+d\zeta} = (n_{wol} - c_{Ca} - c_{Ca})V + \beta c_{Ca} , \tag{IV.6a}$$

and

$$(J_{Si})_{x=\zeta} - (J_{Si})_{x=\zeta+d\zeta} = (n_{wol} - c_{Si} - c_{Si})V + \beta c_{Si} . \tag{IV.6b}$$

where c_i and V represent the difference in concentration of component i and that in volume of wollastonite, respectively, per unit time in the lamina. Because the term βc_i is negligibly small, by comparing the above equations with (IV.5a and b'):

$$V = \frac{(dc_{Si} / dc_{Ca})_{x=\zeta} - (dc_{Si} / dc_{Ca})_{x=\zeta+d\zeta}}{k_{Ca}(dc_{Si} / dc_{Ca})_{x=\zeta+d\zeta} - k_{Si}} \left(\beta^n \frac{\partial c_{Ca}}{\partial x} \right)_{x=\zeta} , \tag{IV.7}$$

where

$$k_{Ca} = (n_{wol} - c_{Ca} - c_{Ca})/\overline{D}_{Ca} \text{ and } k_{Si} = (n_{wol} - c_{Si} - c_{Si})/\overline{D}_{Si} .$$

Note that k_{Ca} and k_{Si} are positive because the molar density of wollastonite (n_{wol}) is always larger than the saturated concentrations (c_i+c_i) by more than three orders of magnitude (Table 1 in [2]) and that the term (dc_{Si}/dc_{Ca}) should be negative throughout the zone. Hereafter, the front at where calcite is replaced by wollastonite is set on the right hand side (i.e., ($\partial c_{Ca} / \partial x$) > 0). Then, since the saturation curve is convex toward the origin in Fig. 2,

$$(dc_{Si}/dc_{Ca})_{x=\zeta} < (dc_{Si}/dc_{Ca})_{x=\zeta+d\zeta} .$$

Hence,

$$V > 0 , \qquad\qquad (IV.8)$$

that is, wollastonite continues to precipitate through the zone. Because of the constant porosity at the replacement fronts, the porosity in the zone should decrease.

The fundamental transport equation (II.5) can be rewritten as follows:

$$\frac{\partial \rho_i}{\partial t} = \overline{D}_i \beta^n \frac{\partial^2 c_i}{\partial x^2} + \overline{D}_i \frac{\partial \beta^n}{\partial x} \frac{\partial c_i}{\partial x} . \qquad\qquad (IV.9)$$

As can be supposed by the above equation, when the porosity in the replaced zone increases toward the front where calcite is replaced (($\partial \beta / \partial x$) > 0) as discussed above, the effect of varying porosity on metasomatism is the same as that of *reverse*-infiltration (that is, a pore solution migrates in an opposite direction to the front progress). Consequently, the decrease in porosity through the zone may bring about the *true* steady state profiles (($\partial c_i / \partial x$) = 0) near the replacement front.

Numerical analyses
The formation process of wollastonite skarn at a boundary between limestone and quartzite has been numerically analyzed for the cases of the ratio of intrinsic diffusion coefficients ($\overline{D}_{Si}/\overline{D}_{Ca}$) of 1 (Fig. 3) and 0.1 (Fig. 4). Volume flux of pore solution due to porosity change has been ignored because of its small effect.

Since the porosity at the replacement front is extremely large as compared with that of unreplaced limestone or quartzite, concentration of a diffusing component in pore solution of the front becomes lower than that of the unreplaced rock by (III.2). Accordingly, the component diffuses into the front from the rock, with the resultant increase in porosity. However, the effect of this porosity change on the replacement process can be negligible (Fig. 3).

Although the porosity decreases throughout the replaced zone (Figs. 3 and 4), it cannot become zero because the degree of supersaturation becomes high as it becomes small. The concentration of Ca (and also Si which can be obtained by (III.2) and Fig. 1b) is almost constant (i.e., the *true* steady state, ($\partial c_i / \partial x$) = 0) near the front, whereas it changes sharply at the place of minimum porosity. The reason is that the pore solution of each side of the place tends to homogenize because of extremely small diffusivity at the place. The decrease in porosity in the zone thus causes the decrease in diffusivity at the front, and hence hinders the front progress.

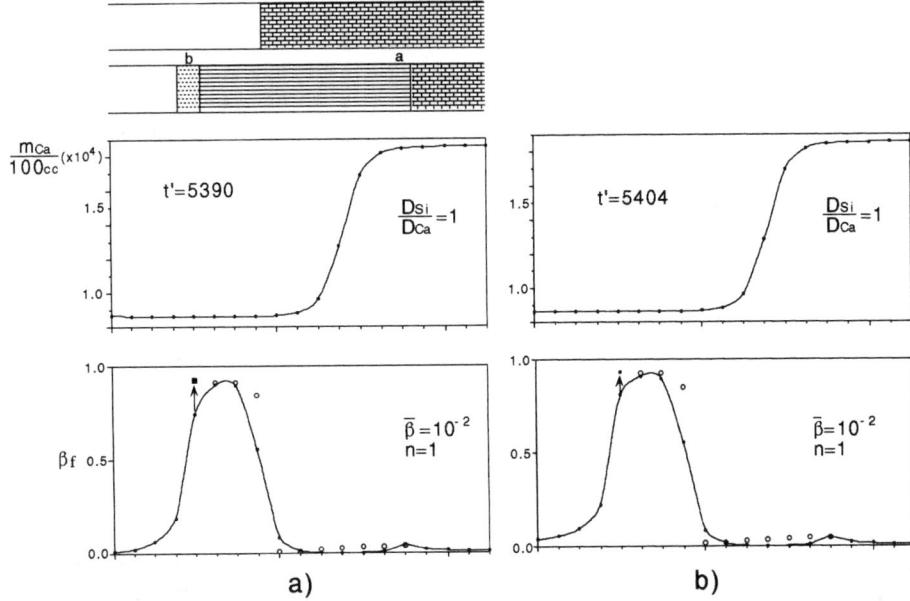

Fig. 3. Profiles of concentration (c_{Ca}) and porosity (β) for the case of $\bar{D}_{Si} / \bar{D}_{Ca} = 1$. Composition of pore solution of the fourth slab before each front has been fixed to an initial composition (i.e., a moving boundary condition) in (a), and that of the eleventh slab has been fixed in (b). Computed time is represented in dimensionless unit ($t' = 2(D_{Si}t)^{1/2}/x'$, where x' represents characteristic distance). A front slab b is not replaced completely. Each open circle represents porosity of a slab at when it was just replaced (i.e., that *was* front porosity). Closed square shows front porosity estimated by (IV.1).

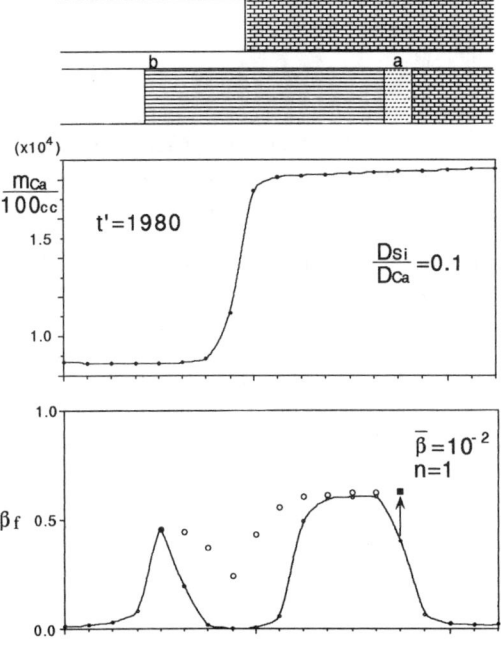

Fig. 4. Profiles of concentration and porosity for the case of $\bar{D}_{Si} / \bar{D}_{Ca} = 0.1$. Composition of the fourth slab before each front has been fixed. A front slab a is not replaced completely.

SUMMARY AND CONCLUSIONS

The formation process of wollastonite skarn at a boundary between limestone and quartzite, a typical bimetasomatism, has been examined, considering the effect of varying porosity on the process itself. Assumptions made are constant volume and local equilibrium, essential in analyzing metasomatism [1]. It has been concluded that the "salting out" effect causes a decrease in porosity in the replaced zone, with the resultant decrease in diffusivity at the replacement front. This means that the zone cannot develop so fast as the case for so-called steady state diffusion ($\partial c_i / \partial x$ is constant but not equal to zero through the zone) with constant porosity and diffusion coefficient.

This may also be true for metasomatic replacement of any reaction mineral. The reason for this statement lies in the fact that so far as local equilibrium is achieved, the mineral should be precipitated by the "salting out" effect. The result may also be applicable to "wall-rock alteration" even when a fissure solution is somewhat undersaturated with a reaction mineral [2]. Therefore, the porosity change, acting as "braking mechanism", is an essential feature of metasomatism by reaction minerals. In other words, the metasomatic replacement is a "self-controlled" process [2].

Acknowledgments
The author is grateful for the guidance given him by Dr. T. Nishiyama of Kyushu University. Dr. M. Watanabe of Hiroshima University kindly read the manuscript and his constructive criticism is gratefully acknowledged. Mr. T. Okudaira, also of Hiroshima University helped the author make computerize. Dr. S. Nakashima of Tokyo University is also thanked for helpful suggestions.

REFERENCES

1. D. S. Korzhinskii. *Theory of metasomatic zoning* (translated by Fean Agrell). Oxford Clarendon Press. 162p (1970).
2. K. Hoshino. Metasomatism as a self-controlled process, *Jour. Sci. Hiroshima Univ. Ser.c*, 9, 419-434 (1992).
3. A. W. Hofmann. Chromatographic theory of infiltration metasomatism and its application to feldspars, *Am. Jour. Sci.*, 272, 69-90 (1972).
4. R. C. Fletcher and A. W. Hofmann. Simple models of diffusion and combined diffusion-infiltration metasomatism. In: *Geochemical transport and kinetics* . Hofmann, A. W., Giletti B. J., Yorder, Jr., H. S. and Yund, R. A. (Eds). pp. 243-286 Carnegie Inst. Washington, (1973).
5. J. D. Franz and H. K. Mao. Bimetasomatism resulting from intergranular diffusion: I. A theoretical model for monomineralic reaction zone sequences, *Am. Jour. Sci.*, 276, 817-840 (1976).
6. _____ and _____. Bimetasomatism resulting from intergranular diffusion : II . Prediction of multimineralic zone sequences, *Am. Jour. Sci.*, v. 279, 302-323 (1979).
7. G. W. Fisher and A. C. Lasaga. Irreversible thermodynamics in petrology . In : *Kinetics of geochemical processes*. pp. 171 - 209 Lasaga, A. C. and Kirkpatrick, R. J. (Eds). Mineralogical Society of America (1983).
8. K. Fujimoto. Factors to control the width of a partially altered zone, Mining Geol., 37, 45-54 (1987). (in Japanese with English abstract)
9. A. Lerman. Maintenance of steady state in oceanic sediments, *Am. Jour. Sci.*, 275, 609 - 635 (1975).
10. K. Skagius and I. Neretnieks. Porosities and diffusivities of some nonsorbing species in crystalline rocks, *Water resources research*, 22, 389-398 (1986).

11. S. Nakashima, H. Kita, T. Suzuki and K. Nishiyama. Diffusion of elements in rock pore water and the pore distribution, *Mining Geol.*, **39**, 65-66 (1989). (Abstract in Japanese)
12. R. A. Berner. Diagenetic model of dissolved species in the interstitial waters of compacting sediments, *Am. Jour. Sci.*, **275**, 88-96 (1975).
13. D. Norton. Metasomatism and permeability, *Am. Jour. Sci.*, **288**, 604 - 618 (1988).
14. P. Ortoleva, E. Merino, C. Moore and J. Chadam. Geochemical self-organization I : Reaction-transport feedbacks and modeling approach, *Am. Jour. Sci.*, **287**, 979-1007 (1987).
15. _____, _____ and _____. Geochemical self-organization II : The reactive-infiltration instability, *Am. Jour. Sci.*, **287**, 1008-1040 (1987).
16. H. Kita, T. Iwai and S. Nakashima. Diffusion coefficient measurement of an ion in pore water of granite nd tuff, *Jour. Japan Soc. Engin. Geol.* (1989). (in Japanese)
17. K. Hoshino, M. Watanabe and A. Soeda. Zonal arrangement of the vein skarns at the Akemidani No. 5 ore body, Fujigatani mine, Yamaguchi prefecture, southwestern Japan, *Mining Geol.*, **32**, 443 - 456 (1982).
18. J. V. Walther and H. C. Helgeson. Description and interpretation of metasomatic phase relations at high pressures and temperatures: I. Equilibrium activities of ionic species in nonideal mixtures of CO_2 and H_2O, *Am. Jour. Sci.*, **280**, 575-606 (1980).

Ostwald ripening and crystal size distribution (CSD) of garnet in various metamorphic rocks

Kazuhiro Miyazaki

Geological Survey of Japan, Higashi 1-1-3, Tsukuba, 305 Japan

Abstract This paper discusses the importance of Ostwald ripening during metamorphism. Crystal size distributions (CSDs) of garnets from high pressure regional metamorphic rocks and low pressure contact metamorphic rocks are examined, and compared with the theoretical distribution (LSW distribution) for Ostwald ripening. CSDs at high grade part of the high pressure regional metamorphic rocks are consistent with the LSW distribution, while those at low grade part are slightly different from the LSW distribution. Mean diameter and the number of grains systematically increases and decreases with increasing grade, respectively. Reported thermal history of the high pressure regional metamorphic rocks shows that duration of peak metamorphism at high grade part is longer than that at low garde part. Then it seems that Ostwald ripening progressed in these high pressure regional metamorphic rocks, and progression of Ostwald ripening at high grade part was more complete than that at low grade part. Discrepancy between patterns of CSDs exist also in the low pressure contact metamorphic rocks. CSDs at the place adjacent to reentrant of intrusive pluton are consistent with the LSW distribution, while those at the place adjacent to apophysis of the intrusive pluton are quite wider than the LSW distribution. Reported model calculation of thermal history in contact metamorphic aureole around rough shaped pluton suggests that metamorphic rocks at the place adjacent to apophysis of pluton cool more rapidly than those at the place adjacent to reentrant of pluton. Then it seems that Ostwald ripening progresses at the place adjacent to reentrant of the intrusive rock in low pressure contact metamorphic rocks. Measured CSDs in this paper and analyses of previously published CSDs suggest that Ostwald ripening occurs in various metamorphic rocks from high to low pressure metamorphic belts and from regional to contact metamorphic terranes.

Introduction

There are two explanations for producing CSDs of garnets in metamorphic rocks, such as nucleation and growth kinetics (Jones and Galway, 1966; Kretz, 1973; Carlson, 1989) and Ostwald ripening (Cashman and Ferry, 1988; Miyazaki, 1991). Ostwald ripening takes place following initial nucleation and growth. Miyazaki (1991) pointed out that heating rate decides which mechanism affects observed CSDs; nucleation and growth kinetics is dominant when heating rate is large, and Ostwald ripening is dominant when heating rate is small. He concluded that CSDs of garnets from high pressure and low temperature (high P/T) metamorphic rocks are consistent with the latter case.

In this paper, CSDs of garnets from high *P/T* regional metamorphic rocks (Sanbagawa metamorphic rocks) and from low *P/T* contact metamorphic rocks (Wagakunisan metamorphic rocks) were examined. Thermal history of the Sanbagawa metamorphic rocks was reported by Takasu and Dallmeyer (1990). Then these metamorphic rocks are suited for researching relation between thermal history and progress of Ostwald ripening. We will also discuss relations between thermal history and progress of Ostwald ripening in the Wagakunisan metamorphic rocks. 3D numerical simulation of thermal history of contact aureole around rough shaped pluton was presented by Kerrick et al. (1991). Their result is used for evaluating thermal history of the Wagakunisan metamorphic rocks. Finally, we will analyze previously published CSDs to discuss the importance of Ostwald ripening in various metamorphic rocks.

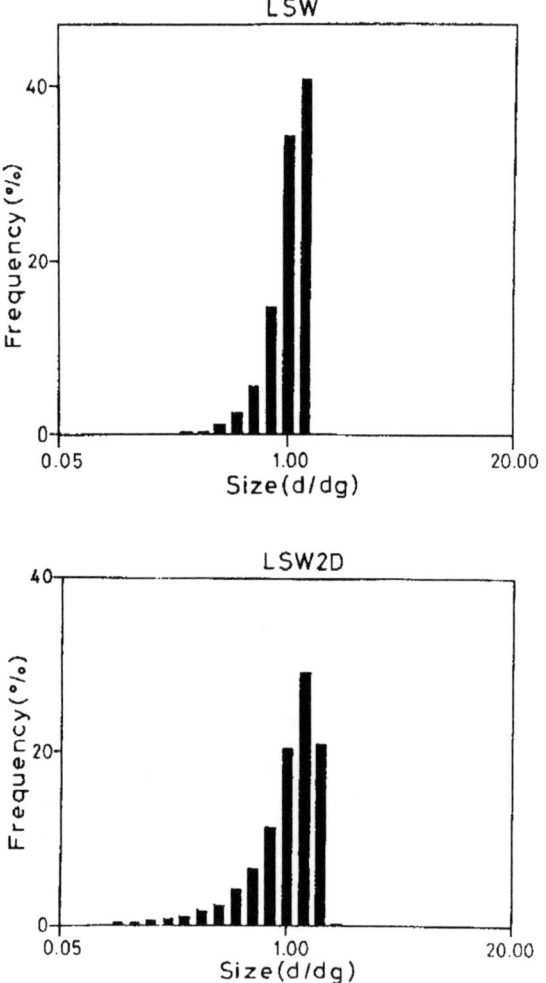

Figugre 1. Calculated frequency histogram of LSW distribution(LSW) and the apparent distribution in thin section of LSW distribution(LSW2D).

Steady-state size distribution during progress of Ostwald ripening

When mean concentration of component of precipitate phase is high, most crystals grow. As the mean concentration decreases with increasing mean radius of precipitate phase in a closed system, larger crystals grow while smaller crystals dissolve. This phenomenon is named Ostwald ripening. The total number of moles of the precipitating phase in the system is kept constant and the total interfacial energy decreases as Ostwald ripening continues. Lifshitz and Slyozov (1961) and Wagner (1961) theoretically formulated the time-evolution of the distribution function of crystal size in closed system in which Ostwald ripening operate.

Define $F(R,t)$ as the crystal size distribution function, where R is the radius of garnet. $F(R,t)dR$ represents the number of crystals per unit volume at time t in a size class R to $R+dR$.

This function satisfies the following continuity equation:

$$\partial F(R,t)/\partial t = -\partial(F(R,t)\ (dR/dt))/\partial R, \qquad (1)$$

where dR/dt is diffusion limited radial growth rate of garnet. This growth rate is derived from Gibbs-Thomson relation and steady-state diffusion of garnet component as follows (Lifshitz and Slyozov, 1961):

$$dR/dt = \{(v\ D\ C_{eq\infty}\ A)/R\}\ (1/R_c - 1/R), \qquad (2)$$

here v is the molar volume of garnet, D is diffusivity, $C_{eq\infty}$ is the equilibrium concentration of imaginary garnet with infinite radius, A is a constant related to the density of interfacial energy (see Miyazaki, 1991), and R_c denotes the critical radius. This critical radius is defined as follows (Lifshitz and Slyozov, 1961):

$$R_c = (C_{eq\infty}\ A)/(C_m - C_{eq\infty}), \qquad (3)$$

where C_m is mean concentration of garnet component in the system. Crystals with radii larger than R_c grow while crystals with radii smaller than R_c dissolve. We assume the closed system. Then the system must satisfy the following mass balance equation:

$$C_o = C_m + (4\pi)/(3\ v) \int F(R,t)\ R^3\ dR, \qquad (4)$$

where C_o denotes the total moles of the garnet component in a unit volume of the system. Lifshitz and Slyozov (1961) obtained analytical solution of size distribution function $F(R,t)$ which satisfies continuity equation (eq.(1)) and mass balance equation (eq.(2)) as follows:

$$F(R,t) = 3\ n(t)\ P_{lsw}(R/R_m(t))\ (R/R_m(t))^2\ R_m(t)^{-1}, \qquad (5)$$

where $n(t)$ is the number of garnet per unit volume, $P_{lsw}(R/R_m(t))$ is the LSW scaling function (see Lifshitz and Slyozov, 1961 or Miyazaki, 1991) and $R_m(t)$ is the mean radius of garnet. The number of garnet per unit volume ($n(t)$) is given by

$$n(t) = 0.22\ Q_o\ R_m(t)^{-3}, \qquad (6)$$

where Q_o is total supersaturation given by

$$Q_o = C_m - C_{eq\infty} + q/v, \qquad (7)$$

where q denotes the sum of the volume of precipitated garnet crystals per unit volume. The mean radius ($R_m(t)$) is given by

$$R_m(t)^3 = (4/9)\ v\ D\ C_{eq\infty}\ A\ t. \qquad (8)$$

Figure 1 shows a histogram of $P_{lsw}(R/R_m(t))$. We will compare this LSW distribution with the measured CSDs. The CSDs are measured on thin section. Then we must convert the LSW distribution in real space to an apparent one on thin section. The apparent LSW was obtained by Miyazaki (1991). Figure 1 also shows a histogram of the apparent LSW distribution.

For derivation of equations (1)-(8), we assume that garnet precipitate form a uniform medium. When garnet precipitate from complex medium consisting of intergranural medium (IGM) and minerals, concentration C_i and diffusivity D must be converted. Miyazaki (1991) converted C_i and D for the IGM-mineral system (he committed some errors when deriving equations of the growth rate and the mean radius; factor φ was erroneously dropped from the equations of the growth rate and the mean radius in Miyazaki (1991)). Conversions of C_i and D are as follows:

$$C_i = \phi C'_i,$$

$$D = \tau \phi D_{igm},$$

where C'_i is concentration of garnet component in IGM, φ is the volume ratio of IGM to IGM + minerals, τ is tortuosity, and D_{igm} is instrsic diffusivity of garnet component in IGM.

Table 1. Petrological description of samples discussed in the text.

Sample	Location	Mineralogy	Mineral zone	P (kbar)	T (°C)	Reference
AS407D	Asemi.R.	chl,bt	Olg-bt	10-12	610	(1)(3)(4)
AS404A	Asemi.R.	chl,bt	Olg-bt	10-12	610	(1)(3)(4)
AS408D	Asemi.R.	chl,bt	Olg-bt	10-12	610	(1)(3)(4)
AS303F	Asemi.R.	chl	Gar	10-12	400-460	(1)(3)(4)
AS303A	Asemi.R.	chl	Gar	10-12	400-460	(1)(3)(4)
Ma40	Wagakuni	Kf,bt	Crd	1.5-3.0	>400	(2)
Ma053	Wagakuni	Kf,bt	Crd	1.5-3.0	>400	(2)

All samples contain qz, pl, mu, gar and gph. Reference: (1) Banno and Sakai (1989); (2) Miyazki et. al. (1992); (3) Isozaki and Itaya (1990); (4) Takasu and Dallmeyer (1990).

Methods of investigation

Measured CSDs were compared with the theoretical CSD to investigate the effect of Ostwald ripening. CSDs were measured for the following metamorphic rock samples: (1) pelitic schists from the oligoclase-biotite and garnet zones of the Sanbagawa metamorphic rocks, at the Asemi River, Ehime Prefecture, Japan; and (2) pelitic and psammitic hornfels from the cordierite zone of the Wagakunisan metamorphic rocks, at Ibaraki Prefecture, Japan. Detailed sample descriptions are presented in Table 1.

The central part of Shikoku Island, Japan, exposes the Sanbagawa metamorphic rocks of high pressure intermediate facies group in the sense of Miyashiro (1961). The metamorphic event has been dated as 116 Ma (Isozaki and Itaya, 1990; Minamishin et. al., 1979). These metamorphic rocks are divided into the following four zones based on mineral assemblages of pelitic schists such as, chlorite, garnet, albite-biotite and oligoclase-biotite zones (Banno and Sakai, 1989). Metamorphic grade increases from the chlorite zone to the oligoclase-biotite zone. The temperature-pressure conditions during the metamorphism were of 400° -460 °C, 10-12 kbar for the garnet zone and similar pressure and 610° C for oligoclase-biotite zone (Isozaki and Itaya, 1990; Takasu and Dallmeyer, 1990; Banno and Sakai, 1989). The thermal structure and P-T-t path of the Sanbagawa metamorphic rocks in central Shikoku are well researched by Isozaki and Itaya (1990), Takasu and Dallmeyer (1990) and Banno and Sakai (1989). Then these metamorphic rocks are suited for research on the effect of duration and metamorphic temperature for Ostwald ripening. The Sanbagawa metamorphic rocks in central Shikoku consist mainly of pelitic schists and basic schists with small amounts of psammitic schists and quartz schists. Pelitic schists contain euhedral garnets (0.037-0.306 mm in diameter) set in a matrix consisting of quartz, plagioclase (albite or oligoclase), muscovite, chlorite, graphite

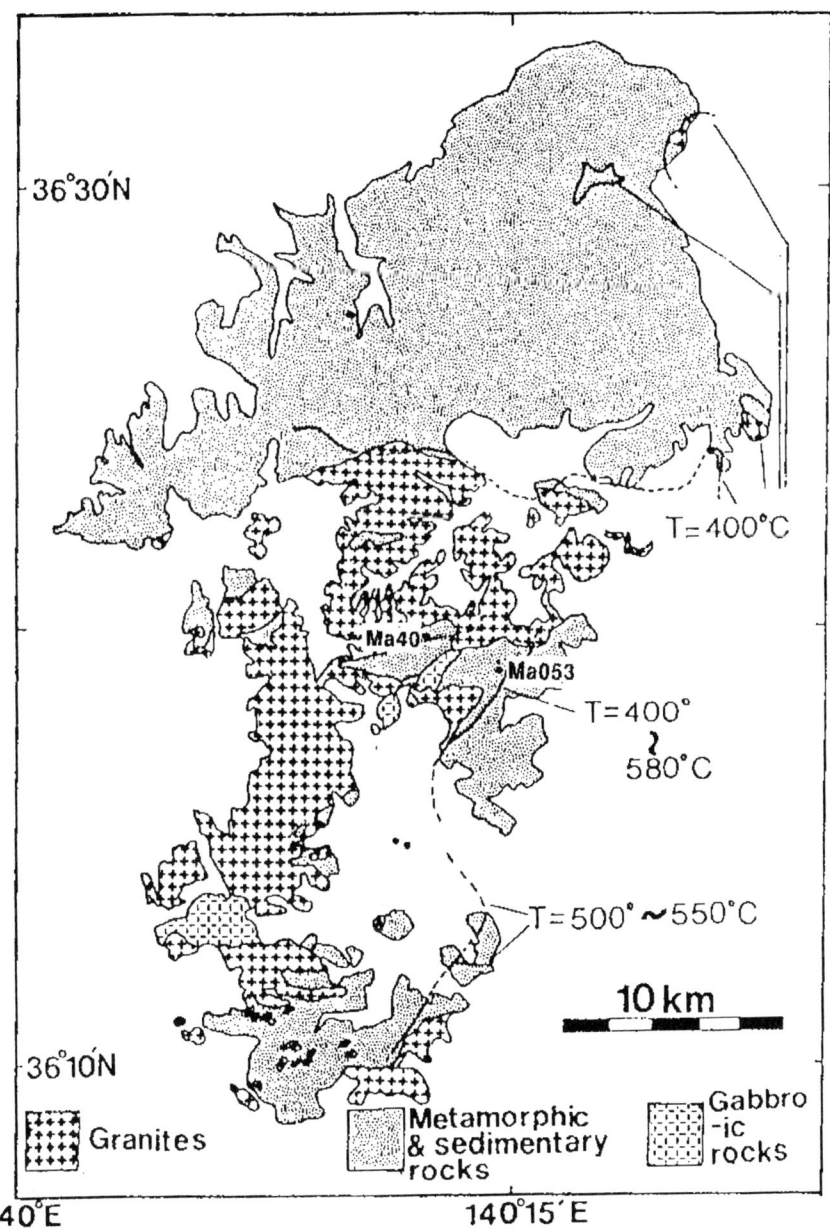

Figure 2. Location of samples used for CSD analyses in the Wagakunisan metamorphic rocks. (Fig.7. of Miyazaki et al., 1992).

and/or biotite. (001) planes of phyllosilicates are parallel to the schistosity. Samples which have homogeneous distribution of matrix minerals were selected. In such samples, the spatial distribution of garnets are also homogeneous. Lower-grade equivalents to the garnet and oligoclase-biotite zones indicate that the origin of garnet would be the continuous reaction of chlorite+quartz to garnet+water. This continuous reaction produced compositional zoning of garnet.

The eastern end of Honshu Island, Japan, exposes the Wagakunisan metamorphic rocks of low pressure facies group in the sense of Miyashiro (1961). The Wagakunisan metamorphic rocks are distributed around the Tsukuba Mountains, Ibaraki Prefecture. These metamorphic rocks are recognized as the eastern end of the Ryoke metamorphic rocks and as contact metamorphic rocks around the Tsukuba granites (Shiba,1979, 1982; Miyazaki et. al., 1992). Contact metamorphic event has been dated as 60 Ma (Kawano and Ueda, 1966; Shibata, 1968; Arakawa and Takahashi, 1988). The temperature-pressure conditions during the contact metamorphism were above 400° C, and 1.5-3.0 kbar for the cordierite zone (Miyazaki et. al., 1992). The Wagakunisan metamorphic rocks consist mainly pelitic and psammitic hornfels with small amounts of marble and meta-chert. Pelitic and psammitic hornfels contain euhedral garnet (0.055-0.066 mm in diameter) set in the matrix of quartz, plagioclase, K-feldspar, biotite, and muscovite. The matrix and garnet appear to be modally homogeneous on a millimetre scale. Sample localities are shown in Fig.2. Samples Ma40 and Ma053 were collected at a point near reentrant and at a point near apophysis of the granitic pluton, respectively. 3D numerical simulation of thermal history of contact metamorphic rocks around rough shaped pluton was presented by Kerrick et al. (1991). This result is used for evaluating thermal history at different part of contact aureole on the Wagakunisan metamorphic rocks. Lower-grade equivalents to the garnet-bearing samples indicate that origin for garnet would be the continuous reaction of biotite+muscovite+quartz to garnet+K-feldspar+water.

Crystal sizes of garnets were measured on thin sections cut perpendicular both to the schistosity and to the lineation of metamorphic rocks. The diameters of garnets inside an inscribed circle were measured under the microscope. The mean diameter (d_g), exponential of standard deviation (SD_g), skewness (Sk), and histograms of the scaled distributions were obtained from the measured data. The mean diameter d_g is defined as an exponential of the mean of log d (diameter), and is equivalent to the geometrical mean:

$$d_g = \exp[(1/N) \sum \log[d_i]] \tag{9}$$

where N denotes the measured number. The exponential of standard deviation SD_g is that of log d, and represents the width of distribution scaled by d_g as follows :

$$\sigma^2 = (1/N) \sum \{\log [d_i/d_g]\}^2 \tag{10}$$

and

$$SD_g = \exp[\sigma]. \tag{11}$$

Skewness (Sk) is

$$Sk = -(1/N) \sum \{(\log[d_i/d_g])^3/\sigma^3\} \tag{12}$$

The diameter scaled by mean diameter was used for constructing frequency diagrams. Each class of histograms was chosen such that the interval between each class was equal on a log scale. In regime of Ostwald ripening, change of pattern of CSD of crystals below mean size is more rapid and larger than that of crystals above mean size. We can get detail structure of CSD of crystals below mean size by using log scale plot. This is the reason why we use the log scale plot. Histograms of the apparent LSW distribution are shown in Fig.1. This apparent distribution will be compared with the measured distributions.

Results of CSDs

Table 2. Mean diameter(d_g), SD_g, Sk, measured number(N), number per unit area (D.G.) and f of garnets in the Sanbagawa metamorphic rocks. The values of d_g, SD_g and Sk are calculated from equations(9), (10), (11) and (12) in the text.

Sample	d_g(mm)	SD_g	Sk	N	D.G.(cm^{-2})	f (%)
AS407D	0.306	1.67	1.16	202	47	5.0
AS404A	0.220	1.66	0.76	204	49	2.7
AS408D	0.191	1.72	0.70	203	94	4.3
AS303F	0.037	1.52	0.03	203	480	0.7
AS303A	0.037	1.63	0.02	203	286	0.5

Table 3. Mean diameter(d_g), SD_g, Sk, measured number(N), number per unit area (D.G.) and f of garnets in the Wagakunisan metamorphic rocks. The values of d_g, SD_g and Sk are calculated from equations(9), (10), (11) and (12) in the text.

Sample	d_g(mm)	SD_g	Sk	N	D.G.(cm^{-2})	f (%)
Ma40	0.055	1.58	0.54	202	284	0.9
Ma053	0.066	2.04	0.49	201	118	0.9

Sanbagawa metamorphic rocks
Table 2 summarizes the mean diameter (d_g), exponential of standard deviation (SD_g), skewness (Sk), measured number (N), number per unit area (D.G.) and areal fraction (f) of garnet in the samples from the Sanbagawa metamorphic rocks. The mean diameter d_g increases with metamorphic temperature as follows; At lower garnet zone d_g is 0.037 mm and at oligoclase-biotite zone d_g ranges from 0.191 mm to 0.306 mm. The value of SD_g (1.52 to 1.72) has the same as that of the narrow-type of the Nagasaki metamorphic rocks. The skewness (Sk) values at garnet zone ranges from 0.02 to 0.03 and that at oligoclase-biotite zone ranges from 0.70 to 1.16. D.G. ranges from 286 cm^{-2} to 480 cm^{-2} at garnet zone and from 94 cm^{-2} to 47 cm^{-2} at oligoclase-biotite zone. The f value ranges from 0.5 % to 0.7 % at garnet zone and from 2.7 % to 5.0 % at oligoclase-biotite zone.

Theoretical distribution and measured CSDs of garnets from the Sanbagawa metamorphic rocks are shown in Fig.3. Measured CSDs at oligoclase-biotite zone have long tails in the region less than the mean diameter while the frequencies above the mean diameter decrease rapidly. These features are similar to those of the apparent LSW distribution. The mean diameter d_g increases from garnet zone to oligoclase-biotite zone and D.G. decreases from garnet zone to oligoclase-biotite zone. These systematic changes of d_g, D.G., and CSD pattern are well explained by progress of Ostwald ripening. Measured CSDs at garnet zone are slightly different from the apparent LSW distribution. Although widths of their distribution pattern are similar to that of the apparent LSW, they have more symmetrical pattern in comparison with the apparent LSW distribution. These discrepancy of CSD patterns at garnet zone and at oligoclase-biotite zone will be discussed in later section.

Wagakunisan metamorphic rocks
Table 3 summarizes d_g, SD_g, Sk, D.G., N and f. Measured samples have similar value of d_g (0.055 to 0.066 mm) and f (0.9 %). However SD_g is quite different from each other (1.58 and 2.04). CSD of Ma40 is similar to the apparent LSW distribution, although that of Ma053 is quite wider than the LSW distribution (Fig.4). CSD of Ma053 has very long tail in the region less than the mean diameter. Samples of Ma40 and Ma053 were collected from different sites in

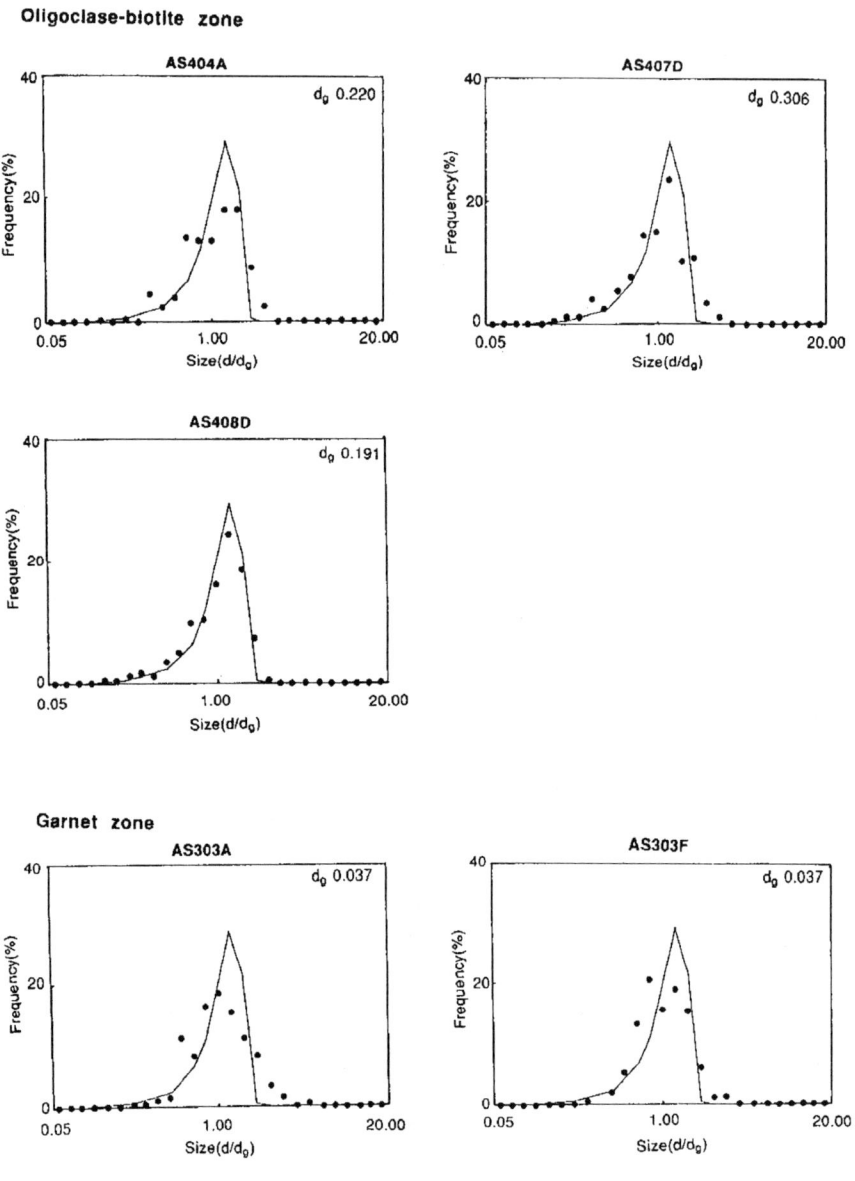

Figure 3. CSDs of garnets in the Sanbagawa metamorphic rocks. Solid lines are two-dimensional theoretical CSDs predicted by the LSW theory.

contact aureole: Ma40 from a point near reentrant and Ma053 from a point near apophysis of the granitic pluton, respectively. Distances of these two points from the nearest pluton are almost the same. However morphologies of the plutons adjacent to the sampling points are different. The relation between progress of Ostwald ripening and thermal history at these different sites in contact aureole will be discussed in later section.

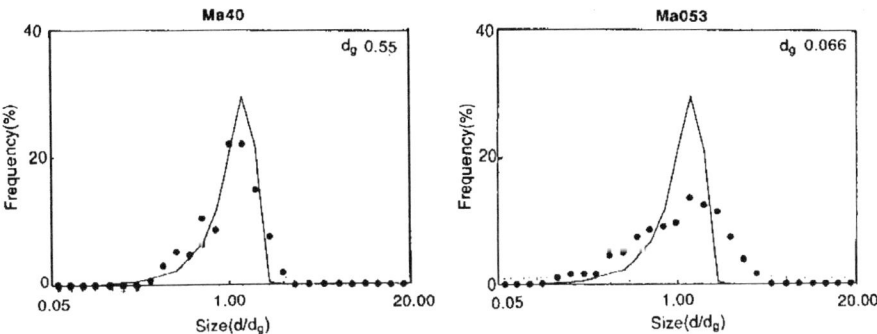

Figure 4. CSDs of garnets in the Wagakunisan metamorphic rocks. Solid lines are same as in Fig.3.

Discussion

Thermal history and Ostwald ripening
In the Sanbagawa metamorphic rocks, the CSDs of garnets at the oligoclase-biotite (higher grade part) zone are very similar to the theoretical CSD (the LSW distribution). Mean diameter and number of garnets per unit volume systematically increases and decreases with increasing grade, respectively. These observations suggest that Ostwald ripening progressed in these high P/T metamorphic rocks. On the other hand, peaks of CSDs at the garnet zone (lower grade part) are slightly shifted to smaller size in comparison with that of the LSW distribution. We interpret that these discrepancies form the LSW distribution are due to incompleteness of progress of Ostwald ripening. Takasu and Dallmeyer (1990) illustrated the thermal history of the Sanbagawa metamorphic rocks. They show that duration of peak metamorphism increases with increasing grade (see Fig.3 in Takasu and Dallmeyer, 1990). Sakai et al. (1985) suggested that duration of nucleation-growth in the Sanbagawa metamorphic rocks is very short in comparison with the whole crystallization, and that nucleation-growth stage is restricted in early stage of the whole crystallization. Ostwald ripening progress following the initial nucleation-growth (Cashman and Ferry, 1988; Miyazaki, 1991). Then duration of Ostwald ripening at high garde part is longer than that at low garde part (see Fig.3 in Takasu and Dallmeyer, 1990). In addition to this, velocity of progress of Ostwald ripening increases with increasing garde, because of diffusion limited growth rate of Ostwald ripening. Taking account of the duration and the velocity of Ostwald ripening, progress of Ostwald ripening at high garde part is more complete than that at low garde part.

In the Wagakunisan metamorphic rocks, there are two different CSDs. One (Ma40) is consistent with the LSW distribution and the other (Ma053) is quite wider than that. Sampling point of Ma40 is located at a point near reentrant of intrusive pluton and that of Ma053 is located at a point near apophysis of the pluton. Kerrick et al. (1991) presented 3D numerical simulation of thermal history of contact metamorphic rocks around rough shaped pluton (see Figs.40 and 41b in Kerrick et al., 1991). They show that the metamorphic rocks adjacent to apophysis of the pluton have more rapid cooling history than the metamorphic rocks adjacent to reentrant part of the pluton (see Figs.40 and 41b in Kerrick et al., 1991). Cooling rate of pluton and surrounding contact metamorphic rocks depends on size of the pluton: the cooling rate of small pluton is larger than that of large pluton. Volume ratio of intrusive pluton to surrounding contact metamorphic rocks within reentrant part of the pluton is larger than that within apophysis part of the pluton. From analogy of the size-dependent cooling rate, metamorphic rocks adjacent to apophysis of pluton cools more rapidly than those adjacent to reentrant of the pluton. These interpretations and the reported numerical simulation suggest that CSD of Ma053 caused by initial nucleation-growth kinetics has not changed completely to the LSW distribution due to shortage of duration of Ostwald ripening.

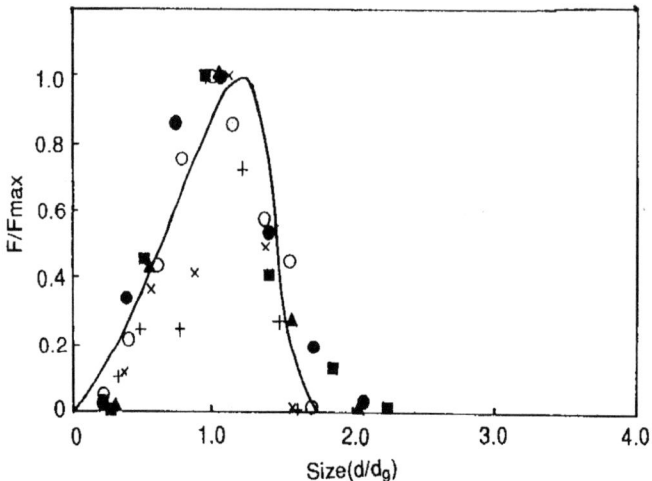

Figure 5. CSDs of garnets in low pressure contact metamorphic rocks in Aradra aureole. Solid line is two dimensional theoritical distribution predicted by LSW theory. CSD data from Jones and Galwey (1964).

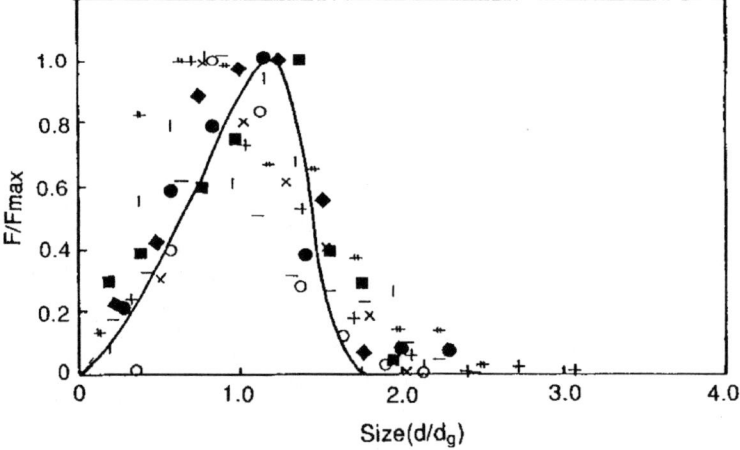

Figure 6. CSDs of garnets in medium pressure regional metamorphic rocks in Mallaing, Scotland. Solid line is the same as in Fig.5. CSD data from Galwey and Jones (1966).

Analysis of previously published CSD
The CSDs in the Sanbagawa metamorphic rocks (oligoclase-biotite zone) and the Wagakunisan metamorphic rocks (site adjacent to to reentrant of pluton), suggest that Ostwald ripening progresses not only in high P/T regional metamorphic rocks but also in low P/T contact metamorphic rocks. One exception described by Cashman and Ferry (1988) is the contact metamorphic rocks from the Isle of Skye, in which the rapid cooling prevent the progress of Ostwald ripening and preserve the initial CSD caused by nucleation and growth. Several authors reported many CSDs from various metamorphic regions (Cashman and Ferry, 1988;

Galwey and Jones, 1966; Jones and Galwey, 1969). Cashman and Ferry (1988) shows CSDs of garnets in Waterville Formation which is a Buchan type metamorphic rocks. These metamorphic rocks suffered from low pressure intermediate facies group of metamorphism in the sense of Miyashiro (1961); the metamorphic pressure is 3.5kbar and the metamorphic temperature ranges from 400° to 550° C (Cashman and Ferry, 1988). Following observations suggest that Ostwald ripening has occurred in their samples: 1) grain size distribution histograms for garnets from the chlorite, garnet, and sillimanite zones of the Waterville Formation, normalized to maximum frequency and average size are similar to steady state distributions predicted by LSW theory (see Fig.10 in Cashman and Ferry, 1988), 2) the total number of crystals decreases with increasing metamorphic grade and 3) the average crystal size increases as metamorphic garde increases (see Fig.6 in Cashman and Ferry, 1988). They concluded that metamorphic garnets have annealed (Ostwald ripening) following volume diffusion kinetics. The change in number and mean size of crystals with increasing grade is qualitatively the same as would be expected from progressive degrees of annealing (Ostwald ripening), because annealing time in high grade zone is longer than in low grade zone. Thus the effects of annealing (Ostwald ripening) should be greater in higher grade than in lower grade samples. These Cashman and Ferry (1988)s interpretations are consistent with our discussions mentioned above. Figures 5 and 6 show CSDs which are normalized to maximum frequency and average size using crystal size distribution reported by Jones and Galwey (1964) and Galwey and Jones (1966), respectively. Jones and Galwey (1964) measured the size of garnets from Aradra low pressure contact metamorphic rocks of which metamorphic pressure is 2.0-5.0 kbar (Pattison and Tracy, 1991). Galwey and Jones (1966) measured the size of garnet from Barrovian type medium pressure regional metamorphic rocks near Malling, Inverness in England of which metamorphic pressure is 5.5-7.1 kbar (Pattison and Tracy, 1991). These CSDs are similar to the LSW distribution (Fig.5 and 6). These previously published CSDs and measured CSDs in this paper suggest that Ostwald ripening occurs in various type of metamorphic rocks in natural system from high P/T to low P/T metamorphic rocks and from regional to contact metamorphicterranes.

Conclusions

In high P/T Sanbagawa metamorphic rocks, CSDs, mean size, and D.G. change systematicaly with increasing grade. CSDs at high grade part in these metamorphic rocks are consistent with the LSW distribution. In the low P/T Wagakunisan (contact) metamorphic rocks, CSD in metamorphic rocks adiacent to reentrant of the pluton is also consistent with the LSW distribution. These obsevations suggest that Ostwald ripening progresses not only in high P/T regional metamorphic rocks but also in low P/T conact metamorphic rocks. This interpretation is also consistent with analysis of previously published CSDs.

Acknowledgements
I would like to express my sincere thanks to Dr. T. Nishiyama of Kyushu University for his encouragement, discussion and critical reading of the manuscript. Dr. T. Nakajima of Geological Survey of Japan is thanked for his help in preparing the samples.

References

Aakawa and Y. Takahashi. Rb-Sr ages of granitic rocks fromthe Tsukuba district, Japan, *Assoc. Petrol. Econ. Geol.* **83**, 232-240 (1988).
S. Banno, C. Sakai and T. Higashino. Pressure-temperature trajectory of the Sanbagawa metamorphism dedued from garnet zoning, *Lithos.* **19**, 51-63 (1986).
S.Banno and C. Sakai. Geology and metamorphic evolution of the Sanbagawa metamorphic belt, Japan, *Geol. Soc. Sp. Publ.* **43**, 519-523 (1989)
W.D. Carlson. The significance of intergranular diffusion to the mechanisms and kinetics of porphyroblast *Contrib. Mineral. Petrol.* **103**, 1-24 (1989).
K.V. Cashman and J.M. Ferry. Crstal size distribution (CSD) in rocks and the kinetics and dynamics of

crystallozation III. Metamorphic crystallization, *Contrib. Mineral. Petrol.* **99**, 401-415 (1988).
A.K. Galwey and K.A. Jones. Cryatal size distribution of garnets in some analyzed metamorphic rocks from Mallaig, Inverness, Scotland. *Geol. Mag.* **103**, 143-152 (1966).
Y. Isozaki and T. Itaya. Chronology of Sanbagawa metamorphism, *J. metamorphic Geol.* **8**, 401-411 (1990).
K.A. Jones and A.K. Galwey. A study of possible factors concernig garnet formation in rocks from Ardara, Co.Donegal, Ireland, *Geol. Mag.* **101**, 76-93 (1964)
K.A. Jones and A.K. Galwey. Size distribution, composition and growth kinetics of garnet crystals of some metamorhic rocks from west of Ireland, Quart. *J. Geol. Soc. London.* **122**, 29-44 (1966).
Y. Kawano and Y. Ueda. K-Ar dating on the igneous rocks in Japan (IV) Granitic rocks of backbone range in northeastern Janpan and its western district, *Sci. Rep. Tohoku Univ.* **3**, 351-367 (1966).
D.M. Kerrick, A.C.Lasaga and S.P. Raeburn. Kinetics of heterogeneous reactions. In: *Contact metamorphism.* D.M. Kerrick (Eds). Reviews in Mineralogy vol. **26**, pp. 587-671, Chap. 12 (1991)
R. Krezt. Kinetics of crystallization of garnet at two localities near Yellownife, *Can. Mineral.* **12**, 1-20 (1973)
I.M. Lifshitz and V.V. Slyozov. The kinetics of precipitation from supersaturated solid solutions, *J. Phys. Chem. Solid.* **19**, 35-50 (1961)
M. Minamishin, T. Yanagi and M. Yamaguchi. Rb-Sr whole rock age of the Sanbagawa metaorphic rocks in central Shikoku, Japan. In: *Isotope geoscience of Japanese Islands*, 68-71. (in Japanese) (1979)
A. Miyashiro. Evolution of metamorhic belts, *Jour. Petrol.* **2**, 277-311 (1961)
K. Miyazaki. Ostwald ripening of garnet in high P/T metamorphic rocks, *Contrib. Mineral. Petrol.* **108**, 118-128 (1991)
K. Miyazaki, M. Sasada and H. Hattori. Low P/T metamorphic rocks formed at different pressure levels around the Tsukuba Mountains, Japan, *J. Geol. Soc. Japan.* **98**, 713-722 (1992)
C. Sakai, S. Banno, M. Toriumi and T. Higashino. Growth history of garnet in pelitic schists of the Sanbagawa metamorphic terrain in central Shikoku, *Lithos.* **18**, 81-95 (1985)
M. Shiba. Stratigraphy and metamorphic zoning of the Tsukuba metamorphic rocks, Ibaraki Prefecture, *Assoc. Mineral. Petrol. Econ. Geol.* **74**, 339-349 (1979)
M. Shiba. Metamorpic conditions of the Tshukuba metamorphic rocks, *Assoc. Minetal. Petrol. Econ. Geol.* **77**, 345-355 (1982)
K. Shibata. K-Ar age determinations on granitic and metamorphic rocks in Japan, *Geol. Surv. Japan. Rep.* **277**. pp73 (1968)
A. Takasu and R.D. Dallmeyer. $^{40}Ar/^{39}Ar$ mineral age constraints for the tectonothermal evolution of the Sambagawa metamorphic belt, central Shikoku, Japan: a Cretaceous sccretionary prism, *Tectonophysics*, **185**, 111-139 (1990)
D.R.M. Pattison and R.J. Tracy. Phase equilibria and thermobarometry of metapelites. In: *Contact metamorphism*, D.M. Kerrick (Eds), Reviews in Mineralogy vol **26**, Chap. 4, pp.105-206 (1991)
C. Wagner. Theorie der alterung von niederschlagen durch umlosen, *Z. Elektrochemie.* **65**, 581-591

Marbles, calc-silicate rocks and skarns of the Moldanubian unit (Bohemian Massif, Czechoslovakia): Carbon and oxygen isotope and fluid inclusion constraints on formation conditions

K. ŽÁK and P. SZTACHO
Czech Geological Survey, Malostranské nám. 19, CS-118 21 Prague 1, CZECHOSLOVAKIA

Abstract Marbles, calc-silicate rocks and skarns of high-grade metamorphic sequences outcropping near the Central Moldanubian Pluton differ significantly in C- and O-isotopic composition of their carbonate phases. C- and O-isotope systematics of studied carbonates were controlled by several factors. The original $\delta^{13}C$ and partly also $\delta^{18}O$ values of sedimentary precursors were preserved only in the central part of large marble bodies. Processes responsible for a decrease in $\delta^{13}C$ and $\delta^{18}O$ values in calc-silicate rocks and skarns were decarbonation and interaction of carbonates with circulating metasomatic fluids. Very low $\delta^{13}C$ values of scheelite-bearing calc-silicate rocks and of all calcite samples from scheelite-bearing skarns indicate involvement of metasomatic fluids carying organic-matter derived carbonaceous components. Circulation of metasomatic fluid and scheelite deposition occured most probably at a late, low pressure stage of regional metamorphism and/or during the Central Moldanubian Pluton intrusion. The youngest development stage of skarns is characterized by invasion of a low $\delta^{18}O$ fluid.

Keywords: stable isotopes, decarbonation, skarns, scheelite deposits

INTRODUCTION

During regional panning prospects for heavy minerals a widespread occurrence of scheelite was found near the northern part of the Moldanubian Pluton (Moldanubian Unit, Bohemian Massif, Czechoslovakia). This abundant occurrence of scheelite is related to the frequent presence of calc-silicate rocks and skarns within the high-grade rocks of the Moldanubian Unit. Similar scheelite mineralizations were described from Austrian [1] and Bavarian [2] part of Moldanubicum.

The use of C- and O-isotopic composition of carbonate phases from calc-silicate gneisses (erlans, skarnoids) and skarns became routine in studies of metamorphic or metasomatic decarbonation reactions in carbonate-bearing rocks during last two decades (see [3] for review). The magnitude of ^{18}O and ^{13}C depletions which occurs during these reactions is directly linked and can be calculated if the reaction stoichiometry and P, T conditions are known. To decipher reaction processes and conditions, stable isotope data are usually combined with the P, T estimates derived from mineral phase equilibria [4-6]. The combination of stable isotope record with fluid inclusion data is less frequent [7].

In the study area marbles, calc-silicate rocks and lenticullar skarn bodies were studied. A combination of stable isotope and fluid inclusion data is applied to estimate the reaction processes of carbonate phases in these rocks and the formation of scheelite mineralization.

SUMMARY OF GEOLOGICAL AND PETROLOGIC RELATIONSHIPS

The Moldanubian Unit (the easternmost section of internal zone of European Variscan orogen, southern part of the Bohemian Massif) experienced a complicated geological and tectonic development and has a very complex polymetamorphic history. Most recently the development of Moldanubian zone in southern Bohemia is interpreted as representing a polyphase evolution of imbricated crustal and (locally) upper mantle segments. In this concept the traditionally defined Varied Group is overthrusted on the Monotonous Group.

In the study area itself (see Fig.1), the Varied Group is characterized by frequent intercalations of marbles, calc-silicate rocks, rare graphite-containing rocks, quartzites and amphibolites within the sequence of paragneisses.

In constrast, the Monotonous Group is formed dominantly by metapelitic and metapsammitic rocks, mostly silimanite - ($\pm$cordierite) - biotite gneisses and migmatites; and intercalations of other kinds of rocks are rare.

The age of formation of both units is not definitely known yet. At least part of the Varied Group sequence can be of Lower Paleozoic age, nevertheless some ~ 2 Ga-old crystalline basement is indicated as well [8]. The rocks of the Monotonous group are probably not older than Upper Proterozoic [9].

The regional metamorphic development comprises several metamorphic events, each characterized by different P, T conditions. Older metamorphic stage is characterized by lower temperature and medium-pressure conditions and by the mineral association staurolite-kyanite-garnet. This stage can be documented especially in areas more distant from large Variscan plutons (see below). The dominant metamorphic stage in the study area represents the Variscan high temperature and lower pressure metamorphic event, characterized by extensive migmatization of rocks. For the high-grade gneisses of the Austrian part of the Moldanubian zone temperatures range from 700 to 770°C and pressures of 70 - 90 MPa are indicated by [10] using conventional mineral termometers and barometers while the peak metamorphic temperature conditions of skarn parageneses from the eastern part of the Moldanubian Unit were estimated at 600 - 725 °C [11].

Large granitoid bodies outcropping in the study area belong to the Bohemian-Moravian branch of the Variscan Moldanubian Pluton and are called Central Moldanubian Pluton (Fig.1.). Leucocratic peraluminous two-mica S-type granites of this part of the Moldnubian Pluton are generally late-to-post orogenic intrusive types. Rb-Sr dating indicated an age of 316$\pm$7 Ma for this type (i.e. Eisgarn type) in the Austrian part of the Moldanubian Unit [12]. The Central Moldanubian Pluton is a multiple intrusion of mainly anatectic origin, decreasing in age from south to north [13]. The intrusion proceeded from SW to NE rising also to a higher crustal level. For the N part of the Central Moldanubian Pluton a Rb-Sr age of 303$\pm$6 Ma was obtained (with $^{87}Sr/^{86}Sr_{initial}$= .7176), while the peak of Variscan metamorphism of metamorphites in the surroundings was dated by Rb-Sr method at 335$\pm$3 Ma [14]. Geophysical data indicate that the eastern contact of the Central Massif dips very steeply, while the western contact is flatter. The granite is relatively thin in its central parts and dips to the west [15]. The contacts cut discordantly the country rocks of the Monotonous and Varied Groups. The P, T conditions of granite emplacement were estimated at 660-700 °C and 3.5 - 4.0 kbar for varieties with andalusite, and 665-710°C and 4.0 - 4.5 kbar for granites without andalusite [16].

Marbles, calc-silicate rocks (skarnoids) and skarns containing locally subeconomic scheelite mineralization are widespread near the northern part of the Moldanubian pluton [17], see Fig.1. In this paper (in agreement with classification of Einaudi and Burt, [18]) we use the common term "calc-silicate rocks" for more fine-grained rock types formed dominantly by isochemical (except for devolatilization) metamorphic recrystallization of impure carbonate layers. Skarns are a coarse-grained bodies of calc-silicates formed by infiltration and diffusion of metasomatic fluids carrying exotic components. This subdivision is to some degree artificial, since all gradations between these two rock types exist.

Marbles

Lenticular marble bodies with thicknesses up to several tens of meters outcrop close to the town of Ledeč. Marbles usually exhibit variable content of silicates - mostly diopside, phlogopite for example. To study transition metasomatic zones between carbonate and silicate rocks formed dominantly by regional metamorphic events, an active quarry near Bohdaneč was selected (see Fig.1). Here thick lenses of dolomitic marble, located close to amphibolite bodies contain talc, tremolite and phlogopite [19].

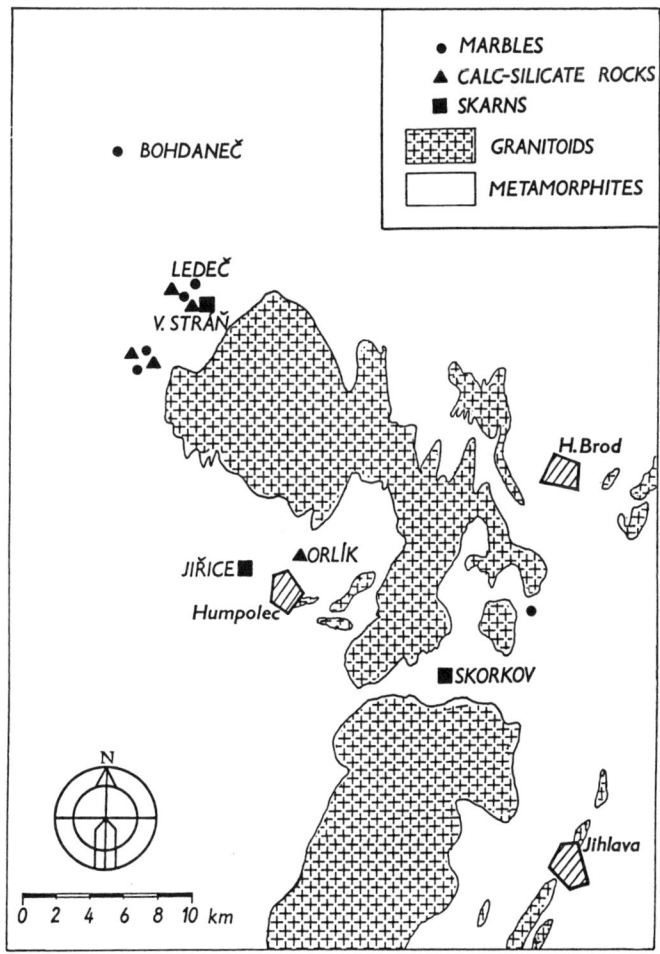

Figure 1. Map of studied marble, calc-silicate rock and skarn localities.

Calc-silicate rocks
Calc-silicate rocks forming lenticular bodies with thicknesses to 1 m are widespread in the study area. Typically, the peripheral zone near the boundary with paragneisses contain clinopyroxene (diopside), plagioclase (bytownite to labradorite), locally Ca-rich garnet and usually secondary hornblende. Wollastonite is rare. Accessory minerals are represented by scheelite, titanite and some sulfides (pyrrhotite, pyrite). The central fine-grained carbonate zone is missing in thinner bodies. At the locality of Orlík near Humpolec the parametamorphic sequence with calc-silicate rocks is spatially connected with lenticular concordant quartz veins containing gold, scheelite, pyrrhotite arsenopyrite, löllingite and microscopic minerals of gold, bismuth and tellurium including maldonite (Au_2Bi) [20,21].

Skarns
Lenticular bodies of skarns with maximum thicknesses to 1 m are characterized, besides the typical coarse-grained mineral assemblage, by proximity to Variscan granitoids. Skarns outcropping in a motorway cut near Skorkov (see Fig.1.) are located within migmatites penetrated by numerous pegmatites. Skarn lenses are characterized by peripheral zone with diopside (Hd 40-50, locally replaced by hornblende), plagioclase and quartz. Titanite, biotite, scheelite, pyrrhotite, chalcopyrite, sphalerite and pyrite are present in minor quantities. Mineral composition of the peripheral skarn zone is similar to the mineral composition of calc-silicate rocks. The internal skarn zone is formed by massive idocrase, Ca-rich garnet and quartz. Less frequent are pyroxene, hornblende, scheelite and Al-rich titanite (up to 9% Al_2O_3). The youngest central skarn zone is represented by coarse-grained calcite, but this zone is developed only locally. (Fig.2).

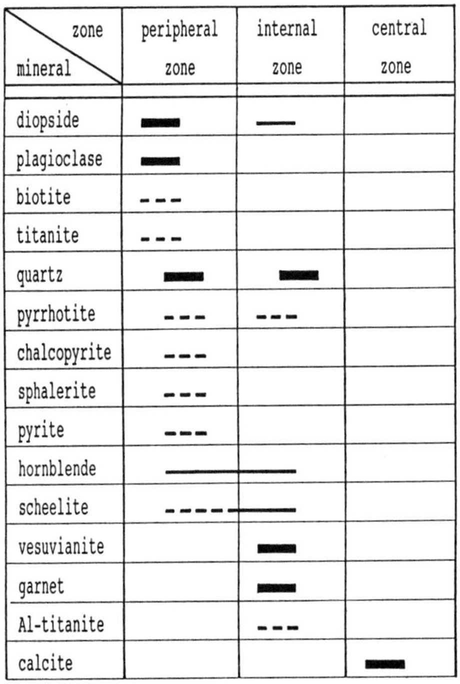

zone / mineral	peripheral zone	internal zone	central zone
diopside	■■■	—	
plagioclase	■■■		
biotite	- - -		
titanite	- - -		
quartz	■■■	■■■	
pyrrhotite	- - -	- - -	
chalcopyrite	- - -		
sphalerite	- - -		
pyrite	- - -		
hornblende		——	
scheelite	- - - -	——	
vesuvianite		■■■	
garnet		■■■	
Al-titanite		- - -	
calcite			■■■

Figure 2. Succesion of skarn zones at Skorkov locality.
■■■ >10 modal %, —— 1-10 modal %, - - - <1 modal %

Skarn lenses outcropping in a motorway cut near Jiřice (Fig.1) are located within cordieritic migmatites that are characterized by diopside (Hd 40-50), Ca-rich garnet, massive wollastonite and scheelite. Similarly, small skarn lenses located in quarry of Velká Stráň near Ledeč (Fig.1) are characterized by idocrase, wollastonite, diopside, scheelite and calcite. Besides of rare coarse-grained skarn lenses numerous bands of calc-silicate rocks containing scheelite are typical for this quarry as well [17].

Magnetite-bearing skarns of uncertain origin are found in this area (Vlastějovice). These may represent metamorphosed accumulations of sedimentary or submarine-exhalative ores [11,22]. Magnetite-bearing skarns were not included in this study.

METHODS

Carbon and oxygen isotope analyses of carbonates were performed using conventional reaction with 100% H_3PO_4 under vacuum. Because most samples are calcitic with only minor dolomitic admixtures no correction of measured $\delta^{18}O$ values was required. Overall analytical precision was $\pm$ 0.1 $^o/_{oo}$.

Fluid inclusions were studied by optical microthermometry on the apparatus Chaixmeca [23].

MODEL CALCULATION OF CARBON AND OXYGEN ISOTOPIC SHIFTS DURING DECARBONATION

As already stressed above the magnitude of ^{18}O and ^{13}C depletions which occur during decarbonation reactions are directly linked and can be calculated if the reaction stochiometry and PT conditions are known. For this calculation two limiting cases are used [3], batch volatilization and Rayleigh volatilization. The batch volatilization represents the system in which all evolved fluid remains together with the rock in a closed system and equilibrates with the rock finally being removed in one step. In contrast, the Rayleigh volatilization involves the continuous exchange and removal of infinitely small aliquots of fluid, each before the volatilization of the next.

Based on the chemistry of marbles in the study area (calcitic with less than 30 % dolomite component) and the mineral assemblages of calc-silicate rocks and skarns, the following reaction was used as a stoichiometric base for the calculations:

$$2\ CaCO_3 + CaMg(CO_3)_2 + 4\ SiO_2 = CaMgSi_2O_6 + 2\ CaSiO_3 + 4\ CO_2$$
(calcite + dolomite + quartz = diopside + wollastonite + CO_2)

The calculations were conducted at temperatures of 400°C (lower temperature limit derived from homogenization temperature of fluid inclusion within youngest mineral of skarn parageneses) and 650°C (minimum temperature of main metamorphic peak) using the equations of [24]. Continuous CO_2 separation from the system (Rayleigh distillation) was selected for this modelling, because continuous separation produces larger ^{13}C and ^{18}O depletions than single-stage CO_2 escape. Calculations were stopped when the mole fraction of residual carbon equalled 0.05, because all samples contained at least 10 modal % of carbonate. Following oxygen and carbon isotope fractionations were used: CO_2-calcite after [25], calcite-diopside after [26], diopside-wollastonite after [27], and calcite-dolomite after [28]. Curves at 400°C and at 650°C do not differ significantly; therefore only the 650°C curve is shown in Fig.3.

RESULTS AND DISCUSSION

$\delta^{13}C$ and $\delta^{18}O$ data of carbonate phases from marbles, calc-silicate rocks and skarns are plotted in Fig.3. together with the model decarbonation curve for open-system decarbonation at 650°C. It is quite clear that carbon and oxygen isotopic compositions of carbonates in marbles and most calc-silicate rocks can be produced from carbonate sedimentary precursors altered by decarbonation reactions. In contrast, all calcites from skarns and some samples from calc-silicate rocks are (with respect to carbon isotopic composition) beyond the limits that can be explained by decarbonation.

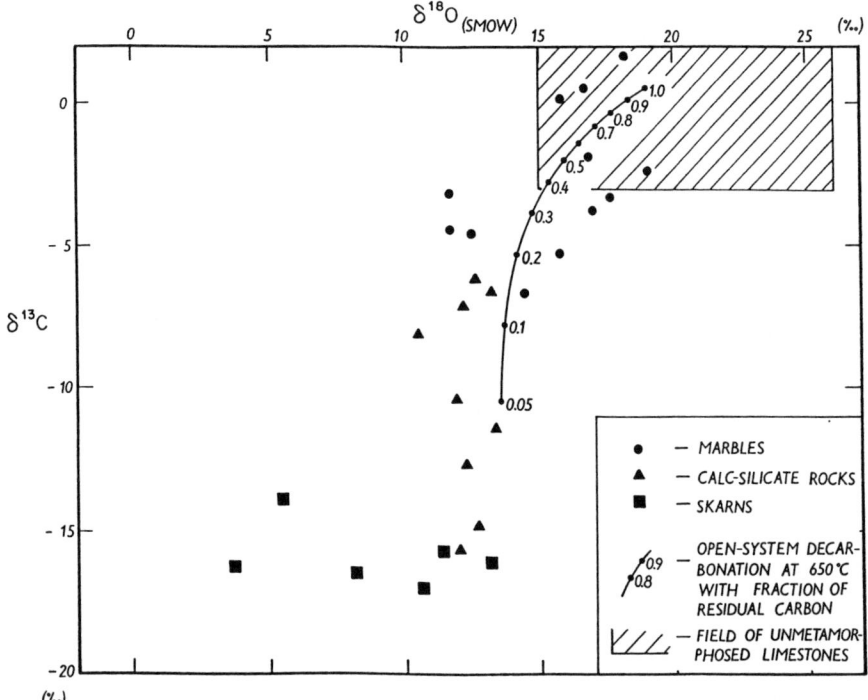

Figure 3. $\delta^{13}C$ and $\delta^{18}O$ data of carbonate phases from studied rock types. The field of unmetamorphosed Upper Proterozoic and Paleozoic carbonates is after Veizer and Hoefs, [29]. Model decarbonation curve represents open system (Rayleigh) decarbonation at 650°C (for explanation see text).

Marbles

For massive bodies of pure marbles, the shifts of carbon and oxygen isotopic compositions to lower δ-values are relatively limited. Decarbonation and isotopic exchange with circulating fluids are minimalized here, because the metamorphic fluid/rock ratios are very low in the interior of pure marble lenses. The highest $\delta^{13}C$ and $\delta^{18}O$ values of the least affected rocks fall into the field of normal marine sedimentary carbonates of Upper Proterozoic and Paleozoic age.

More significant changes of carbon and oxygen isotopic compositions of marbles are restricted to the peripheral contact zones with silicate rocks where local metamorphic exchange of components (bimetasomatic diffusion) between the two rock types have occurred. To study the changes of carbon and oxygen isotopic composition near the contacts of large marble bodies with amphibolite, two short profiles across the boundary were sampled and analysed (Fig.4). Generally the C- and O-isotope systematics seem to be controlled by metamorphic devolatilization. However the observed irregularities (especially in the $\delta^{18}O$ trends) indicate that the system was probably infiltrated by some externally derived fluid, which circulated preferentially in the zones of volume changes connected with decarbonation reactions.

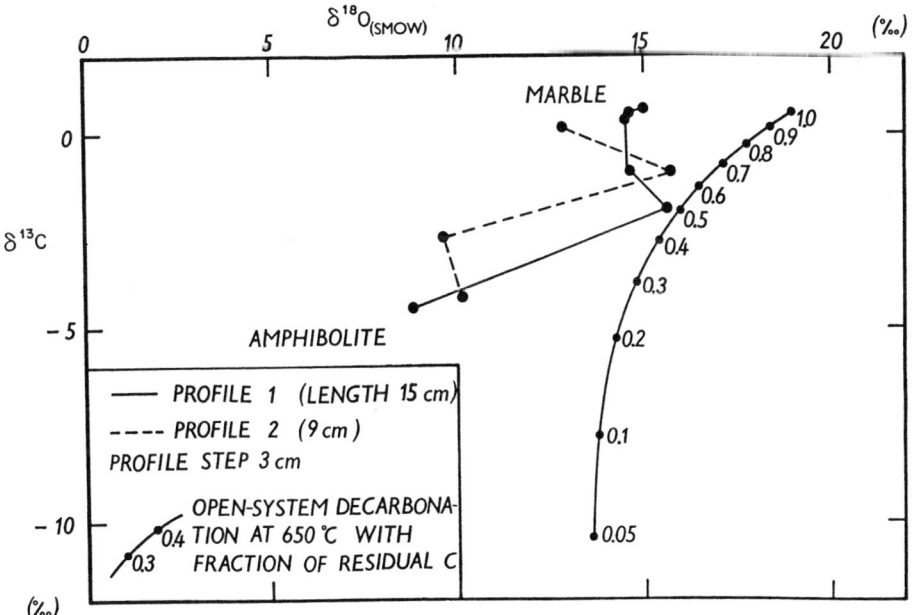

Figure 4. $\delta^{13}C$ and $\delta^{18}O$ changes in two short profiles across marble/ambhibolite bimetasomatic zone.

Calc-silicate rocks
Carbon and oxygen isotopic compositions of carbonates from most calc-silicate rocks plot usually relatively close to the model decarbonation curve (Fig.3). This indicates that decarbonation reactions were probably the dominant factor controlling the isotope systematics. Nevertheless several samples from the Velká Stráň quarry near Ledeč, and all samples from Orlík near Humpolec exhibit more negative $\delta^{13}C$ values than can be explained by decarbonation (i.e. all samples with the $\delta^{13}C$ values below -11 ‰).

At Orlík calc-silicate rocks are associated with lenticular concordant quartz veins containing gold, scheelite, pyrrhotite, arsenopyrite, löllingite and microscopic minerals of gold [20]. In the ore-bearing zone, a transition of CO_2-rich fluids to CH_4 (+ N_2)-rich and water-rich fluids was observed in fluid inclusions of quartz and scheelite [21], with the carbonaceous components probably derived from organic matter. Therefore, the anomalous decrease of $\delta^{13}C$ in calc-silicate rocks at Orlík is a result of infiltration of external fluids containing organic matter-derived components and cannot be explained by decarbonation reactions only.

Gilbert et al. [30] calculated the scheelite solubility at varying PT conditions in chloride solutions controlled by variable host rock mineralogy. The solubility of scheelite is significantly lower in an environment controlled by calc-silicate mineralogy than in micaschist. Moreover, solubility of scheelite decreases with an increasing N_2 component in complex C-H-O-N fluids. For scheelite deposited in calc-silicate gneisses of Montagne Noire, France Gilbert et al. [30] rejected a possibility that scheelite mineralizations are metamorphosed sedimentary-exhalative concentrations. Instead, scheelite was deposited here by hydrothermal solutions at a late, low pressure, stage of regional metamorphism, mostly in idocrase-grossular calc-silicate gneissic host.

For the gold - scheelite ore-bearing zone at Orlík a significant pressure decrease during mineralization was interpreted from a fluid inclusion study [21]. Primary fluid inclusions in scheelite indicate minimum P, T conditions of fluid trapping at 360-380°C and 500 - 700 bars. Calc-silicate mineralogy of the host rocks together with increased N_2 content of infiltrating fluid represent favourable conditions for scheelite deposition. The relationship between scheelite deposition in lenticular concordant quartz veins of the gold-scheelite ore-bearing zone, and scheelite deposition in surrounding calc-silicate rocks is not clear. Similar fluid inclusion data in both scheelite types, together with field evidence (i.e. crosscutting of ore bearing zone by late pegmatite dykes) indicate that both processes could have occurred during hydrothermal fluid movement at a late, low pressure stage of regional metamorphism and/or during granite intrusion.

With this is in agreement the carbon isotopic composition of calc-silicate rock carbonates at localities with more abundant scheelite mineralization (i.e. Orlík and Velká Stráň near Ledeč). These localities indicate influence of organic carbon-derived component transported by hydrothermal fluids. Other localities have less negative $\delta^{13}C$ values (which can be a result of decarbonation only) and contain scheelite more rarely.

Skarns

C and O isotopic composition of carbonates from skarns was studied in numerous papers [4-6,31-36]. Commonly, skarn carbonates do not represent the residual carbonate from decarbonation reactions. Instead, skarn carbonates usually belong to the youngest skarn minerals and are typically formed in isotopic equilibrium with infiltrating metasomatic fluids, which were frequently isotopically equilibrated with adjacent plutonic body.

Petrographical study indicate that typical coarse-grained central calcite zone in studied skarn at Skorkov is the youngest skarn mineral (Fig.2). The carbon isotopic composition of calcite samples from all described skarns is significantly lower than can be reached by decarbonation reactions. This calcite was formed from metasomatic fluids. Very low $\delta^{13}C$ values of both skarn carbonates (Fig.3) and of metasomatic fluids (calculated $\delta^{13}C_{fluid}$ values from -12 to -17 $^o/_{oo}$) indicate dominance of organic matter-derived carbon. Similarly, very low $\delta^{13}C$ values (to -15 $^o/_{oo}$) were found for carbonates from W, Sn and some Cu(-Fe) skarns of Japan [36].

Calcite from the central skarn zone at Skorkov contain two principal types of fluid inclusions, H_2O-CO_2 type and H_2O type. The H_2O-CO_2 type inclusions exhibit features typical of the primary inclusions. The density of CO_2-rich phase ranging from 80 to 137 kg/m^3 indicate relatively low trapping pressure. H_2O-type inclusions have several populations with different homogenization temperatures, with the highest interval ranging 385-390°C.

The oxygen isotopic composition of skarn calcites from all localities ranges from $\delta^{18}O = +3.7$ to $+13.1$ $^o/_{oo}$ (SMOW) (see Fig.3). The lowest $\delta^{18}O$ values, found especially for skarn at Jiřice, indicate participation of fluids with relatively low $\delta^{18}O_{fluid}$ values below ca. $+3$ $^o/_{oo}$ (SMOW). These low $\delta^{18}O$ fluids cannot be of metamorphic or

magmatic origin. Fluids depositing this calcite type are most probably of meteoric origin, and are only partially affected by oxygen isotope exchange with rock during the circulation. Mineral assemblages of these skarns was therefore formed in three sequential steps:
  i. Decarbonation reactions during regional metamorphism.
  ii. Metasomatic processes conected with hydrothermal circulation during late, low pressure stage of regional metamorphism and/or during granite intrusion. However, the metasomatic fluids do not have typical "magmatic" carbon isotopic signature.
  iii. Youngest invasion of low $\delta^{18}O$ fluids probably of meteoric origin.

CONCLUSIONS

Internal parts of larger pure marble lenses show only limited ^{13}C and ^{18}O depletions and their $\delta^{13}C$ and $\delta^{18}O$ values are in the range of normal marine sedimentary carbonates. Only the peripheral zones of marble bodies, where bimetasomatic diffusion during metamorphism took place, exibit ^{13}C and ^{18}O depletions caused dominantly by decarbonation. Nevertheless irregularities especially in $\delta^{18}O$ changes indicate that stable isotope systematics of these transitional zones was disturbed by fluid circulation.

Carbon isotopic composition of carbonates from calc-silicate rock generally reflect more significant ^{13}C and ^{18}O depletions during decarbonation reactions. At localities with the most abundant scheelite mineralization the carbonate $\delta^{13}C$ values are frequently shifted to very low values beyond the limits which can be reached by decarbonation. Significant infiltration of external fluid carrying organic matter-derived carbonaceous species is probable. This fluid movement occurred most probably at a late, low-pressure stage of regional metamorphism.

All skarn calcite $\delta^{13}C$ values are extremely low indicating dominance of organic matter-derived carbon during metasomatic processes. Metasomatic fluids circulated either during a late, low pressure stage of regional metamorphism and/or during granite intrusion. Low $\delta^{18}O$ values of several calcite samples suggest invasion of low $\delta^{18}O$ fluids during final formation stages.

Acknowledgements
This research was funded by Czech Geological Survey, Prague. The authors wish to thank Drs. S. Vrána and Z. Pertold for thoughtful comments and criticism of an earlier version of this manuscript.

REFERENCES

1. A. Beran, R. Göd, M. Götzinger and J. Zemann. A scheelite mineralization in calc-silicate rocks of the Moldanubicum (Bohemian Massif) in Austria, *Mineral. Deposita.* **20**, 16-22 (1985).
2. R. Jung and R. Höll. Wolframvorkommen in Nordost- Bayern, *Erzmetall.* **35**, 142-147 (1982).
3. J.W. Valley. Stable isotope geochemistry of metamorphic rocks. In: *Stable isotopes in high-temperature geological processes. Reviews in Mineralogy.* **16**. J.W. Valley, H.P. Taylor Jr. and J.R. O'Neil (Eds). Chap. 13, pp. 445-489. Mineralogical Society of America, Washigton, D.C. (1986).
4. P.E. Brown, J.R. Bowman and W.C. Kelly. Petrologic and stable isotope constraints on the source and evolution of skarn-forming fluids at Pine Creek, California, *Econ. Geol.* **80**, 72-95 (1985).
5. J.R. Bowman, J.J. Covert, A.H. Clark and G.A. Mathieson. The CanTung E Zone scheelite skarn orebody, Tungsten Northwest Territories: Oxygen, hydrogen and carbon isotope studies, *Econ. Geol.* **80**, 1872-1895 (1985).
6. J.R. Bowman, J.R. O'Neil and E.J. Essene. Contact skarn formation at Elkhorn, Montana. II: Origin and evolution of C-O-H skarn fluids, *Amer J. Sci.* **285**, 621-660 (1985).

7. J. Salmenik. Skarn and ore formation at Seriphos, Greece. Ph.D. Dissertation, Univ. Utrecht. *Geologica Ultraiectina*. No. 40 (1985).
8. J.I. Wendt, A. Kröner, W. Todt, J. Fiala, P. Rajlich, T.C. Liew and J. Vaněk. U-Pb zircon ages and Nd-whole rock systematics for Moldanubian rocks of the Bohemian Massif, Czechoslovakia. In: *Conference on the Bohemian Massif, Abstracts*. Z. Kukal (Ed.). pp. 1-2. Geological Survey, Prague. (1988).
9. A. Kröner, I. Wendt, T.C. Liew, W. Comston, W. Todt, J. Fiala , V. Vaňková and J. Vaněk. U-Pb zircon and Sm-Nd model ages of high-grade Moldanubian metasediments, Bohemian Massif, Czechoslovakia, *Contr. Mineral. Petrology*. **99**, 257-266 (1988).
10. J. Petrakakis. Metamorphism of high - grade gneisses from the Moldanubian zone, Austria, with particular reference to the garnets, *J. Metamorph. Geol.* **4**, 323 - 344 (1986).
11. J. Pertoldová, M. Pudilová and Z. Pertold. Conditions of skarn formation at Pernštejn. In: *New trends and knowledges in Czechoslovak economic geology, Proceedings of conference*. Z.Křibek (Ed.). pp. 43-61. Charles University, Prague (1987). (*In Czech*)
12. S. Scharbert. Rb-Sr Untersuchungen granitoider Gesteine des Moldanubikums in Österreich, *Mitt. Österr. Mineral. Gesell.* **132**, 21-28 (1987).
13. S. Scharbert. Rb-Sr geochronology of intrusive rocks of the Moldanubicum. In: *Conference on the Bohemian Massif, Abstracts*. Z. Kukal (Ed.). p. 78. Geological Survey, Prague (1988).
14. S. Scharbert and M. Veselá. Rb-Sr systematics of intrusive rocks from the Moldanubicum around Jihlava. In: *Thirty years of geological cooperation between Austria and Czechoslovakia, Festive Volume*. D. Mináříková and H. Lobitzer (Eds). pp. 262-272. Federal Geological Survey, Vienna and Geological Survey, Prague (1990).
15. L. Mottlová. The deep structure of the central and Melechov massifs interpreted on the basis of alternative solutions of inverse gravity problems, *Věst. Ústř. Úst. Geol.* **57**, 351-363 (1984).
16. P. Novotný. Identification of peraluminous minerals in the Moldanubian Pluton. In: *Proceedings of IInd conference of mineralogists and petrologists Brno-Blansko 1987*, 101-107. Brno (1987). (*In Czech*)
17. L. Jurák and I. Tenčík. A review of tin and tungsten mineralizations in the area of Bohemian-Moravian Highlands, *Vlast. Sbor. Vysočiny Odd. Věd. Přír.* **6**, 21-28 (1970). (*In Czech*)
18. M.T. Einaudi and D.M. Burt. Introduction - terminology, classification and composition of skarn deposits, *Econ. Geol.* **77**, 745-754 (1982).
19. K. Beneš et al. *Explanations to the geological map of ČSSR 1 . 200 000*. Geofond, Prague (1963). (*In Czech*)
20. J. Litochleb and J. Malec. Paragenesis of gold and bismuth in gold-bearing veins near Humpolec, *Geol. Průzk.* **20**, 152 (1978). (*In Czech*)
21. J. Durišová, P. Sztacho and J. Dubessy. A fluid inclusion study of Au-W stratiform mineralization at Orlík near Humpolec, Czechoslovakia, *European J. of Mineralogy*. **4**, 965-976 (1992).
22. V. Zoubek. Comments to the question of skarns, granulites and south-Bohemian graphite deposits, *Sbor. St. Geol. Úst. Čs. Republ.* **13**, 488-498 (1946). (*In Czech*)
23. B. Poty, J. Leroy and L. Jachimowicz. Un nouvel appareil pour le mesure des températures sous le microscope: L'installation de microthermométrie Chaixmeca, *Bull. Soc. Fr. Minéral. Cristallogr.* **99**, 182-186 (1976).
24. D. Rumble. Stable isotope fractionation during metamorphic volatilization reactions. In: *Characterization of metamoprhism through mineral equilibria. Reviews in Mineralogy*, **10**. J.M. Ferry (Ed.). p. 327-353. Mineralogical Society of America, Washigton, D.C. (1982).
25. T. Chacko, T.K. Mayeda, R.N. Clayton and J.R. Goldsmith. Oxygen and carbon isotope fractionations between CO_2 and calcite, *Geochim. Cosmochim. Acta*. **55**, 2867-2882 (1991).
26. S.W. Kieffer. Thermodynamics and lattice vibrations of minerals: 5. Applications to phase equilibria, isotopic fractionation and high-pressure thermodynamic properties, *Rev. Geophys. Space Phys.* **20**, 827-849 (1982).
27. A. Matthews, J.R. Goldsmith and R.N. Clayton. Oxygen isotope fractionations involving pyroxenes: the calibration of mineral-pair geothermometers, *Geochim. Cosmochim. Acta*. **47**, 631-644 (1983).
28. S.M.F. Sheppard and H.P. Schwarcz. Fractionation of carbon and oxygen isotopes and magnesium between metamorphic calcite and dolomite, *Contr. Mineralogy and Petrology*. **26**, 161-198 (1970).
29. J. Veizer and J. Hoefs. The nature of $^{18}O/^{16}O$ and $^{13}C/^{12}C$ secular trends in sedimentary carbonate rocks, *Geochim. Cosmochim. Acta*. **40**, 1378-1400 (1976).
30. F. Gilbert, B. Moine, J. Schott and J.L. Dandurand. Physical and chemical controls of tungsten deposition in the calc-silicate gneisses from the Montagne Noire, France. In: *Source, transport and deposition of metals, Proceedings of the 25 years SGA meeting*. M.Pagel and J.L. Leroy (Eds). p. 45-48. A.A. Balkema, Rotterdam (1991).
31. Y.H. Shieh and H.P. Taylor Jr. Oxygen and carbon isotope studies of contact metamorphism of carbonate rocks, *J. Petrology*. **10**, 307-331 (1969).

32. R.O. Rye, W.E. Hall and H. Ohmoto. Carbon, hydrogen, oxygen and sulfur isotope study of the Darwin lead-silver-zinc deposit, southern California, *Econ. Geol.* **69**, 468-481 (1974).
33. B.E. Taylor and J.R. O'Neil. Stable isotope studies of metasomatic Ca-Fe-Al-Si skarns and associated metamorphic and igneous rocks, Osgood Mountains, Nevada, *Contr. Mineral. Petrol.* **63**, 1-49 (1977).
34. B.J. Cooke and C.I. Godwin. Geology, mineral equilibria, and isotopic studies of the McDame tungsten skarn prospect, North-central British Columbia, *Econ. Geol.* **79**, 826-847 (1984).
35. A.W. Rose, D.C. Herrick and P. Deines. An oxygen and sulfur isotope study of skarn-type magnetite depositsof the Cornwall type, Southeastern Pennsylvania, *Econ. Geol.* **80**, 418-443 (1985).
36. H. Shimazaki, H. Shimizu and T. Nakano. Carbon and oxygen isotopes of calcites from Japanese skarn deposits, *Geochem. J.* **20**, 297-310 (1986).

FLUID-CONTROLLED GRANULITE FORMATION IN THE EAST GONDWANA

M. SANTOSH[1] and M. YOSHIDA[2]

[1] *Centre for Earth Science Studies, P.B. 7250, Akkulam, Thuruvikkal Post, Trivandrum 695 031, India*
[2] *Department of Geosciences, Faculty of Science, Osaka City University, Osaka 558, Japan*

Abstract--Orthopyroxene-bearing anhydrous granulites (charnockites) constitute many segments of the East Gondwana crustal fragments. Field structures associated with patchy- and vein-type charnockites (incipient charnockites) in southern India and Sri Lanka are reminiscent of fluid pathways along which CO_2-rich fluids advected, and locally buffered the water activity to low levels, thereby stablizing the granulite assemblage. Fluid inclusion studies indicate contrastingly higher abundance of CO_2 in the charnockite as compared to the precursor gneiss. In many localities, CO_2 infiltration has resulted in the precipitation of coarse graphite crystals. The regional granulites in these terranes display a concomitant increase in the density of trapped CO_2-rich fluids with mineral thermobarometric gradient across amphibolite to granulite progressions. The carbon isotopic compositions of fluid inclusions and graphites are consistent with a "juvenile", sublithospheric source for the CO_2, which was transferred to higher levels of the crust through structural pathways or magmatic conduites.

Keywords: Charnockite, fluid inclusions, CO_2, graphite, carbon isotopes, East Gondwana

INTRODUCTION

The East Gondwana is composed of India, Sri Lanka, Antarctica and Australia. Many segments of these continental fragments expose granulite facies rocks. These rocks are thought to represent exhumed sections of the Earth's middle to lower crust, and have received considerable attention in respect of petrogenetic studies covering deep crustal processes and the crust-mantle interaction (e.g., Newton, 1990; Santosh and Yoshida, 1992a; Yoshida and Santosh, 1993). Of particular interest is orthopyroxene-bearing anhydrous granulites (charnockites) which represent formation under very low water activities. Fig. 1 shows the pressure-temperature (P-T) fields of equilibration of granulite facies assemblages in four principal terranes of the East Gondwana. P-T conditions estimated from mineral phase equilibria in these rocks vary between 600-800°C and 4.5 to 8 kbar. Fig. 1 shows that in all these terranes, the partial pressure of water was quite low, generally around 0.1 to 0.3.

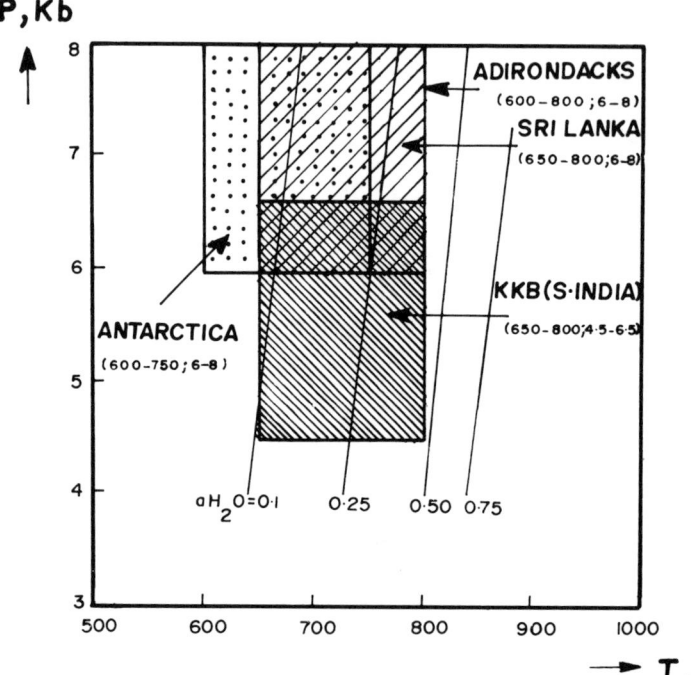

Fig. 1 Pressure-temperature diagram showing the fields of epuilibration of charnockites in the Lützow-Holm Bay region of East Antarctica (Santosh and Yoshida, 1992a), Kerala Khondalite Belt in southern India (Chacko et al., 1987) and Sri Lanka (Burton and O'Nions, 1990). Also shown is the P-T field for the Adirondack granulites in N.Y. State, U.S.A. (after Lamb et al., 1987) for comparison.

In the mid to lower crustal environment, low water activities can result from a variety of processes. Three important mechanisms are often cited: (1) extraction of water in anatectic melts, leaving a dry granulite residue (Fyfe, 1973); (2) vapor-absent metamorphism of contact-aureole rocks which have been pre-baked by shallow intrusives (Lamb and Valley, 1984); and (3) dilution of pore fluids by streaming of CO_2 (Newton et al., 1980). Considerable debate exists on the relative importance of each mechanism in stabilizing the anhydrous mineral assemblages which characterize granulites (Newton, 1990; Valley, 1992; Santosh, 1992).

In this paper, we present evidence from a number of lines which converges to indicate that granulite formation in the East Gondwanian crustal fragments was predominantly fluid-controlled.

FIELD EVIDENCE

By far, the most compelling evidence for the role of fluids in granulite petrogenesis has been provided by the field structures associted with arrested charnockite formation in southern India and

Fig. 2 Field photograph (in Kurunegala, Sri Lanka) showing arrested charnockite formation in a dense network of interconnected veins and "dykes" within upper amphibolite facies gneisses (after Yoshida and Santosh, 1993). The apparently isolated patches are thought to represent cross-sections of the interconnected tubular fluid channels.

Sri Lanka (Hansen et al., 1987; Raith et al., 1989; Santosh and Yoshida, 1992b; Yoshida and Santosh, 1993). In many quarry sections are exposed upper amphibolite facies gneisses which show conversion along structurally-controlled zones into coarse veins and patches of orthopyroxene-bearing granulite (Fig. 2). Many of these veins show shear-derived characteristics as judged from the distortion of foliation of the host gneisses. The mode of occurrence and the trend of the shear veins filled with incipient charnockite indicate that charnockite formation post-dated the regional metamorphism (Yoshida and Santosh, 1987). The intimate relationship between gneiss and granulite in the scale of a few decimeters precludes any significant temperature gradients, and invokes a fluid-controlled transformation process. The field structures associated with incipient charnockites are strongly reminiscent of fluid channels. The veins are thought to form an interconnected network, with the isolated patches representing the cross-sections of the tubular channels.

Granulite facies rocks also form regional charnockite massifs, covering vast expanses of mappable units in the absence of associated amphibolite facies lithologies. Although the role of fluids in their petrogenesis is not unequivocally demonstrable from field structures, the transcrustal shear systems which dissect these major crustal blocks, as in southern India, could represent deep-seated conduits along which substantial volumes of fluids were transferred. Additional evidence comes from the fluid inclusion characteristics across regional amphibolite to granulite progressions, as discussed in a later section.

MINERAL TRANSFORMATIONS

Metamorphic charnockites in East Gondwana have resulted by the conversion of either amphibole-bearing orthogneisses, or garnet- and biotite-bearing paragneisses. Based on the igneous or sedimentary parentage of the precursor rocks, charnockites can be classified into ortho- and para-types. While the vast majority of the regional charnockites ("massive" charnockites) come under the category of orthocharnockites, the local transformation products ("incipient" charnockites) occur both within orthogneisses and paragneisses. In the

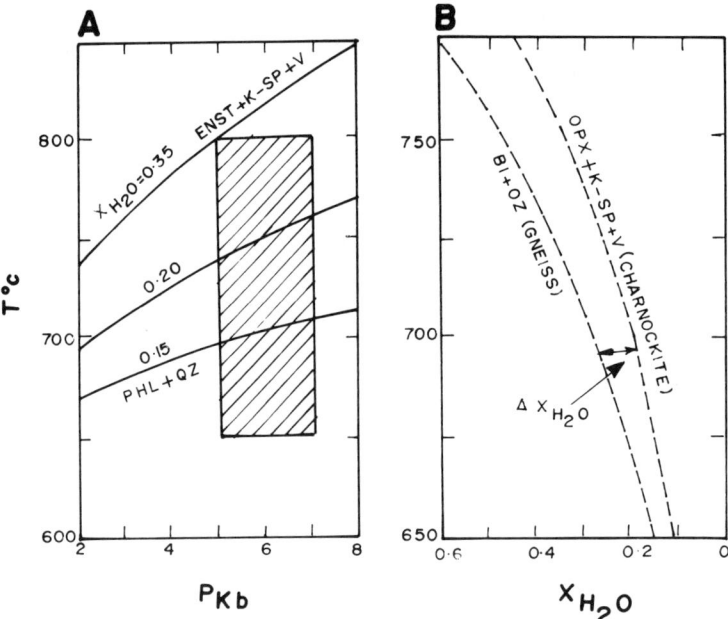

Fig. 3 Water activities estimated for charnockite formation. The shaded region in "A" represents the range of temperatures under which the south Indian granulites equilibrated. The shift in the mol fraction of water required to dehydrate the gneiss to charnockite at the Mannantala incipient charnockite quarry as estimated by Santosh et al. (1990) from mineral phase equilibria is shown in "B".

orthocharnockite localities, the orthopyroxene forms through the breakdown of calcic amphibole by reaction with biotite and quartz. In the paracharnockites, garnet, biotite and quartz in the gneisses react to form orthopyroxene, alkali feldspar and plagioclase. Several studies have documented the temperature and pressure conditions of charnockite formation in the East Gondwanian charnockites from computations based on experimental mineral phase equilibria thermo-barometers (e.g., Hansen et al., 1987; Raith et al., 1989; Burton and O'Nions, 1990; Santosh et al., 1990; Santosh and Yoshida, 1992a). None of the studies detected any significant pressure-temperature gradients across gneiss-incipient charnockite reaction fronts. This has been taken to indicate that the breakdown of hydrous minerals in the precursor gneisses, and the stabilization of orthopyroxene in the charnockite is a fluid-controlled process, where the buffering of water activity to low values (Fig. 3) was achieved by the dilution of pore fluids with an externally-influxed non-aqueous fluid, generally CO_2.

FLUID INCLUSIONS

Notwithstanding the debate over the peak- versus post-metamorphic entrapment of CO_2-rich fluid inclusions in granulites (cf. Lamb et al., 1987; Santosh et al., 1991), charnockites in the East Gondwana fragments carry abundant CO_2 trapped within various minerals which are thought to represent the traces of a synmetamorphic fluid phase (Hansen et al., 1987; Touret and Hansteen, 1988; Santosh et al., 1990; Santosh and Yoshida, 1992a,b). In most cases, the high density of the trapped CO_2 phase is consistent with entrapment during the high P-T conditions defined by mineral phase equilibria, and the isochores penetrate the mineral P-T windows. In the East Antarctic charnockites, a systematic variation in fluid densities was documented within the domain of individual samples, with garnet > feldspar > quartz. Along the pressure gradient from north to south in this terrane, the fluid densities show an increase from c. 0.98 g/cm^3 to c. 1.10 g/cm^3 (Fig. 4). This strengthens other evidence for a synmetamorphic entrapment of CO_2 in the East Antarctic charnockites (Santosh and Yoshida, 1992a).

The most compelling evidence for an external introduction of CO_2-rich fluids is provided by the CO_2 abundance measurements in adjacent gneiss and charnockite samples from the arrested charnockite localities through a step-wise thermal decrepitation and gas extraction method (Jackson et al., 1988; Santosh et al., 1991). Up to four fold increase in the volume of CO_2 has been recorded while passing from gneiss to charnockite in some localities (Fig. 5).

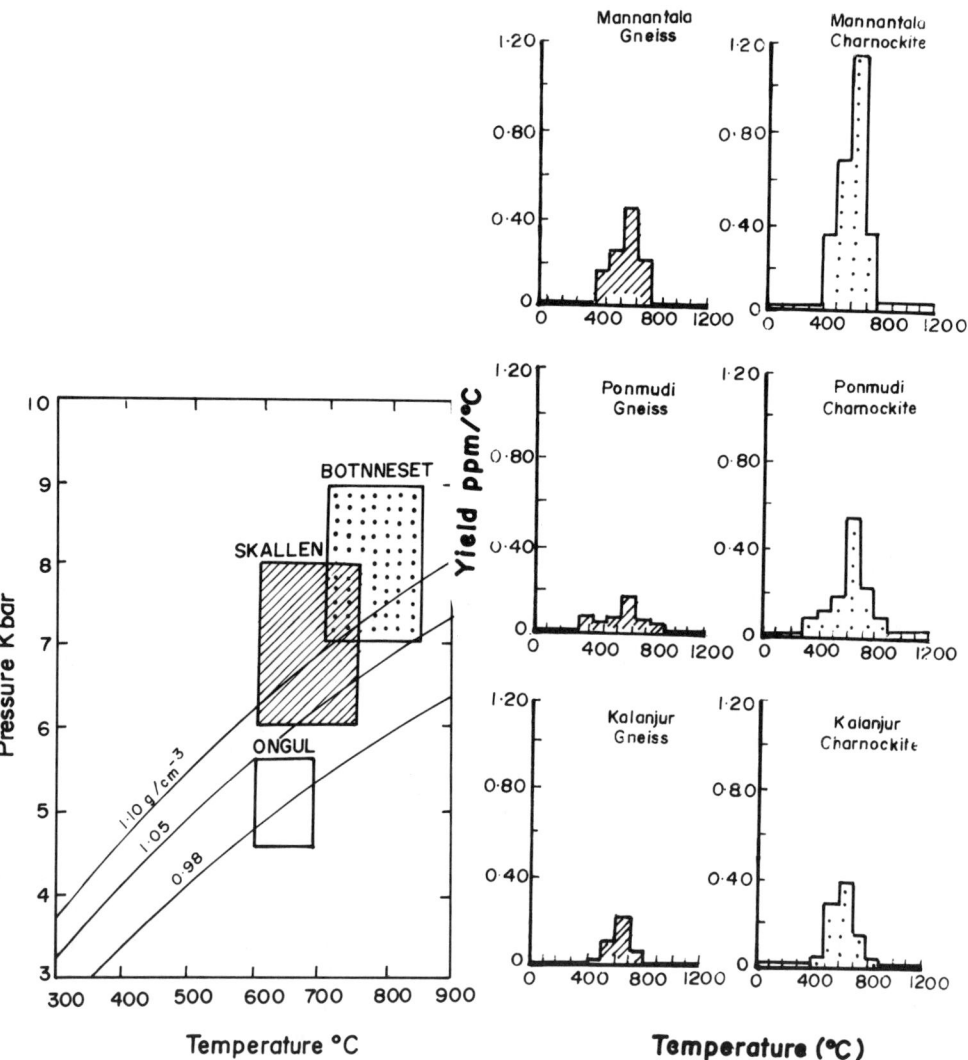

Fig. 4 P-T diagram showing mineral phase equilibria thermo-barometric estimates (shaded boxes) and fluid inclusion isochores (thin curves with density values) for the amphibolite to granulite progression in the Lützow-Holm Bay region of East Antarctica (after Santosh and Yoshida, 1992a). The data show a concomitant increase in fluid density with pressure and temperature, suggesting synmetamorphic entrapment of CO_2 in the East Antarctic charnockites.

Fig. 5 The abundance of CO_2 released from fluid inclusions by step-wise heating of pure mineral grains from gneiss-incipient charnockite zones of southern India (after Santosh et al., 1991). Note the increase in abundance while passing from the gneiss to the charnockite at the peak release interval (500-800 degree centigrade).

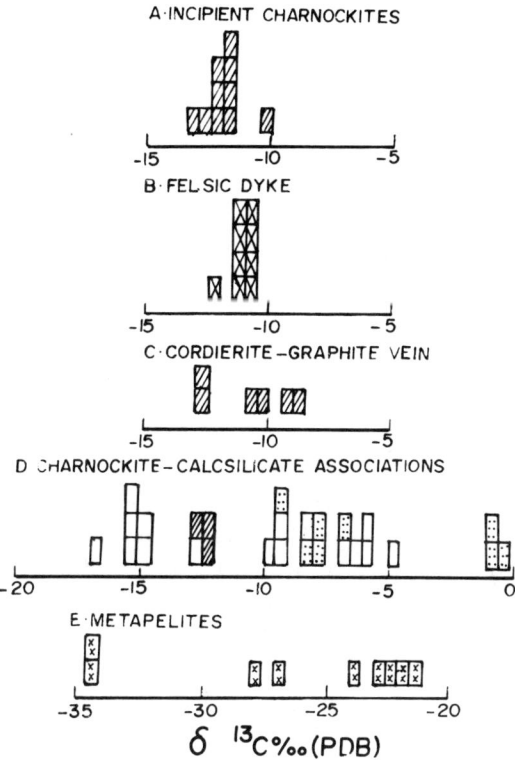

Fig. 6 Histograms showing the $\delta^{13}C$ range in graphites from various associations in the Kerala Khondalite Belt in the chanockite-calcsilicate associations. Open: graphite from charnockite; hatched: graphite from calcsilicate; stipled: calcite from calcsilicate. (after Santosh and Wada, in prep.).

CARBON STABLE ISOTOPES OF FLUID INCLUSIONS

The carbon stable isotopic composition of CO_2 released from inclusions by thermal decrepitation defines $\delta^{13}C$ values in the range of -8 to -10 per mil in the incipient charnockites (Jackson et al., 1988; Santosh et al., 1990). Exchange with crustal carbon has resulted in isotopic alteration and shift to lighter carbon values in some cases. Jackson et al.'s (1988) study recorded almost 2 per mil variation while passing from the gneiss to charnockite, with the latter showing heavier-carbon enriched fluids. An isotopic traverse across a gneiss-incipient charnockite boundary in southern India revealed that during charnockite formation, the mineral reaction front and the carbon isotope front are decoupled, with the isotope front transposed to a few decimeters into the host gneisses. This has been incorporated into a model of CO_2 advection across shears and faults, and the charnockite formation has been correlated to structurally-enhanced zones of fluid flow (Santosh et al., 1990).

Carbon isotope data on fluids in granulites from the other parts of East Gondwana are sparse; a limited data-set from East Antarctica (Santosh and Yoshida,1992a) show $\delta^{13}C$ values in the range of -5 to -6 per mil. These values, as well as those from the incipient charnockite zones are comparable with the isotopic composition of carbon derived from "mantle" settings such as diamonds, carbonatites, CO_2 trapped within Mid Ocean Ridge Basalts etc. (Javoy et al., 1986). Hence it is postulated that the ultimate source of CO_2 associated with charnockite formation could be from degassing of subcontinental lithosphere.

STABLE ISOTOPIC STUDIES IN GRAPHITES

Graphite occurs in the East Gondwana granulites in a number of associations. Disseminated graphite flakes occurring along the compositional layers in metapelites may represent the conversion of organic material to structurally-ordered carbon during the high grade metamorphism of the pelitic and psammo-pelitic sediments. In most cases, their carbon stable isotopic composition is characterized by $\delta^{13}C$ values less than -25 per mil (Fig. 6), comparable with biogenically-derived carbon (Santosh and Wada, unpub. data). Coarse graphite flakes are associated with arrested charnockite patches, felsic pegmatite dykes, cordierite + quartz veins, and charnockitized gneisses adjacent to calcsilicate layers, as in several localities in southern India. The remarkably high $\delta^{13}C$ values displayed by graphites from the structurally-controlled zones (-8.6 to -13.4 per mil), together with the narrow spread in individual localities (±1.2 per mil) correlate with precipitation from isotopically homogeneous CO_2-rich fluids which either infiltrated along structural pathways like shears/faults, or were transferred through magmatic conduits. Some of the graphites from charnockite-calcsilicate associations show extreme ^{13}C enrichment ($\delta^{13}C$ up to -4.8 per mil). Changing fluid regimes are detected from up to 5 per mil variation in $\delta^{13}C$ among graphites sampled adjacent to calcsilicate horizons, and within single crystal of coarse graphite in association with cordierite-rich veins (Santosh and Wada, in prep.). Steep isotopic gradients across fluid channels, and the preservation of biogenic carbon signature in disseminated graphite flakes within adjacent metapelites ($\delta^{13}C$ = -21.3 to -34.4 per mil) suggest that CO_2 influx was strongly channelized in such examples. These features provide evidence for the precipitation of graphite in reduced crustal lithologies which were infiltrated by CO_2-rich fluids.

MECHANISM OF FLUID TRANSFER

If CO_2 was indeed instrumental in buffering the water activity and aiding charnockite formation, then the source characteristics and transfer mechanism of the CO_2-rich fluids are topics of importance. The available stable isotope data suggest that most of the CO_2-rich fluids were externally-derived, from deep sources of "juvenile" nature, although local carbonate reservoirs like decarbonation of interlayered cabonate lithologies were also significant in specific cases. CO_2 exsolved from underplated basaltic magmas, which were channelized along deep-seated faults or shears is considered as one possibility. Several alkaline and subalkaline granite and syenite plutons and K-basalts are distributed throughout the reactivated areas of East Gondwanian crustal fragments, probably representing extensive fault-controlled magmatism of mantle derivation. A magmatic conduit for juvenile CO_2 is hence considered possible. The several felsic dykes, some of which are in intimate association with incipient charnockites (e.g., Farquhar and Chacko, 1991), may represent the cryptic pathways of magmas which probably transported CO_2 to higher crustal levels and caused the incipient charnockitic alteration of the gneisses. These features can be incorporated into a geodynamic model (cf. Yoshida and

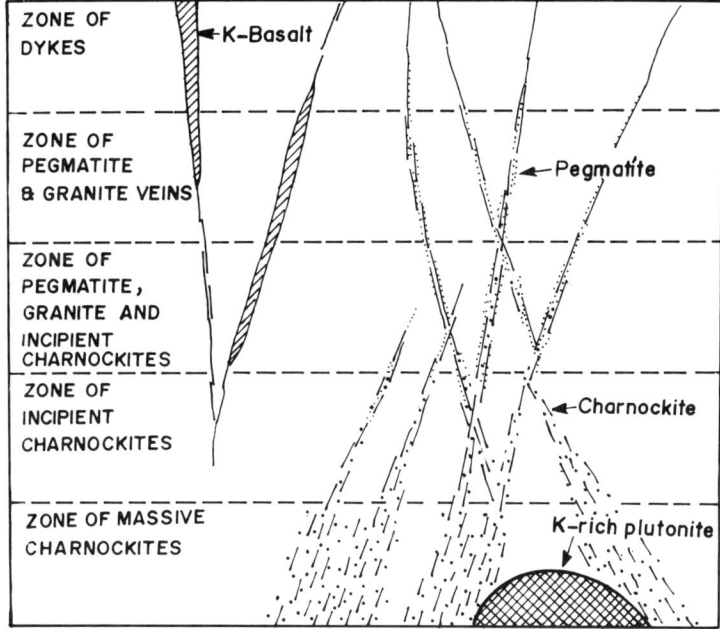

Fig. 7 Schematic model of incipient charnockite formation in association with fractures, CO_2 fluid streaming, and alkaline-subalkaline magmatic rocks (after Yoshida and Santosh, 1993).

Santosh, 1993), which links magmatism, CO_2 infiltration and charnockite formation (Fig. 7).

ACKNOWLEDGEMENTS

The senior author thanks the Department of Science and Technology (Govt. of India) for project support to study fluid processes in the deep crust, and the Director, CESS for facilities. This work is also a contribution to the Monbusho International Scientific Research Program No. 04041090 lead by M. Yoshida and IGCP288 lead by R. Unrug.

REFERENCES

Burton K.W. and R.K. O'Nions. The timescale and mechanism of granulite formation at Kurunegala, Sri Lanka, *Contrib. Mineral. Petrol.* **106**, 66-89 (1990).

Chacko T., G.R.R. Kumar and R.C. Newton. Metamorphic P-T conditions of the Kerala (S. India) Khondalite belt, a granulite facies supracrustal terrain, *Jour. Geol.* **95**, 343-358 (1987).

Farquhar J. and T. Chacko. Isotopic evidence for involvement of CO_2-bearing magmas in granulite formataion, *Nature.* **354**, 60-63 (1991).

Fyfe. W.S. The granulite facies, partial melting and the Archean crust, *Philos. Trans. R. Soc. London, Ser. A.* **273**, 457-461 (1973).

Hansen E.C., A.S. Janardhan, R.C. Newton, W.K.B.N. Prame and G.R.R. Kumar. Arrested charnockite formation in southern India and Sri Lanka, *Contrib. Mineral. Petrol.* **96**, 225-244 (1987).

Jackson D.H., D.P. Mattey and N.B.W. Harris. Carbon isotope compositions of fluid inclusions in charnockites from southern India, *Nature.* **333**, 167-170 (1988).

Javoy M., F. Pineau and H. Delorme. Carbon and nitrogen isotopes in the mantle, *Chem. Geol.* **57**, 41-62 (1986).

Lamb W.M. and J.W. Valley. Metamorphism of reduced granulites in low-CO_2 vapor-free environment, *Nature.* **312**, 56-58 (1984).

Lamb W.M., J.W. Valley and P.E. Brown. Post-metamorphic CO_2-rich fluid inclusions in granulites, *Contrib. Mineral. Petrol.* **96**, 485-495 (1987).

Newton R.C. Fluids and shear zones in the deep crust, *Tectonophys.* **182**, 21-37 (1990)

Newton R.C., J.V. Smith and B.F. Windley. Carbonic metamorphism, granulites and crustal growth, *Nature.* **288**, 45-50 (1980).

Raith M., S. Hoernes, E. Klatt and H.J. Stahle. Contrasting mechanisms of charnockite formation in the amphibolite to granulite terrane of the Nilgiri Hills (southern India): characterisation of high-grade metamorphism. In: *Granulites and crustal evolution.* D. Veilzeuf and Ph. Vidal (Eds). pp. 339-365, Kluwer Acad. Pub. (1989).

Santosh M. Carbonic fluids in granulites: cause or consequence? *J. geol. Soc. India.* **39**, 375-399 (1992).

Santosh M., M.B.W. Harris, D.H. Jackson and D.P. Mattey. Dehydration and incipient charnockite formation: a phase equilibria and fluid inclusion study from South India, *Jour. Geol.* **98**, 915-926 (1990).

Santosh M., D.H. Jackson, M.B.W. Harris and D.P. Mattey. Carbonic fluid inclusions in south Indian granulites: evidence for entrapment during charnockite formation, *Contrib. Mineral. Petrol.* **108**, 318-330 (1991).

Santosh M. and M. Yoshida. A petrologic and fluid inclusion study of East Antarctic charnockites from the Lützow Holm Bay: evidence for fluid-rich metamorphism in the deep crust, *Lithos.* (1992a in press).

Santosh M. and M. Yoshida. The role of CO_2 in charnockite formation at Kurunegala, Sri Lanka, *J. Geol. Soc. Sri Lanka 4.* (1992b in press).

Touret J.L.R. and H.T. Hansteen. Geothermobarometry and fluid inclusions in a rock from the Doddabetta charnockite complex, southwest India, *Rend. Soc. Ital. Mineral. Petrol.* **43**, 65-82 (1988).

Valley J.W. Granulite formation is driven by magmatic prosesses in the deepcrust, *Earth Science Rev.* **32**, 145-146 (1992).

Yoshida M. and M. Santosh. Charnockite "in the breaking" and "making" in Kerala, South India: tectonic and microstructural evidences, *J. Geosci. Osaka City Univ.* **30**, 23-49 (1987).

Yoshida M. and M. Santosh. A tectonic perspective of incipient charnockites in East Gondwana, *Precamb. Res.* (1993 in press).

SANDSTONE PETROLOGY IN RELATION TO TECTONICS

Editors
F. Kumon and K.M. Yu

CONTENTS

Preface   67

Effect of provenance, sorting and weathering on the geochemistry of fluvial sands from different tectonic and climatic environments
S.B. Kroonenberg   69

A Precambrian Andean-type flysch: Petrology and geochemistry of Yangzhanling flysch in southern Anhui Province, China
B. Xia, X. Ma, H. Lu, Z. Fang, and J. Zhou   83

Petrofacies analysis and tectonic evolution of Zagroside flysch suites from northeastern Iraq
B. Al-Qayim   97

Permian to mid-Triassic evolution of sandstone composition in a complex back-arc extensional to foreland basin: the Bowen Basin, eastern Queensland, Australia
C.R. Fielding, P. de Caritat, J.C. Baker, and M.M. Wilkinson   109

Compositional changes of the sandstone in relation to the evolution of a mobile belt - an example of the Paleozoic to Mesozoic sandstones of North Japan
K. Okami, M. Ehiro, and S. Koshiya   119

Modal and chemical compositions of the representative sandstones from the Japanese Islands and their tectonic implications
F. Kumon and K. Kiminami   135

Serpentinite protruded into fore-arc region: implications of detrital chromian spinels in Cretaceous sandstones of the Kanto Mountains, Japan
K.-I. Hisada and S. Arai   153

Sedimentation and paleoenvironment in the uppermost Lower to Middle Miocene basin on the Japan Sea coast, Southwest Japan
T. Matsumoto   165

Coexistence of provenance-reflected shallow-marine and deep-marine turbidite sandstones - sedimentation at the eastern margin of the Niigata Neogene backarc basin, northeast Japan
S. Tokuhashi   173

Petrography of turbidite sandstones in Niigata basin, northernmost part of Fossa Magna, central Japan
M. Tateishi, A.A.A. El Habab, and M. Shimazu   183

Preface

The relationship between sandstone composition and tectonic setting has been a controversial subject during the last several decades. Recently, W.R. Dickinson and his colleagues proposed a model of the relationship on the basis of the refined modal analysis method (Gazzi-Dickinson method) and plate tectonics, and made a milestone on this problem. Geochemical approach on the relationship have been also advanced vastly by Bhatia and others. The diagrams proposed offer useful key to clarify the tectonic situations of ancient sandstones. These diagrams, however, are not almighty, and the routine use of the discrimination diagrams lead to the recognition of error populations as pointed by G.H. Mack.

The mineralogical and chemical composition of the sandstone is controlled by the various factors such as provenance, relief, climate, weathering, sorting, transportation mechanism, sedimentary environment and diagenesis. Most sedimentologists believe that the provenance factor is most effective in usual cases and the provenance factor is strongly influenced by the tectonic setting, especially plate movement. This is the reason why the sandstone composition enable to distinguish its tectonic setting. The other factors, however, should be taken in count seriously. For example, the strong weathering is capable of wiping out completely the signature of source rocks in less than 50,000 years (Kroonenberg, in this proceeding). Coastal environment modifies the original composition toward quartz-rich sand. The goal of sandstone petrology is to evaluate quantitatively the effect of each factor in sandstone and to build up the models which is capable of distinguishing it.

Sandstone petrologists have been made various efforts for this purpose. Along this line, 15 papers were presented in Symposium II-2-3 titled as "Sandstone petrology in relation to tectonics", 29th IGC held in Kyoto, on August 25, 1992. Revised nine papers of them and an additional paper which was canceled in the symposium are included in this volume. Kroonenberg synthesized the chemical composition of fluvial sands in various conditions, and proposed a new diagram, SAM diagram, which enable to distinguish tectonic setting. Kumon and Kiminai also proposed two discrimination diagrams for magmatic arc provenance. One is based on traditional modal composition and the other is on chemical composition. Xia et al. elucidated the nature of the Precambrian flysch in China mainly based on chemical composition. Fielding et al. clarified the tectonic development of a Permian backarc basin in Australia, and Al-Qayim discussed the tectonic evolution of the late Cretaceous in Zagros orogenic belt in Iraq, mainly on the basis of modal composition of sandstone. Okami et al. showed the tectonic history of Northeast Japan viewed from sandstone compositions. Matsumoto elucidated the geologic development of the Miocene a backarc basin along Sea of Japan. Hisada and Arai indicated the important role of detrital spinels for researching a forearc basin history. Tokuhashi showed a good usage of heavy mineral assemblage for basin analysis. Tateishi et al. demonstrated the usefulness of chemical analysis of heavy minerals in sandstone.

The editors heartily thank Professor Emeritus Keiji Nakazawa of Kyoto University, Dr. Philip A. Jarvis, a guest researcher of Geological Survey of Japan, and Dr. K. Hisatomi of Wakayama University for their kind reviewing and advice.

    Kang-Min Yu
      Department of Geology, Yonsei University, Seoul, Korea
    Fujio Kumon
      Department of Geology, Shinshu University, Matsumoto, Japan

Effect of provenance, sorting and weathering on the geochemistry of fluvial sands from different tectonic and climatic environments

SALOMON B. KROONENBERG

Department of Soil Science and Geology, Agricultural Universiy, P.O. Box 37, 6700 AA Wageningen, the Netherlands.

Abstract: Major element chemistry of 500 samples of Quaternary sands from seven different drainage basins in Costa Rica, Colombia, Russia, Spain, France and the Netherlands has been plotted in Silica-Alkali-Mafic (SAM) diagrams to show the variability due to lithology, sorting, weathering in each basin and its impact on the interpretation of its tectonic setting. It appears that the composition of most Quaternary fluvial sediments does not depart greatly from those of published average suites of marine sandstones from similar provenances. However, sorting in mixed sediments derived from strongly contrasting source rocks such as basalt and granite can give rise to a wide range of chemical compositions, which may lead to erroneous tectonic interpretations. Tropical weathering is capable of wiping out the signature of source rocks altogether in less than 50,000 years.

Key words: Silica-Alkali-Mafic plot, bulk chemistry, major elements, variability, igneous line, maturity, Quaternary, river terraces

INTRODUCTION: THE SAM DIAGRAM

The geochemistry and petrograpy of sands and sandstone reflect basically three factors: (1) provenance, (2) transport and deposition, and (3) post-depositional processes. There is general consensus that provenance outweighs the other processes. Therefore, much effort has been made to relate modal and chemical composition of sediments to tectonic regime in the same way the geochemistry of igneous rocks is related to it.

When modal compositions are used, they are usually plotted in triangular diagrams to this end [1, 2, 3]. The main diagram QFL (Figure 1) differentiates proportions of quartz, feldspars and lithic fragments in sands. Other triangular diagrams differentiate between monocrystalline and polycrystalline quartz, K-feldspars and plagioclase, and sedimentary/metamorphic and volcanic fragments [1].

How can modal composition give clues about tectonic settings such as magmatic arcs, continental blocks and recycled orogen? This is basically because of the weathering behaviour of the pertaining lithologies. Acid plutonics produce mainly individual quartz and feldspar grains, their relative proportions depending upon the degree of chemical weathering in the source area. Schistose rocks and shales produce micaceous lithic fragments, and volcanics produce lithic fragments of volcanic groundmass, and varying proportions of individual phenocrysts. The tectonic interpretation is based on the occurrence of these rock types in specific settings. Interpretations are not always unambiguous, however: sediments from granites in active orogenic belts such as the Peruvian Coastal Batholith, uplifting old basements area or glaciated granitic Precambrian shields are not likely to differ very much in composition. The advantage of modal studies is that they enable distinction of individual components from the source area such as rock fragments and heavy minerals in sediments deposited far from the hinterland.

The geochemical study of sandstones has other advantages, apart from the rapidity of analysis and easy statistical handling, as has been shown by numerous authors [see 4, 5, 6 for major sandstone suites, and 7, 8 for Quaternary river sands]. The most straightforward distinction between different tectonic environments is the type of magmatism. Andesitic magmatism characterizes island arcs, andesitic to rhyolitic magmatism continental margin volcanic arcs, basaltic magmatism is typical for hot spots, mid-oceanic ridges and continental rift valleys. In petrographic studies, the composition of individual volcanic lithic grains in sediments cannot be distinguished because of the fine grain size of the volcanic groundmass. Geochemical composition of sediments derived from volcanic areas can give direct clues as to tectonic setting. Volcanic sands in beach barriers have been traced back by their geochemistry to individual eruptions in the source area [9].

Therefore I designed a diagram to plot chemical composition in such a way that (1) the geochemical differentiation of magmatic source rocks becomes evident, (2) the effect of sorting, and (3) the maturity of the sediments as a result of pre- or post-depositional chemical weathering or recycling of sediment, is to be seen. This SAM diagram (Figure 2) shows corner points formed by SiO_2 (divided by 20 to give a better data spread in the diagram; S), K_2O+Na_2O (Total Alkali; A), and $MgO+FeO+TiO_2$ (M). An earlier version of the diagram was published with the alkali and mafic corners interchanged [8], but I now prefer the present configuration, to facilitate comparison with Dickinson & Suczek's diagram [1].

Magmatic differentiation produces depletion in ferromagnesian elements by fractional crystallization of olivine, pyroxenes and amphiboles, and enrichment in both SiO_2 and alkalies in alkali feldspars and quartz. The SiO_2/K_2O+Na_2O ratio remains roughly constant during differentiation, as can be seen from the standard TAS diagram for the nomenclature of volcanic rocks [10]. Sands from pure basaltic to pure rhyolitic sources should therefore move from near the mafic corner to a point somewhere between the silicic-alkalic corners. This "igneous line" is at once a check for the immaturity of the sediment. In-situ weathering or increasing maturity by recycling will lead to a departure from the main igneous line towards the silicic corner. Al and Ca are excluded from the diagram, because these elements occur in both ferromagnesians and feldspars. Other diagrams can be constructed to show variation in these elements (see below, Figures 11 and 12).

SOURCES AND SINKS

One may study the sediment composition from two ends, either from the marine sink upstream, or from the upland source downstrean.

On the one hand, there is much attention for sandstone suites in large marine basins such as continental margins. These basins are of same order of magnitude as the tectonic provinces in the source area one wishes to distinguish. Good correlations between tectonic setting and sandstone chemistry are found in such a way, as for instance by Bhatia [4] and Sawyer [5] for major greywacke series. The modal and chemical signatures of such sandstones average the input from innumerable small terrestrial drainage basins of greatly differing lithologies. However, very little is known about the pathways in which the characteristics of source areas become preserved in sinks.

Therefore, studies of the pathways of recent sands in large rivers are most rewarding, as has been advocated by Potter [11] and since then by many others [12, 7]. But also here, as the emphasis is on linking sediment composition with tectonic environment, sources of variability have often been neglected by studying only sediments of specific grain sizes and averaging the chemical composition of large numbers of samples.

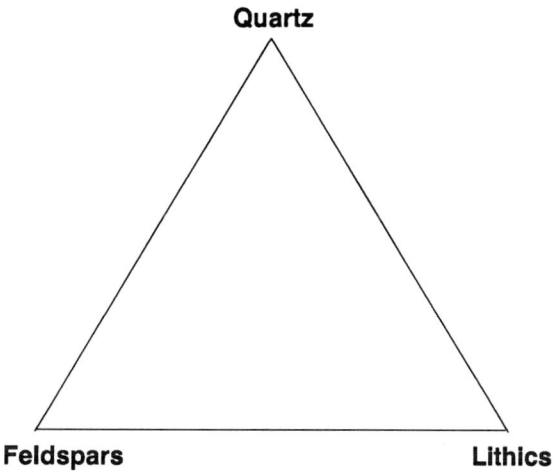

Figure 1. QFR plot for modal composition of sandstone, after [1].

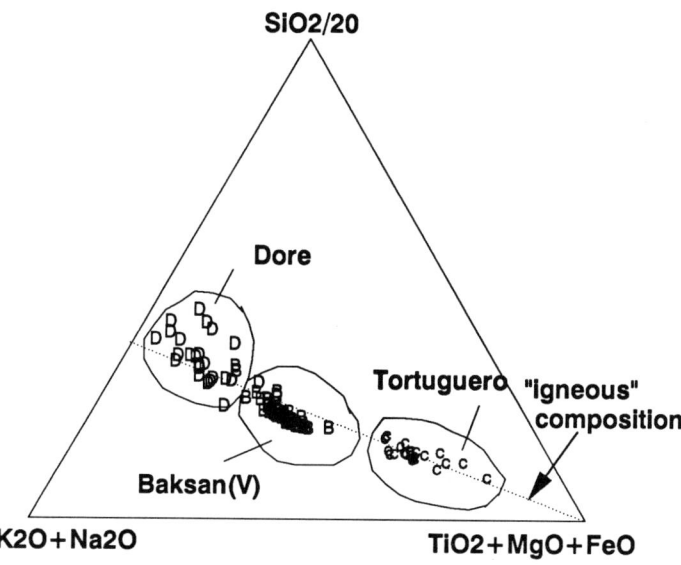

Figure 2. SAM plots for Recent sediments derived from three types of igneous rocks.

Table 1. Sampled river basins and their settings.

River (area)	Source area	Lithologies	Basin	Climate
Tortuguero (Costa Rica)	Island arc	Andesite	Retroarc	Trop
Caquetá (Colomb. Amazon)	Cordilleran	And/met/sedim	Retroarc	Trop
Guadalhorce (Spain)	Collisional	Serp/met/sedim	Intramont.	Med.
Baksan (Caucasus, Russia)	Collisional	Dacite or bsmt.	Upland riv.	Temp
Allier (C. Mass., France)	Up.basement	Basalt+granite	Rift Valley	Temp
Dore (Centr.Mass.,France)	Up.basement	Granite	Rift Valley	Temp
Maas (S. Netherlands)	Up.basement	Schist	Epicontin.	Temp

In this paper I will give a few examples of how differentiation by sorting and post-depositional weathering can strongly modify the signature from the source rocks, and how great the variability of sediment chemistry can be within single drainage basins. I will show the results of major element XRFS analyses of 500 sediment samples (mainly sands) from seven different drainage basins in different tectonic and climatic settings (Table 1). Data from each individual basin and minor element data are partly reported elsewhere, partly unpublished.

THE SIGNATURE OF PROVENANCE

To assess the effect of magmatic differentiation on sand composition, I will first study fluvial sands from three lithologically "pure" environments from well-defined tectonic settings (Figure 2).
(1) The Tortuguero River in the Atlantic Zone of Costa Rica drains the Quaternary stratovolcanoes Irazú and Turrialba, belonging to the great Central American andesitic island arc, uplifted in the Pliocene. Its Holocene sands (both from the present river bed and from sub-Recent low terraces) consist essentially of andesitic rock fragments and plagioclase and pyroxene phenocrysts, without much admixture [2, 9, 12]. Chemically they are perfectly comparable to andesites.
(2) The upper part of the Baksan River and its tributaries drain the glaciated dacitic El'brus stratovolcano (5,644 m) in the Kabardino-Balkarian Republic of the Russian Federation [14]. The Caucasian volcanism is the result of the northwards subduction of oceanic Tethys crust and subsequent continental collision. The volcanogenic sediments of the present river bed of the upper Baksan River consist essentially of dacitic lithic fragments with minor amounts of acid plagioclase phenocrysts. Other tributaries drain granitic and metamorphic terrains (not shown in this figure).
(3) The Dore River in the French Central Massif drains the uplifted, essentially granitic Hercynian basement of the Massif Central. Recent uplift and incision have shed large amounts of fresh granitic sediment in the river bed, largely undiluted by other components [15, 16, 17]. The scatter of compositions is slightly larger than in the volcanic group.

The three purely magmatic groups plot roughly on a straight line, the "igneous line" mentioned above. Sands from purely basaltic or acid volcanic environments are still lacking. It is to be noted that, while magmatic differentiation related to tectonic setting is well expressed, no distinction can be made between plutonic and volcanic sources without petrographic data. This illustrates the complementarity of both methods.

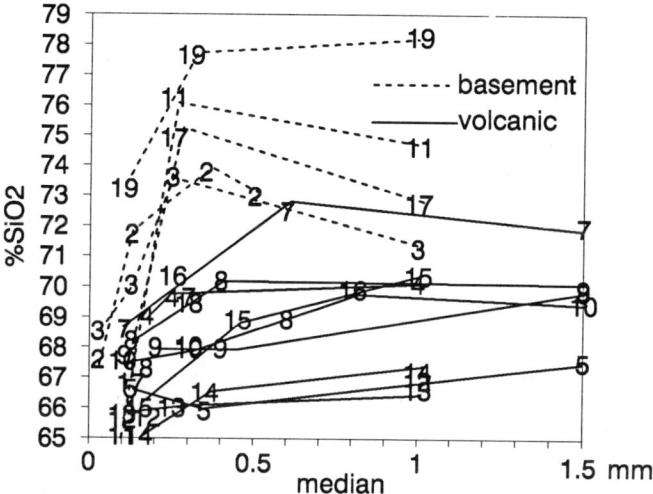

Figure 3. Relation of SiO$_2$ content to median in sands of basement and volcanic provenances from the Baksan drainage basin, N. Caucasus, Russia.

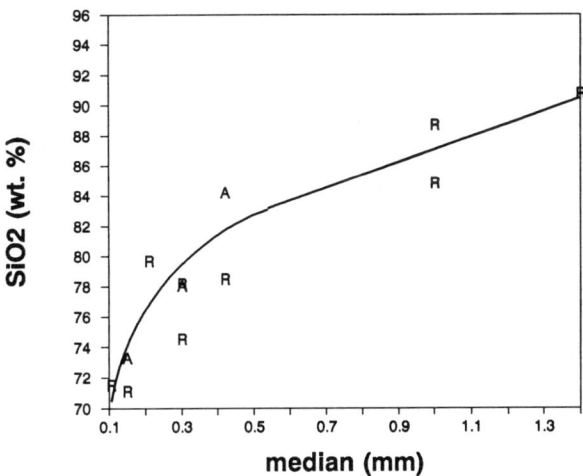

Figure 4. Relation of SiO$_2$ to median in Recent sands of mixed provenance, Caquetá River, Colombian Amazons.

THE EFFECT OF SORTING

Considerable compositional differentiation occurs by sorting during transportation and deposition. When sorting occurs in sediments in their marine sinks, as in graded bedding in turbidites, it is very difficult to evaluate how far this suppresses or enhances the signatures of specific source rocks, and hence modifies the signal of the tectonic setting. In Quaternary fluvial sands the way in which sorting and postdepositional processes affect the signature of the lithology can be studied in a much more straightforward way.

Geochemical and mineralogical differentiation by sorting is caused by (1) differences in primary grain size of specific minerals, such as zircon, (2) by differences in density, such as heavy minerals, and (3) differences in shape, such as micas [18, 19, 20, 21]. I studied sorting by analysing several samples of different granulometry from the same sampling site. I prefer this to separation of a single sample into different grain size fractions to avoid the effect of density sorting of heavy minerals: density sorting leads to different mineralogical compositions in one grain size class in dependence of the overall grain size in sediments: the fine sand fraction washed from gravel may contain more gold than the same fraction washed from pure sands, to give just an extreme example.

Modern sediments from different tributaries of the Baksan River in the Caucasus show that the degree of chemical differentiation by sorting is very much dependent on the type of rock in the drainage basin (Figure 4). Sediments from tributaries draining the El'brus volcano show flat differentiation curves (SiO_2 to median), because they consist of dacitic fragments of all grain sizes. Sediments from tributaries draining basement rocks (granites, gneisses and schists), however, are strongly differentiated, because physical weathering produces mainly individual mineral grains of greatly differing size and hydraulic behaviour (quartz, feldspars, micas). The intermediate curves are of mixed tributaries.

Differentiation by sorting in mixed sediments is even greater if the original sediments are further apart in composition. The Caquetá River, one of the tributaries of the Amazon, drains the Eastern Cordillera of the Colombian Andes, a cordilleran-type orogenic belt consisting of uplifted basement rocks, Paleozoic and Mesozoic sediments and andesitic volcanoes [22]. Deposition takes place in the retro-arc foreland basin in a tropical rainforest climate. Recent sediments contain abundant volcanic rock fragments, next to much quartz, feldspars and mica mainly of crystalline provenance. Figure 4 shows the steep differentiation curve (SiO_2 to medium) of the Recent sediments due to in-situ sorting, measured along a stretch of 250 km. When plotted in the SAM diagram (Figure 5) the most mafic sediments are close to the andesitic composition of the volcanic source rocks, but the most quartz-rich ones are considerably displaced towards the siliceous corner, outside the "igneous line". This is due to the fact that quartz predominates in coarse sand fractions, while volcanic rock fragments and individual non-quartz mineral grains are restricted mostly to the fine to medium sand fractions.

THE EFFECT OF WEATHERING

While the issue of downstream weathering during fluvial transport is far from solved [2, 3], also post-depositional weathering and diagenesis alter sediment composition in a way which may obscure the provenance signal. Weathering may be studied by comparing the composition of modern sediments with those in fluvial terraces of the same river. Fluvial terraces are old floodplains now beyond the reach of present floods as a result of uplift and/or incision of the river. Progressive weathering in a sequence of terraces of known age enables calculation of weathering rates (which may have varied according to past climatic conditions).

Late Pleistocene river terraces of the Caquetá River in the Colombian Amazons, dated

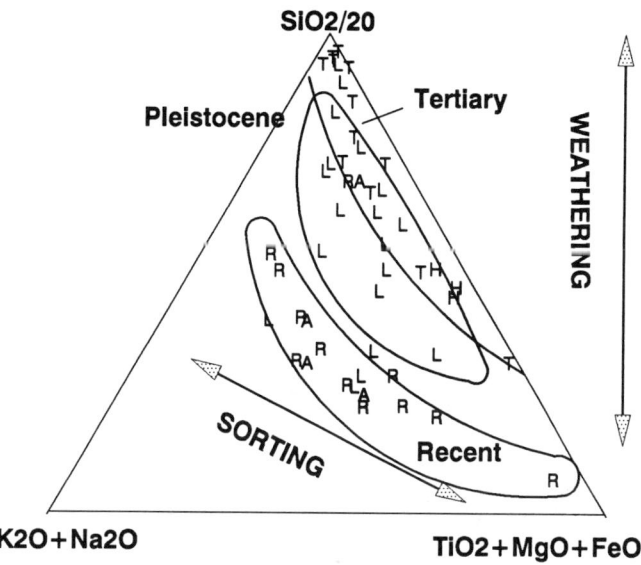

Figure 5. Chemistry of Recent, Pleistocene and Tertiary sands, Caquetá River, Colombian Amazons.

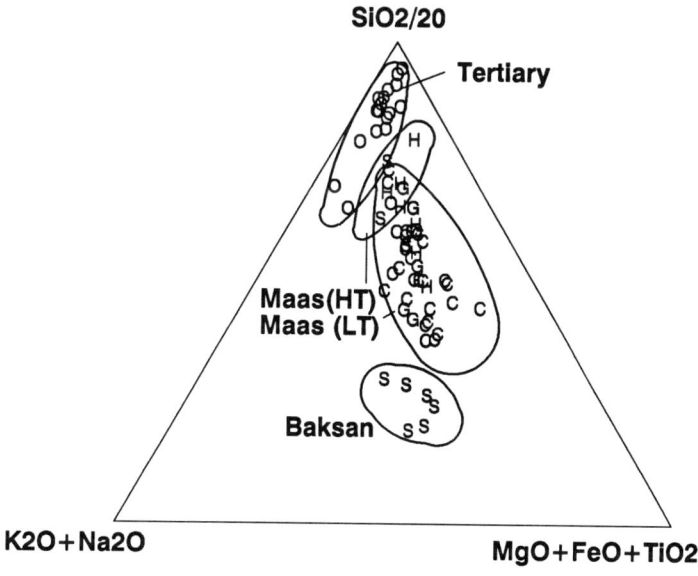

Figure 6. Chemistry of Recent, Pleistocene and Tertiary sands from uplifted low-grade metamorphic basement (Maas, the Netherlands), and low-grade metamorphic rocks in Alpine fold belt (Baksan basin, Caucasus).

30,000 – 40,000 BP have already lost so much of their less stable constituents that their chemical composition approaches that of quartz arenites [22] (Figure 5). It is striking that the data points for weathered fluvial sediments have not moved more towards the right hand side of the diagram in their course to silicic corner: apparently Fe and Ti are relatively enriched by the loss of K, Na, and Mg by weathering. River terraces in temperate climates show less advanced weathering and more influence of lithological variations, but still weathering rates can be calculated using ratios of sensitive elements such as CaO/TiO_2 [23, 8, 16].

EFFECT OF MIXING

River terraces are also useful in another way. They enable establishing how far fluvial sediments have changed in composition in time as a result of events in the drainage basin such as volcanic eruptions, tectonic uplift, glaciation, river capture etc. The River Maas originates in NE France and traverses the uplifted schistose Hercynian Ardennes Massif in Belgium before entering the Netherlands at the border of the North Sea Basin. The terraces of the River Maas in the Netherlands [24], show an increase in siliceous components (both chemically and petrographically) with age (Figure 7). These differences are not due to in situ postdepositional weathering, but reflect uplift in the hinterland: old terraces still contain large amounts of pre-uplift deeply weathered quartz-rich Tertiary sands, younger terraces contain increasing amount of schist particles set free by uplift and dissection of fresh bedrock in the Ardennes Massif [25].

The river terraces of the Allier River in the French Central Massif show the combined effect of sorting, mixing, weathering and downstream changes in composition. The Allier river, a major tributary to the Loire, occupies the Limagne Rift Valley, filled by Oligocene detritic and calcareous rocks and bordered by Hercynian and older granitic to gneissic basement, crowned by numerous Miocene to Recent volcanic centres, largely of basaltic composition. Basalt and gneiss/granite are the main source rocks of Allier sediments [23, 16]. Figure 8

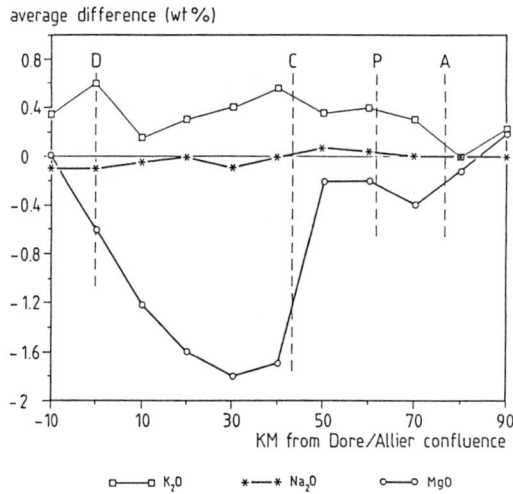

Figure 7. Longitudinal profile of average differences (wt%) of K_2O, Na_2O and MgO concentration between Holocene and Late Weichselian sands (average Holocene minus average Weichselian). After [16]. The Allier flow direction is from right to left. The main tributaries are indicated by vertical lines and symbols: A=Allagnon, P=Couze Pavin, C=Couze Chambon, D=Dore.

shows how chemical composition has changed both in alongstream direction and in time. Na_2O content in present-day sediments shows little change at major confluences, and is comparable to that in Late Glacial terraces. Sediments from the Late Glacial terrace are much richer in MgO and much poor in K_2O than present-day sediments after the confluence with the Couze Chambon River (C in Figure 8). This is due to a large influx of basalt-rich sediments by deglaciation of the Mont Dore mountain, a major composite volcano in the Allier basin. Further details are given by Veldkamp [16].

SYNTHESIS OF ALL BASINS

All data from the discussed examples are plotted together in Figure 8. The samples from Costa Rica, Caucasus, Allier and Dore (French Central Massif) all plot around the igneous line. Clusters for each basin are elongated along the igneous line, suggesting sorting of ferromagnesians and feldspars. The Allier River is not indicated as a separate cluster, as its data points occupy almost the whole "andesitic" to "granitic" range. That is because its source rocks are basalt and granite, and sorting and mixing have given rise to almost every possible sand composition between these two end members. Departures towards the siliceous corner are especially clear in the Dore, which because of its granitic character contains much more coarse-grained quartz. Sediments from mixed fold belts such as the Caquetá and the Guadalhorce, are definitely situated above the igneous line, suggesting enrichment in quartz from preexisting sedimentary rocks. The Guadalhorce River in the Betic Cordillera of southern Spain (not further discussed in this paper, Aseyeva, in prep.) contains several peridotite bodies in its basin, and therefore is displaced more towards the mafic corner, whereas igneous

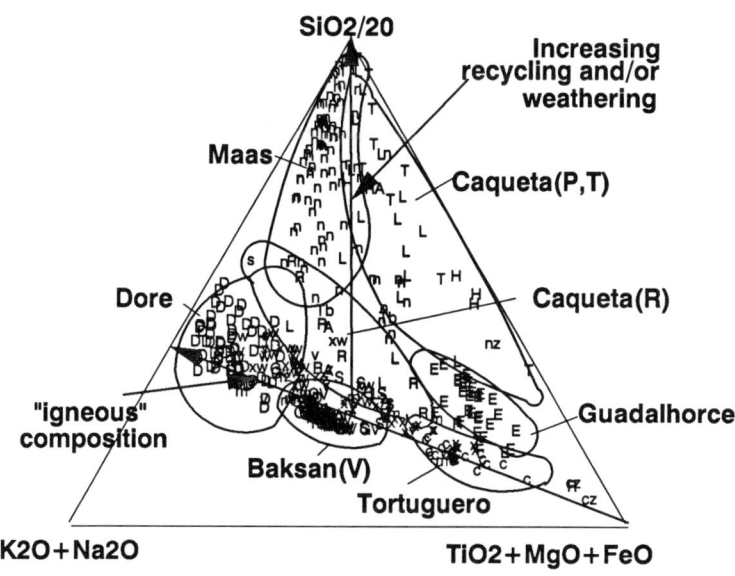

Figure 8. SAM plots of data from all seven basins.

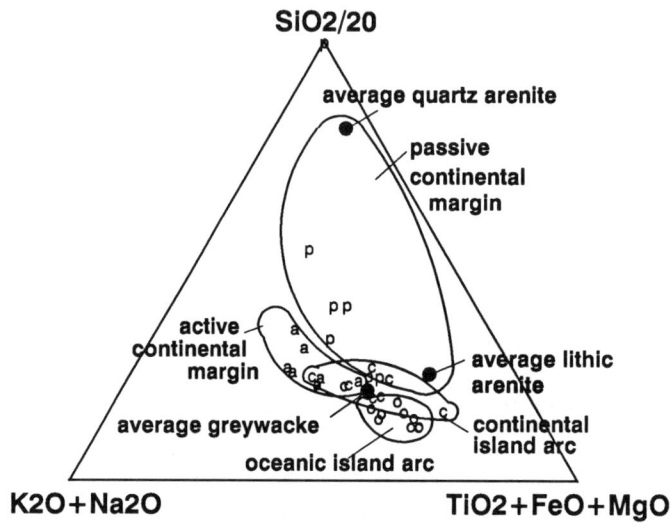

Figure 9. SAM plots of sandstone suites cited from Bhatia [4].

rocks in the Caquetá basin are andesitic to granitic. Considerable displacement towards the silicic corner is shown by the Maas sediments, due to admixture of quartz-rich Tertiary sands, and by the Pleistocene river terraces of the Caquetá, due to weathering.

The examples cited above show that the SAM diagram is adequate in portraying sediment composition in relation to lithology, sorting and weathering. In spite of a great variability of sediment composition within a single drainage basin, individual basin can be well characterized by their position in the SAM diagram. For calcareous and clayey rocks separate $SiO_2/20$ $-Al_2O_3/4-CaO$ and $CaO-K_2O-Na_2O$ diagrams can be constructed. For the present data set also these diagrams show good separation of clusters (Figures 10 and 11).

SANDS FROM SOURCE, SANDS IN SINKS

A comparison of our data with a SAM plot of geochemical data for different suites of marine sandstones in their tectonic setting from Bhatia [4] (Figure 9) shows that the variation within one basin is larger than the variability of the averages of a particular group of average sandstone suites. Nevertheless, there is a fair degree of correspondence. The Costa Rican and Caucasian volcanic samples plot around the fields of the oceanic and/or continental island arcs, the quartz-rich sediments are similar to the passive continental margin rocks. The Dore sediments from the uplifted granitic basement plot closest to the active continental margin cluster, but then any sand of granitic provenance would do so, irrespective of its tectonic setting. No allowance is made in Bhatia's plot for mafic-rich rocks from ophiolitic active continental margins as the Guadalhorce basin.

Geochemistry of fluvial sands

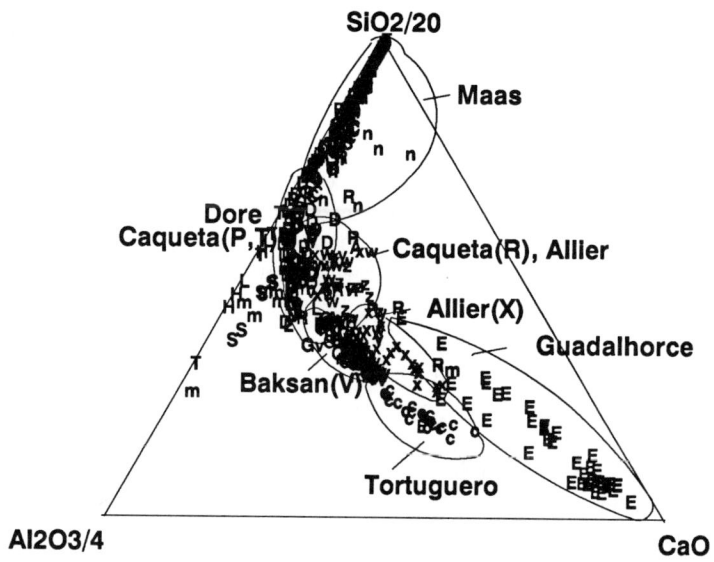

Figure 10. Silica–aluminum–calcium plots of all samples from the seven basins.

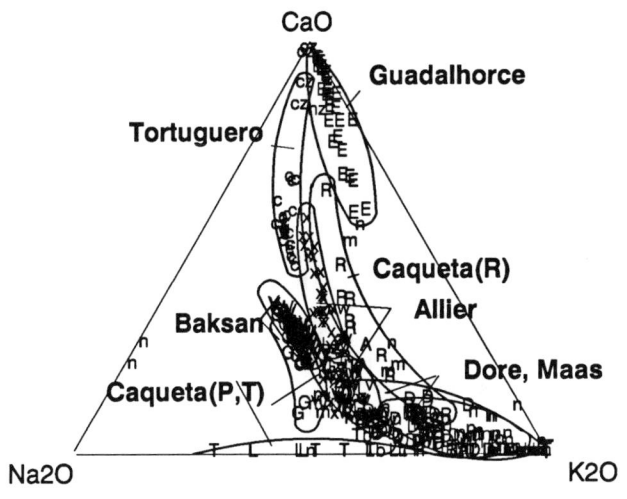

Figure 11. Calcium–sodium–potassium plots of all samples from the seven basins.

CONCLUSIONS

(1) SAM diagrams can be used to adequately portray bulk geochemical composition of sedimentary rocks as related to lithology, sorting and post-depositional weathering.
(2) In order to assess the variability of sediment composition in a drainage basin several samples of different granulometry should be taken at each sampling site. Weathering studies in terrace chronosequences should be included.
(3) In spite of the large compositional variability induced by sorting, weathering and mixing, basic chemical characteristics of most Quaternary fluvial sediments do not depart too much from those of published average suites of marine sandstones from similar provenances. Tropical weathering, however, is capable of wiping out the signature of source rocks in less than 50,000 years.

ACKNOWLEDGEMENTS

This paper is based on data collected together with many colleagues, including E.N. Aseyeva, V.V. Batoyan, M.W. van de Berg, M. de Cleen, A.N. Gennadiev, M.C. Hoorn, A.T.J. Jonker, N.S. Kasimov, A. van der Koppel, M.L. Moura, A. Nieuwenhuyse, R.J.M. van Seeters, C. Slomp, H. Tiemensma, A. Veldkamp. They are all thanked for their cooperation. A.J. Kuijper, BSc is thanked for carrying out the XRFS analyses. Mr. Ari Linna of Turku University, Finland, and Fujio Kumon of Shinshu University are thanked for their critial comments.

REFERENCES

1. W.R. Dickinson and C. A. Suczek. Plate tectonics and sandstone compositions, *AAPG Bull.* **63**, 2164–2182 (1979).
2. M.K. Johnsson. Tectonic versus chemical–weathering controls on the composition of fluvial sands in tropical environments, *Sedimentology*. **37**, 713–726 (1990).
3. P.G. De Celles and F. Hertel. Petrology of fluvial sands from the Amazonian foreland basin, Peru and Bolivia, *Geol. Soc. Amer. Bull.* **101**, 1552–1562 (1989).
4. M.R. Bhatia. Plate tectonics and geochemical composition of sandstone, *Jour. Geology*. **91**, 611–627 (1983)
5. E.W. Sawyer. The influence of source rock type, chemical weathering and sorting on the geochemistry of clastic sediments from the Quetico Metasedimentary Belt, Superior Province, Canada, *Chem. Geology*. **55**, 77–95 (1986).
6. A.B. Ronov. *The sedimentary shell of the earth*, Izd. Nauka, Moskva 79p.(1980). (in Russian).
7. R.L. Cullers, A. Basu, and L. J. Suttner. Geochemical signature of provenance in sand–size material in soils and stream sediments near the Tobacco Root batholith, Montana, USA, *Chem. Geology*. **70**, 335–348 (1988).
8. S.B. Kroonenberg. Geochemistry of Quaternary fluvial sands from different tectonic regimes, *Chem. Geology*. **84**, 88–91 (1990).
9. A. Nieuwenhuyse and S.B. Kroonenberg. Volcanic origin of Holocene beach ridges along the Atlantic coast of Costa Rica, *29th IGC, Abstract*. 318, Kyoto (1992).
10. M.J. Le Bas, R.W. Le Maitre, A. Streckeisen and B. Zanettin. A chemical classification of volcanic rocks based on the total alkali–silica diagram, *Jour. Petrology*. **27**, 745–750 (1986).
11. P.E. Potter. Petrology and chemistry of modern big river sands, *Jour. Geology*. **86**, 423– 449 (1978).
12. E. Franzinelli and P. E. Potter. Petrology, chemistry and texture of modern river sands, Amazon river system, *Jour. Geology*. **91**, 23–39 (1983).

13. R.J.M. Van Seeters. A study on the geomorphology, mineralogy and geochemistry of the Toro Amarillo-Tortuguero and the Chirripó-Matina river system in the Atlantic Zone of Costa Rica, *MSc thesis, Wageningen Univ.*, 50 p. (1992).
14. G.K. Tushinski. *The Glaciation of Mount Elbrus*, Izd. Moskva Univ. 346p. (1968). (in Russian).
15. S.B. Kroonenberg, M.C. Hoorn, M.L. Moura, and A. Veldkamp. Variability in bulk geochemistry of fluvial terrace sands, consequences for the study of weathering chronose quences, *Pedologie.* **40**, 19–31 (1990).
16. A. Veldkamp. Quaternary river terrace formation in the Allier Basin, France. A recontruction based on sand bulk geochemistry and 3-D modelling, *PhD thesis, Wageningen Univ.* 172p. (1991).
17. A. Veldkamp and S.B. Kroonenberg. The application of bulk sand geochemistry in Quaternary research. A methodological study of the Allier and Dore terrace sands (Limagne, France), *Applied Geochemistry.* **8**, 177–187 (1993).
18. W.W. Rubey. The size distribution of heavy minerals within a water-laid sandstone, *Jour. Sed. Petrology,* **3**, 3–29 (1933).
19. N.M. Strakhov. *Principles of Lithogenesis (2).* Oliver & Boyd (1969).
20. A. Basu. Petrology of Holocene fluvial sand derived from plutonic source rocks: implications for paleoclimatic interpretation, *Jour. Sed. Petrology.* **46**, 694–709 (1976).
21. V.A. Kuznetsov. *Geochemical correlations in river valleys.* Nauka i Tehnika, Minsk 288p. (1984). (in Russian).
22. S.B. Kroonenberg and M.C. Hoorn. Bulk geochemistry of Tertiary and Quaternary fluvial sands in the Colombian Amazons, *Chem. Geology.* **84**, 92–95 (1990).
23. S.B. Kroonenberg, M.L. Moura and A.T.J. Jonker. Geochemistry of the sands of the Allier river terraces, *Geologie & Mijnbouw.* **67**, 75–89 (1988).
24. M.W. van den Berg and A. Veldkamp. 3-D modelling of Quaternary fluvial dynamics in a climo-tectonic dependent system: a case study of the Maas record (Maastricht, the Netherlands), *Global and Planetary Change.* (in press)
25. C. Slomp. Geochemie en mineralogie van tertiaire en kwartaire sedimenten in Zuid-Limburg: een methode voor de bepaling van achtergrondgehalten aan zware metalen, *MSc thesis, Agric. Univ., Wageningen/State Geol. Survey.* 53p. (1990). (in Dutch).

(Manuscript received 25 August 1992, accepted 17 February 1993.)

A Precambrian Andean-type flysch: Petrology and geochemistry of Yangzhanling flysch in southern Anhui Province, China

BANGDONG XIA, XUEMIN MA, HONGBO LU, ZHONG FANG and JINCHENG ZHOU

Department of Earth Sciences, Nanjing University, Nanjing 210008, China

abstract: The sandstones in the Late Proterozoic Yangzhanling flysch are mainly lithic graywacke. In most circumstances, feldspars are abundant and quartz is moderate. Importantly, the sandstones are rich in lithoclasts of volcanic rocks in some places. The deposition of the flysch was accompanied with the extrusion of intermediate to acidic magmas. The features of light and heavy minerals and lithoclasts indicate that the source materials of the flysch are intermediate-acidic volcanic rocks, granites, epimetamorphic rocks and sedimentary rocks. The framework modal composition of the sandstones and the geochemical signatures of major elements and REE show that the flysch sedimentary basin was an interarc basin which developed under an Andean-type tectonic setting.

Key words: sandstone, petrology, geochemistry, Yangzhanling flysch, Precambrain, Andean-type, East China

INTRODUCTION

Flysch is a sedimentary assemblage of which tectonic significance is great. Flysch is formed under various tectonic settings. Flysch was firstly divided into six types by Reading [1], i.e., Atlantic, Japan Sea, Island arc, Andean, Mediterranean and Californian types. He described the general macroscopic geological features of the six types of flysch. Flysch was also divided into three types by Crook [2], according to quartz content, SiO_2 value and K_2O/Na_2O ratio of flysch sandstone, i.e., Atlantic, Andean, and magmatic island arc types. Chinese flysch was classified as three assemblage orders by Xia[3], i.e., volcanic-rich rocks order (including both lithic graywacks and quartz graywacke subdivisions), volcanic-free rocks order (including three subdivision: non-carbonate, carbonate-rich terrigeneous clastics and carbonate) and volcanic-sedimentary rocks order. The three orders are considered to occur in interplate, or active continental margin basin, faulted depression basin of intraplate, and rift basin, respectively. The above-mentioned flysch classification are of great significance. They may be used for identifying various tectonic settings to form flysch and studying plate-tectonic evolution. However, the report of detailed studies about various flysches are rare. Moreover, the research on Precambrian Andean-type flysch has not almost been found up to now.

Authors have studied the sediments of the Late Proterozoic Yangzhanling Formation (or Niuwu Formation) in southern Anhui. Now, it can be confirmed that the deposit of the Yangzhanling Formation is a rather typical Andean-type flysch.

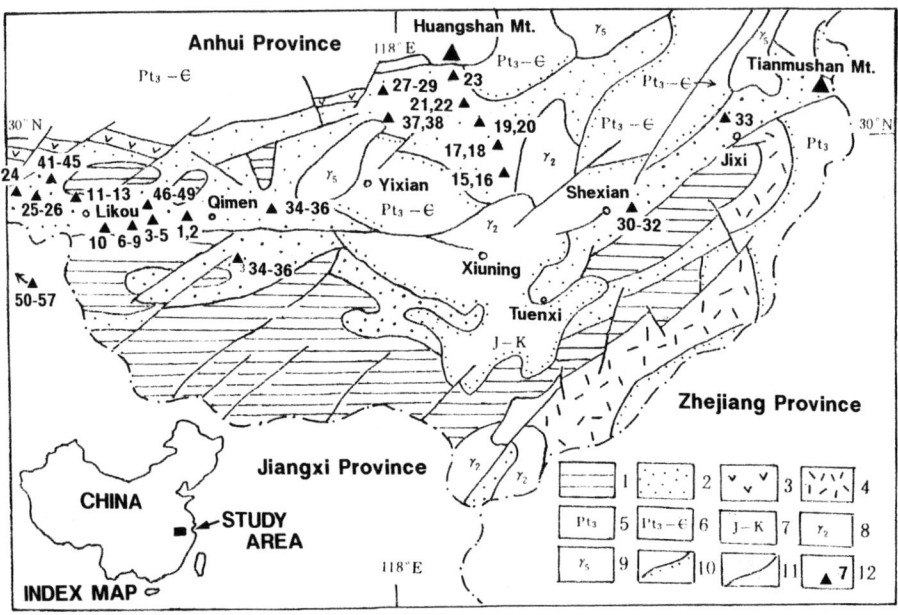

Figure 1. Schematic geological map of southern Anhui Province, China.
1: Xikou Group, 2: Yangzhanling Formation and Danjia Formation (lower and middle parts of Likou Group),
3: Puling Farmation (upper part of Likou Group), 4: Jingtan Formation (upper part of Likou Group),
5: Sinian System, 6: Sinian to Cambrian System, 7: Jurassic to Cretaceous Systems, 8: granites,
9: granodiorites, 10: Unconformity, 11: Faults, 12: Sample location

GEOLOGICAL SETTING

As a main part of the low-metamorphic basement constituted the Yangtze Plate, the Middle Proterozoic strata (Pt_2) are widely spread in the southern Anhui Province, Eastern China with a thickness more than 15,000 meters. The Pt_2 strata has been divided into the Xikou Group and the Likou Group, and they were unconformably overlain by Sinian system (Pt_3) (Figure 1). The Xikou Group consists of slightly-metamorphosed flysch-like formation, including slate, phyllite, metamorphic siltstone, sandstone and some intermediate to acid volcanic rocks. The thickness is about 10,000 meters. The Shuangqiaoshang Group (Pt_2) is considered to be contemporaneous with the Xikou Group located in a neighboring area (north-east Jiangxi Province). Its Rb-Sr whole rock isochron age is 1,401 Ma [4]. Other datings are 963 Ma and 913 Ma [5]. The Likou Group is also a suite of clastic rocks with some volcanic rocks, and is subdivided into three parts: lower part, the Yangzhanling flysch; middle part, a molasse named as Dengjia Formation, only locally occurring in this area, and upper part, volcanic rocks named as Puling(or Jingtan) Formation [6]. The Likou Group was intruded by the Xiuning granite and Shexian granite, of which the K-Ar biotite ages are 935 Ma and 877 Ma, respectively [4].

The thickness of the Yangzhanling flysch is over 2,800 meters. Grey-green and blue grey rhythmic alternations of sandstone and slate show characteristic sedimentray structures such as graded texture, sole marks and synsedimentary deformation etc. [7, 8]. Some crystalloclastic, vitric and volcanic detritus are contained in the Yanzhanling flysch sandstones. In the middle to upper parts of the Yangzhanling flysch, some bedded volcanic rocks and tuffite are interbedded within the flysch deposits. For example, a sequence of meta-rhyolite, meta-tuff yielding gravels and tuffite about 200 meters in thickness are present in the flysch deposits in Jixi County; three layers of volcanic rocks are present in the formation in Huangcun, Wujiatan and Shexian counties. They are composed mainly of andesite, rhyolite, tuff and tuffite etc., and the total thickness is over 100 meters [9]. Some synchronous andesite-prophyrite and dacite are also discovered in the Tianmushan region.

The Dengjia Formation of post-flysch deposits is composed of coarse clastic rocks, several hundred metres thick, and occur within a series of great lenticular bodies. A few volcanic rocks are alternatively contained in the deposits. The grains of the clastic rocks are coarser in the lower and finer in the upper. An unconformity is supposed between the Dengjia Formation and the underlain strata. The Dengjia Formation must be an orogenic molasse [10]. The Pulling or Jingtan volcanic rocks overlying the Dengjia Formation was a product under a volcanic arc or back-arc extensional tectonic setting [11]. The volcanic rocks developed in the south of the basin (Jingtan Formation) (Figure 1) are andesite, dacite and volcaniclastic rocks of calc-alkaline series which belongs to a volcanic arc type. The synchronous eruptive products located in the north of the basin (Pulling Formation) are composed of tholeiite-spilite assemblage, formed under a back-arc extensional tectonic setting.

PETROLOGY OF FLYSCH SANDSTONES

Flysch sandstones are lithic graywacke. The modal compositions of 26 specimens are listed in Table 1. In framework grains, quartz is 25 to 78%, average 53.8%. Maximum content of lithoclasts is 48%, average 27.3%. Maximum amount of feldspars is as high as 40%; mimimum, only a several percents and average 18.9% (Table 1).

The clastic quartz may be distinguished as monocrystalline and polycrystalline quartz. Some of monocrystalline quartz are idiomorphic. The basal plane is hexagonal. Some quartz grains exhibit arc or embayed edge. They must be volcanic in origin. Some monocrystalline quartz grain with metasomatic perforated texture might be derived from granites. The amount of polycrystalline quartz is about 1/4 of the total quartz. Some polycrystalline quartz grains are aggregates of different-sized five or more quartz crystals. The boundaries of quartz grains are either straight or sutured. According to Voll [12], these polycrystalline quartz are mainly derived from metamorphic rocks. Phyllite, slate, argillaceous rock and silicic rock are very common in rock fragmetns. Importantly, igneous rock fragments are abundant in some places, for example, in Huangshan (see the samples numbered "m-" in Table 1) located around the northern margin of the flysch basin, where volcanic fragments attain up form a half to two thirds of all rock fragments. The components of the volcanic rock fragments are mainly andesite, andesitic basalt, dacite, rhyolite, volcanic glass, associated with granites. The surfaces of volcanic rock fragments look clean; their edges and corners are clear. The most of the volcanic lithoclasts could be products of synchronous volcanic activites, which seem to be especially important in the northern margin of the flysch basin.

The plagoclases are intermediate to acidic (An = 20 to 45 %), and the potash feldspars include orthoclase, and a few of perthite and microcline. Most of them must have been derived from intermediate-acidic intrusive and extrusive rocks, judging from the feature of hypautomorphic crystal form and/or the presence of crystalloclastics .

Table 1. Modal composition of framework grains and matrix of the sandstones form the Yangzhaling flysch in southern Anhui Provine, China.

Number of order	Number of thin section	Framework Composition (%)						Mica	Chlorite	Matrix (%)
		Quartz		Feldspar		Lithoclast				
		MQ	PQ	KF	PG	VL	SL			
1	D−1	45	10	4	14	/	26	<1		25
2	D−11	50	14	4	12	/	17	3		20
3	D−33	45	15	2	8	/	30	<1		30
4	D−40	50	16	2	6	/	26			25
5	D−48	41	20	2	6	/	27	3		28
6	D−50	30	19	6	12	<1	33			30
7	D−52	30	15	5	25	/	25			25
8	D−58	60	7	1	7	5	20	<1		28
9	D−60	40	5	5	25	5	20	<1		25
13	L−18	35	10	5	20	5	20	5	<1	30
16	M−10	15	10	10	15	35	13	1	1	35
18	M−22	22	6	5	20	32	15		<1	40
19	M−29	30	15	5	10	10	30		<1	30
20	M−36	30	10	5	15	20	15	3	2	40
21	M−37	30	8	10	30	12	10	<1		30
22	M−43	25	10	5	25	15	15	5	<1	22
23	M−46	30	8	10	30	10	12			25
24	YJ−11	58	5	5	15	/	15	2		25
25	YJ−22	44	18	6	9	/	22	1	<1	23
26	YJ−30	60	15	5	10	/	10		<1	20
27	P−8	50	15	2	5	/	28			15
28	P−12	60	18	2	8	/	12			10
30	C−13	40	5	5	15	<1	35			35
31	C−15	45	15	3	12	<1	25			30
32	C−21	52	10	3	10	<1	20	3	2	25
33	Z−10	28	18	2	8	<1	40	2	2	25

MQ: monocrystalline quartz, PQ: polycrystalline quartz, KF: potash feldspars, PG: plagioclase, VL: volcanic rock fragments, SL: sedimentary rock fragments

The sandstones are poorly sorted, and the grains are angular to subrounded. The matrix ragnes from 20 % to 40 % in amount, and is mainly argillaceous. There is a small amount of siliceous and calcareous cement. The principal textural coefficients of rocks are 3.69 to 7.32, showing submature to immature [13].

All of above descriptions suggest that the flysch is the products of rapid transportation and accumulation. Epimetamorphic rocks, intermediate to acidic volcanic rocks and granites were widely distributed in the source area. There were some sedimentary rocks in the area, too.

The determination of heavy mineral components and contents from eleven sandstone samples have been made by us, there may be four kinds of rock types in the source area, i.e., granodiorite, andesite, metamorphic rocks and sedimentary rocks. The characteristic heavy mineral associations from these specimens are listed in Table 2.

Table 2. Compositions of heavy minerals of the flysch sandstones in southern Anhui Province, China.

Number of order	Number of Samples	Epidote	Hornb-lende	Garnet	Pyroxene	Biotite	Hematite and ferro-hydrite	Tourma-line	Chlorite	Zircon	Apatite	Titanite	Magne-tite	Pyrite
11	L—6	few	13.99	/	11.26	8.52	118.43	4.36	13.62	66.85	16.35	0.73	5.09	193.27
13	L—18	1.40	5.16	few	0.97	4.19	42.88	3.55	7.63	38.15	9.99	1.83	/	168.4
16	M—10	164.40	43.35	/	/	14.74	142.29	14.17	286.28	115.08	123.02	17.01	8.05	11.91
18	M—22	/	2.70	/	/	38.42	107.44	2.40	33.01	42.02	15.01	1.2	3.60	23.41
20	M—36	/	10.34	few	few	49.34	1098.21	/	79.42	31.02	54.51	0.28	3.76	38.06
29	P—20	/	8.65	/	/	2.08	834.60	/	65.74	79.93	25.61	few	/	2598.27
32	C—21	/	0.23	0.69	0.23	2.08	35.75	1.38	/	30.22	20.07	few	3.46	4.15
33	Z—10	4.72	24.78	/	/	26.55	782.60	2.66	28.02	22.71	18.88	0.89	few	17.40
34	F—7	1.02	10.86	/	3.05	5.11	222.86	15.94	24.72	75.98	44.44	0.34	/	22.05
35	F—30	/	41.96	/	/	40.54	212.66	27.74	67.21	260.31	126.25	0.36	7.47	80.73
36	F—38	few	0.23	few	/	few	101.70	0.32	0.64	53.75	15.13	0.13	/	1.29

It is clarified that some heavy minerals exhibit various features showing diverse sources. For example, some zircon grains are rounded and semirounded; some are idiomorphic crystals of macrograins (0.125 mm in width) or fine-crystalline grains (0.03 mm in width).

A definite tectonic zone always has a unique rock association. From above data, it may be inferred that the characteristic rock association of the source rocks of the Yangzhanling Formation are composed of sedimentary rocks, intermediate to acidic volcanic rocks and granitoids. The presence of this rock association in the source area suggests that the source area possesses the tectonic characters of an Andean-type orogenic belt. According to the QFL diagram of Dickinson and Suczek [14] which is used for judging the provenance of sandstone, the data-points of the sandstone of Yangzhanling flysch are plotted in the domain of the recycled orogen and magmatic arc provenances (Figure 2). This means that the source area of the Yangzhanling flysch has the characters of both recycled orogenic belt

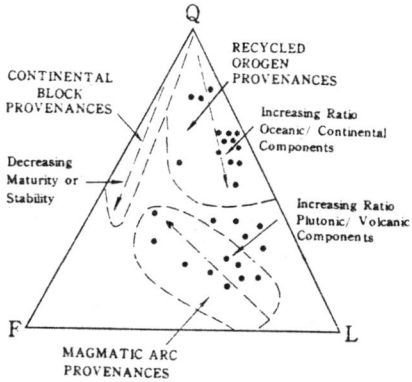

Figure 2. QFL plots of the Yangzhanling flysch sandstones (after [14]).

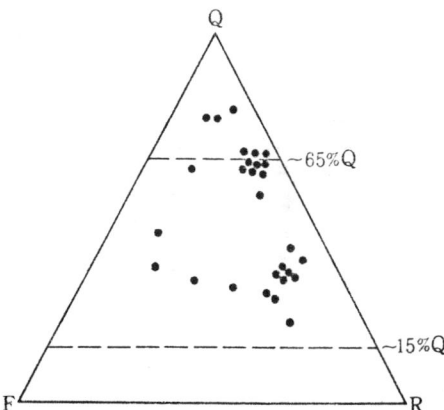

Figure 3. QFR plots of the sandstones from the Yangzhanling flysch (after [2]).

and magmatic arc. This is just one of the features of a modern Andean-type tectonic zone. In the diagram proposed by Crook [2], the most points of framework composition of the Yangzhanling flysch are plotted in the domain of "moderate quartz" (Figure 3). According to Crook [2], the tectonic setting corresponding to the domain of "quartz-intermediate" is also just modern Andean-type zone.

THE GEOCHEMISTRY OF SANDSTONE IN THE FLYSCH

1. The geochemistry of major elements

The chemical analysis of the sandstone from the Yangzhanling flysch (27 samples) are shown in Table 3. The samples show moderate SiO_2 (average 70.19%), relatively high Al_2O_3 (average 14.25%) and low K_2O/Na_2O ratio (average 0.84). According to the chemical classification of Pettijhon [15], twenty four samples are classified as graywacke, two samples as lithic sandstone, and one as arkose. The result is basically concordant with the result of modal analysis on thin section. The quartz content (50%), SiO_2 value (70.19%) and K_2O/Na_2O ratio (< 1) of Yangzhanling flysch are all coincident with those of an Andean-type tectonic setting classified by Crook [2].

The plate tectonic settings forming sandstones were divided into four types by Bhatia [16], i.e., ocean island arc (OIA), continental island arc (CIA), active continental margin (AM) and passive continental margin (PM). Each type has peculiar geochemical parameters.

In the flysch sandstones, (Fe_2O_3+MgO) is 6.37, TiO_2 0.74, Al_2O_3/SiO_2 0.20, K_2O/Na_2O 0.84, $Al_2O_3/(CaO+Na_2O)$ 4.79 in average (Table 3). These are nearly same as the model for CIA and AM. According to the discrimination diagram proposed by Bhatia [16], the fifteen flysch sandstones are located in AM, seven in CIA and five in PM (Figure 4). CIA distinguished by Bhatia [16], means a volcanic arc with normal continental crust or thined one; it is a mass split from continental crust. Such a sedimentary basin could be situated in a forearc, back-arc or inter-arc. The source rocks are felsic volcanic rocks. It is different from AM which is a kind of continental margin with very thick crust, including an Andean-type margin basin, retro-arc basin and California-type strike-slip basin.

Roser and Korsch [17] proposed a diagram to discriminate three types of tectonic settings, e.g., island arc (ARC), active continental margin (ACM) and passive continental margin (PM), based on SiO_2/Al_2O_3 VS. K_2O/Na_2O and K_2O/Na_2O VS. SiO_2 parameters of sandstones and shales. In the discrimination diagrams, 26 sandstones of the Yangzhanling flysch are plotted in ACM area, only one in PM area (Figure 5). So that, the tectonic setting of the Yangzhanling flysch should be identified as a magmatic arc occurring in the active margin, e.g., the Andean-type conntinental margin in this paper. It should be particularly noted that from available chemical models, the tectonic settings of the flysch under consideration are all judged as an active Andean-type tectonic belt.

2. REE Geochemistry

Rare earth element (REE) analysis was performed for eight sandstones from the Yangzhanling flysch measured by ICP method. The values are shown in Table 4. The schematic chondrite-normalized REE patterns indicate LREE negative slope, flat patterns of HREE and intermediate negative anomaly of Eu (Figure 6). Total REE values of the Yangzhanling flysch are nearly same as the average value of the crust 165.35 ppm [18]. But, LREE/HREE ratio are much larger than the average value of the crust, 2.5 to 3.0 [19]. This fact

Table 3. Chemical compositions of the sandstones from the Yangzhanling Formation, southern Anhui Province, China.

Number of order	Number of samples	SiO_2	TiO_2	Al_2O_3	Fe_2O_3	FeO	MnO	MgO	CaO	Na_2O	K_2O
3	D-33	76.32	0.64	12.34	1.72	2.77	0.09	1.20	0.22	2.42	1.82
8	D-58	69.06	0.89	16.67	3.36	3.39	0.08	1.62	0.17	1.88	2.66
13	L-18	71.82	0.71	14.76	0.95	4.18	0.07	1.44	0.46	2.57	2.48
15	M-3	69.19	0.80	15.58	1.07	4.71	0.09	1.76	0.85	3.70	2.42
21	M-37	65.68	0.83	17.39	1.12	4.16	0.09	1.99	0.99	4.24	2.83
27	P-8	68.39	0.63	15.20	5.36	1.13	0.04	1.60	1.03	0.90	4.89
34	F-7	74.42	0.60	13.26	1.07	3.09	0.03	1.09	0.13	3.87	1.53
35	F-30	72.58	0.68	13.75	1.99	3.08	0.07	1.49	0.44	3.06	2.14
36	F-38	72.17	0.89	16.20	2.12	1.33	0.02	0.67	0.11	2.91	3.10
41*	H-C-1	68.98	0.55	14.06	1.16	4.21	0.11	1.40	0.11	2.15	2.15
42	H-C-2	68.68	0.71	14.90	1.25	4.48	0.06	1.34	0.13	2.25	2.00
43	H-C-3	75.30	0.69	9.90	1.28	2.08	0.15	0.42	0.15	1.98	1.40
44	H-C-4	67.82	0.44	12.81	2.92	3.43	0.14	1.34	0.75	1.50	1.90
45	H-C-5	73.32	0.91	10.41	1.96	2.68	0.15	0.89	0.34	2.25	1.30
46	H-C-6	72.05	0.42	11.46	2.05	2.68	0.17	0.70	0.88	2.60	0.38
47	D-C-1	73.06	0.49	12.26	0.80	3.04	0.06	0.52	0.89	2.25	1.75
48	D-C-2	68.85	0.81	11.59	1.51	2.46	0.23	1.13	0.17	2.60	1.50
49	D-C-3	72.89	0.71	12.74	1.28	3.21	0.07	1.11	0.22	2.30	1.80
50	HK-1	71.24	0.33	12.21	1.33	2.94	0.13	0.30	0.39	3.30	1.25
51	HK-2	65.39	0.66	15.80	1.84	3.85	0.11	0.13	0.44	3.10	2.85
52	HK-3	67.42	0.73	15.29	2.80	3.22	0.09	0.15	0.63	3.40	2.45
53	HK-4	66.79	0.80	16.45	1.58	3.41	0.06	1.30	0.71	3.18	3.00
54	HK-5	69.86	0.79	13.90	1.12	3.22	0.11	0.27	0.38	3.55	2.30
55	HK-6	64.06	0.71	17.59	1.55	3.40	0.08	0.32	0.90	2.35	3.80
56	Z-C-1	68.66	0.85	13.09	0.59	4.69	0.10	1.21	0.27	2.70	1.53
57	Z-C-2	63.80	0.76	16.15	0.45	4.89	0.07	0.33	0.50	2.15	2.50
58	Z-C-12	70.80	0.70	11.97	0.67	5.45	0.10	0.95	0.09	2.65	1.10
Average		70.19	0.74	14.25	1.77	3.31	0.06	1.13	0.49	2.71	2.33

* 41-58 from [8]

Table 3. (continued)

P_2O_5	LOI	Total	$Fe_2O_3^* + MgO$	$\dfrac{Al_2O_3}{SiO_2}$	$\dfrac{SiO_2}{Al_2O_3}$	$\dfrac{Al_2O_3}{CaO+Na_2O}$	Fe_2O_3+MgO	$\dfrac{K_2O}{Na_2O}$
0.11	2.62	101.66	5.52	0.16	6.18	4.67	2.92	0.75
0.09	4.24	99.84	8.03	0.24	4.14	8.13	4.97	1.41
0.14	3.09	99.61	6.48	0.21	4.87	4.87	2.39	0.96
0.18	2.46	99.36	7.43	0.23	4.38	3.42	2.83	0.65
0.19	3.26	99.54	7.16	0.26	3.78	3.33	3.11	0.67
0.29	2.72	99.48	7.55	0.22	4.50	7.88	6.96	5.56
0.08	1.96	100.49	5.14	0.18	5.61	3.32	2.16	0.40
0.15	2.54	99.43	6.36	0.19	5.28	3.93	3.48	0.70
0.10	2.54	99.64	3.91	0.22	4.45	5.36	2.79	1.06
0.12	2.71	101.67	2.55	0.20	4.91	5.27	2.55	1
0.16	3.12	100.69	2.59	0.22	4.61	4.31	2.59	0.89
0.17	4.13	100.51	1.70	0.13	9.61	3.02	1.70	0.71
0.13	2.56	99.96	4.26	0.19	5.29	4.28	4.26	1.27
0.12	2.74	99.59	2.85	0.14	7.04	5.01	2.85	0.58
0.14	1.12	101.11	2.75	0.16	6.29	6.32	2.75	0.53
0.09	1.14	100.28	1.32	0.17	5.96	4.23	1.32	0.78
0.20	1.15	101.15	2.64	0.17	5.94	3.49	2.64	0.58
0.21	2.24	99.32	2.39	0.17	5.72	5.21	2.39	0.78
0.19	2.15	99.41	1.63	0.17	5.83	4.11	1.63	0.38
0.11	3.15	99.81	2.97	0.24	4.14	3.16	2.97	0.92
0.10	2.89	99.85	2.95	0.23	4.41	6.10	2.95	0.72
0.17	1.26	100.19	2.88	0.25	4.06	3.25	2.88	0.74
0.13	1.45	100.52	1.39	0.20	5.03	5.44	1.37	0.65
0.09	2.21	99.82	1.87	0.27	3.64	3.20	1.87	1.62
0.28	3.16	99.76	1.80	0.19	5.25	7.48	1.80	0.57
0.07	2.28	100.02	0.78	0.25	3.95	4.18	0.78	1.16
0.09	2.51	100.15	1.62	0.17	5.91	4.32	1.62	0.42
0.15			6.37	0.20	5.21	4.79		0.84

implies that there was a thick crust in the provenance. This is an important characteristics of the Andean-type tectonic zone.

Comparing with the REE discrimination parameters of sandstones from different plate tectonic settings [20], the average value of the Yangzhanling flysch is well concordant to that of the Andean-type (Table 4).

Wang et al. [21] made REE scatter diagram for distinguishing source rock type in provenance. An andesite type is discriminated for the Yangzhanling flysch on REE scatter diagram.

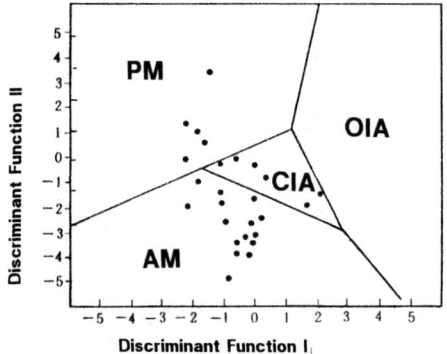

Figure 4. Discrimination diagram of tectonic setting for the sandstones of the Yangzhaling flysch based on a chemical compositon function pair after [16].
PM: Passive continental margin, AM: active continental margin, CIA: continental island arc,
OIA: Ocean island arc

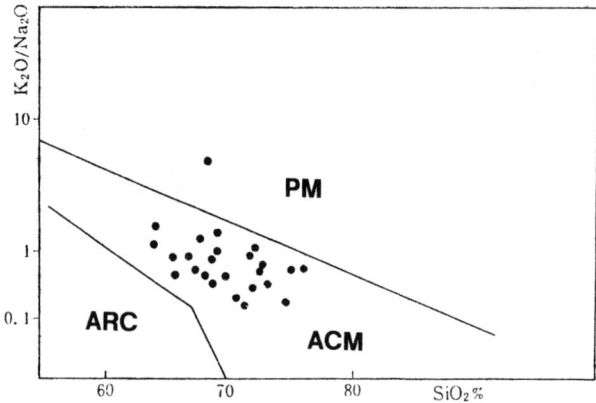

Figure 5. Discrimination diagram of tectonic setting for the sandstones of the Yangzhanling flysch. The diagram is after [17].
PM: passive continental margin, ACM: active continental margin, ARC: island arc

Table 4. REE contents of the flysch sandstones from the Yangzhanling Formation, southern Anhui Provine, China. (ppm)

Number of order	Number of soumples	La	Ce	Pr	Nd	Sm	Eu	Gd	Tb	Dy	Ho	Er	Tm	Yb	Lu	Y	ΣREE	La/Yb	ΣLREE/ΣHREE	Eu/Eu*
2	D—11	36.53	70.06	9.07	33.05	7.57	1.13	7.19	1.20	7.16	1.85	4.21	0.61	4.15	0.56	37.96	184.34	8.80	5.84	0.51
12	L—14	49.62	91.00	11.56	41.77	8.70	1.51	7.34	1.08	6.33	1.79	3.90	0.53	4.10	0.56	33.87	229.79	12.10	7.97	0.62
14	Ls—3	28.89	59.82	7.31	26.87	6.18	1.37	5.88	0.80	5.66	1.35	3.05	0.44	3.06	0.42	26.42	151.09	9.44	6.31	0.75
15	m—3	42.69	76.93	8.94	30.15	5.96	1.29	4.42	0.63	3.06	0.94	1.56	0.20	1.46	0.18	15.05	178.41	29.23	13.33	0.80
17	m—13	24.69	51.32	6.19	21.86	5.11	0.93	4.57	0.75	4.17	1.14	2.68	0.40	2.80	0.37	22.12	126.97	8.82	6.52	0.63
24	YJ—11	39.36	81.97	9.90	35.29	7.23	1.44	5.69	0.83	4.92	1.46	3.07	0.40	3.16	0.43	26.03	200.27	12.39	8.77	0.72
37	0—10	37.10	79.39	9.13	32.43	6.92	1.39	6.02	0.85	5.68	1.50	3.48	0.45	3.49	0.48	30.10	188.31	10.63	7.58	0.70
38	0—14	46.46	92.26	10.44	36.04	7.52	1.67	6.39	0.95	5.27	1.50	2.97	0.36	2.90	0.36	26.55	215.12	13.31	9.39	0.79
Average		38.17	75.34	9.07	32.19	6.90	1.34	5.94	0.89	5.28	1.44	3.12	0.42	3.14	0.42	27.26	184.27	13.09	8.21	0.69
Andean* type Continental margine		37	78														186	12.5	8.1	0.6
Passive* Continetal margine		39	85														210	15.9	8.5	0.56

* from [19]

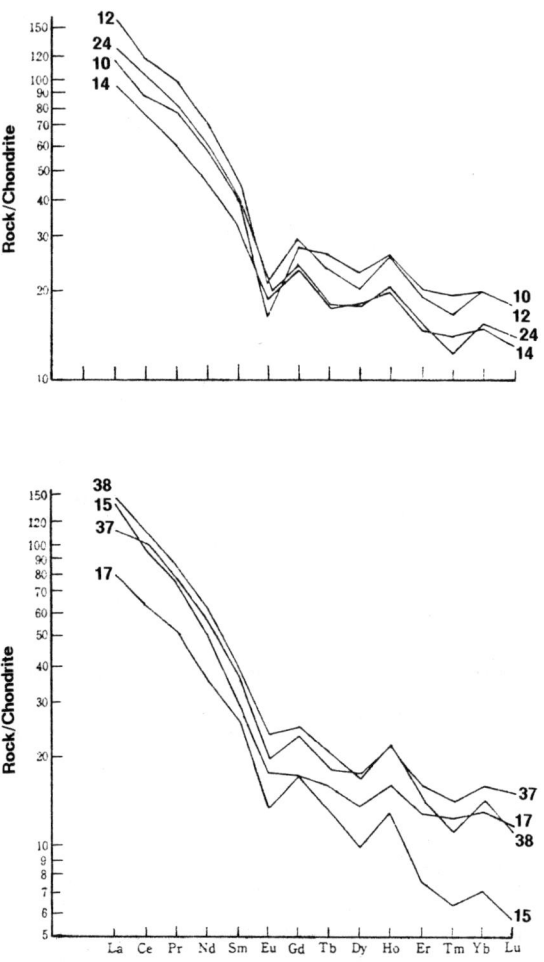

Figure 6. Schematic chondrite-normalized REE patterns of the sandstones from the Yangzhanling flysch. LREE negative slope, flat for HREE, and intermediate negative anomaly of Eu.

CONCLUSIONS

The Late Proterozoic Yangzhanling flysch was one of a typical Andean-type. The formation of the flysch basin may be related to the extension of an active volcanic arc located on mature crust. During the flysch deposit accumulation the volcanic activity was proceeding, and the detritus came mainly from the volcanic arc. This conclusion is coincident with an inference about the presence of a great "Jiangnan" old island arc series [22], in which the studied area is included.

ACKNOWLEGEMENTS

We should give thanks to Dr. Roger C. Brewer for his kindness to discuss the manuscript of this paper. Dr. Fujio Kumon reviewed this paper meticulously. This study was sponsored by the National Natural Science Foundations of China.

REFERENCES

1. H.G. Reading. Global tectonics and the genesis of flysch successions, *24th Intern. Geol. Congr. Proc.* Sect. 6, 59–66 (1972).
2. K.A.W. Crook. Lithogenesis and geotectonics: The significance of compositional variation in flysch arenites (graywackes), In: *Modern and Ancient Geosynclinal Sedimentation*. R.H. Dott Jr. and R.H. Shaver (eds). SEPM, Spec. Pub., no.19, 304–310 (1974).
3. Xia Bangdong. Research on flysch in China and some problems concerned, *Acta Sed. Sinica.* **4**, no.1, 49–65 (1986). (in Chinese with English abstract).
4. Bureau of geology and mineral resoures of Anhui Province. *Regional of geology of Anhui Province*, Geological Memoirs. Series 1, no.5, Geological Publishing House, Beijing, 721p. (1987). (in Chinese with English abstract).
5. Zhou Xinmin and Wang Dezi. The Peraluminous granodiorites with low initial $^{87}Sr/^{86}Sr$ ratio and their genesis in Southern Anhui Province, Eastern China, *Acta Petrol. Sinica*, no.3, 37–45 (1988). (in Chinese with English abstract).
6. Xia Bangdong. Precambrian strata in southern Anhui Province and the metamorphosed volcanic rocks in them. *Jour. Nanjing Univ.(Geology)*. no.1, 89–97 (1962). (in Chinese).
7. Xu Shutong, Lu Jingyuan, Zhang Weiming and Chen Guanbao. Primary tectonic environment and deformation of strata of Proterozoic metamorphics in the Qimen-Shexian region of southern Anhui, *Scientia Geol. Sinica*. no.2, 101–115 (1979). (in Chinese with English abstract).
8. Zhou Hongrui. Late Proterozoic strata and texture-paleogeography in north-eastern Jianxi Province and southern Anhui Province, In: *The Tectonic History of Ancient Continental Margin in South China*. Wang Hongzhen (ed.), Geological Publishing House, Beijing, 173–182 (1983). (in Chinese).
9. Anhui No.332 team for geological survey. *1: 50000 report of regional geological survey at Jixi County*. (1989). (in Chinese) (in press).
10. Xia Bangdong, Fang Zhong, Lu Hongbo and Yu Jinhai. Molasse and global tectonics, *Experimental Petroleum Geol.*. **11**, 314–319 (1989). (in Chinese with English abstract).
11. Liu Shouhe and Xia Bangdong. Late Proterozoic volcanic rocks and their tectonic setting in southern Anhui Province. *Jour. Nanjing Univ.(Earth Science)*. no.1, 43–52 (1990). (in Chinese with English abstract).
12. G. Voll. New work on petrofabrics, *Liverpool and Manchester Geol. Jour.*, **2**, 503–567 (1960).
13. Song Tianrui. A new formula for counting "Predominant textural coefficient" of sandstone in thin sections, *Geological Review*. **25**, 43–47 (1979). (in Chinese with English abstract).
14. W.R. Dickinson and C.A. Suczek. Plate tectonics and sandstone compostions, *AAPG Bull.* **63**, 2164–2182 (1979).
15. F.J. Pettijohn. *Sedimentary rocks*. 3rd edition, Harper and Row Publishers., New York (1975).
16. M.R. Bhatia. Plate tectonics and geochemical composition of sandstones, *Jour. Geology*. **91**, 611–627 (1983).
17. B.P. Roser and R.J. Korsch. Determination of tectonic setting of sandstone-mudstone suites using SiO_2 content and K_2O/Na_2O ratio. *Jour. Geology*. **94**, 635–650 (1986).
18. Li Tong. Chemical element abundances in the Earth and it's major shells. *Geochimica*. no.3, 167–174 (1976). (in Chinese).
19. Liu Yingjun and Cao Liming. *Introduction on element geochemistry*. Geological Publishing House, Beijing (1987). (in Chinese).
20. M.R. Bhatia. Rare earth element geochemistry of Australian Paleozoic greywackes and mudrocks: Provenance and tectonic control, *Sed. Geology*, **45**, 97–113 (1985).

21. Wang Xianjue, Chen Yuwei, Lei Jianquan, Wu Mingqing and Zhao Yiyang. REE geochemistry in sea-floor sediments in the continental shelf of East China Sea. *Geochimica.* no.1, 56-65 (1982). (in Chinese with English abstract).
22. Guo Lingzhi, Shi Yangshen, and Ma Ruishi. Proterozoic plate tectonic movement in south China and resultant and evolution of island arc tectonics, In: *Proceedings of International symposium on Precambrian Crustal Evolution.* Geological Publishing House, Beijing, no.1, 30-37 (1986). (in Chinese with English abstract).

(Manuscript received 25 August 1992; accepted 22 February 1993.)

Petrofacies analysis and tectonic evolution of Zagroside flysch suites from northeastern Iraq

BASIM AL-QAYIM

Department of Geography, College of Arts, University of Baghdad, Baghdad, Iraq

Abstract: The Zagros orogenic belt in northeastern Iraq is associated with flysch-type sediments represented by the Tanjero Formation (Late Campanian to Maastrichtian), and the Kolosh Formation (Paleocene to Early Eocene). Investigations of the stratigraphic associations, lithologic characters and sandstone petrofacies of these sediments from several outcrops in northeastern Iraq revealed their tectonic evolution and association to the adjacent Zagros crush zone.

The petrofacies analysis of sandstones from both units indicates a "recycled orogen" association. The Tanjero Formation is characterized by a thick sequence of proximal turbidites and shows a deposition in a remnant oceanic basin. This basin is developed between the Arabian plate on the one side and the subduction complex to the northeast. The latter is dominated by the complex assemblages of phiolite, radiolarite and metasediments, and supplied most of the detrital sands of the sediments. The basin filling continued as the compressional tectonics progressed. Eventually and by the end of the Cretaceous, the northeastern part of the basim was uplifted, and incorporated into the already exposed subduction complex. This unstable tectonic conditions ultimately leaded to the closing of the basin and the generation of the a new peripheral foreland basin. This basin of the Paleocene time extended over the Arabian plate margin southward of the precursor oceanic basin. It is characterized by distal turbidite sequence of the Kolosh Formation.

Key words: Zagros orogenic belt, sandstone, petrofacies, trench, foreland basin, Iraq, Arabian plate, Zagros

INTRODUCTION

Studies of the relationship between sandstone detrital composition and tectonic setting become increasingly important and widely used [1, 2, 3]. The work of Dickinson and Suczek [3], however, is the most conventional and widely referred to. It furnishes an elaborate model which simply demonstrates the different types of sandstone composition and its relations to different tectonic settings of source areas. This model which inspired the author has been applied to sandstones from northeastern Iraq [4, 5]. This paper further discusses the characters of flysch sediments from northeast Iraq and its association to the adjacent Zagros orogenic belt.

The studied flysch sediments are the Tanjero Formation (Late Campanian to Maastrichtian) and the Kolosh Formation (Paleocene to Early Eocene). Both units represent a part of the imbricated and high-folded zone of northeast Iraq, and extended in a longitudinal belt trending northwest-southeast parallel to the adjacent thrust zone (Figure 1). The high-folded zone is characterized by Jurassic to Paleogene folded and faulted strata of several kilometers

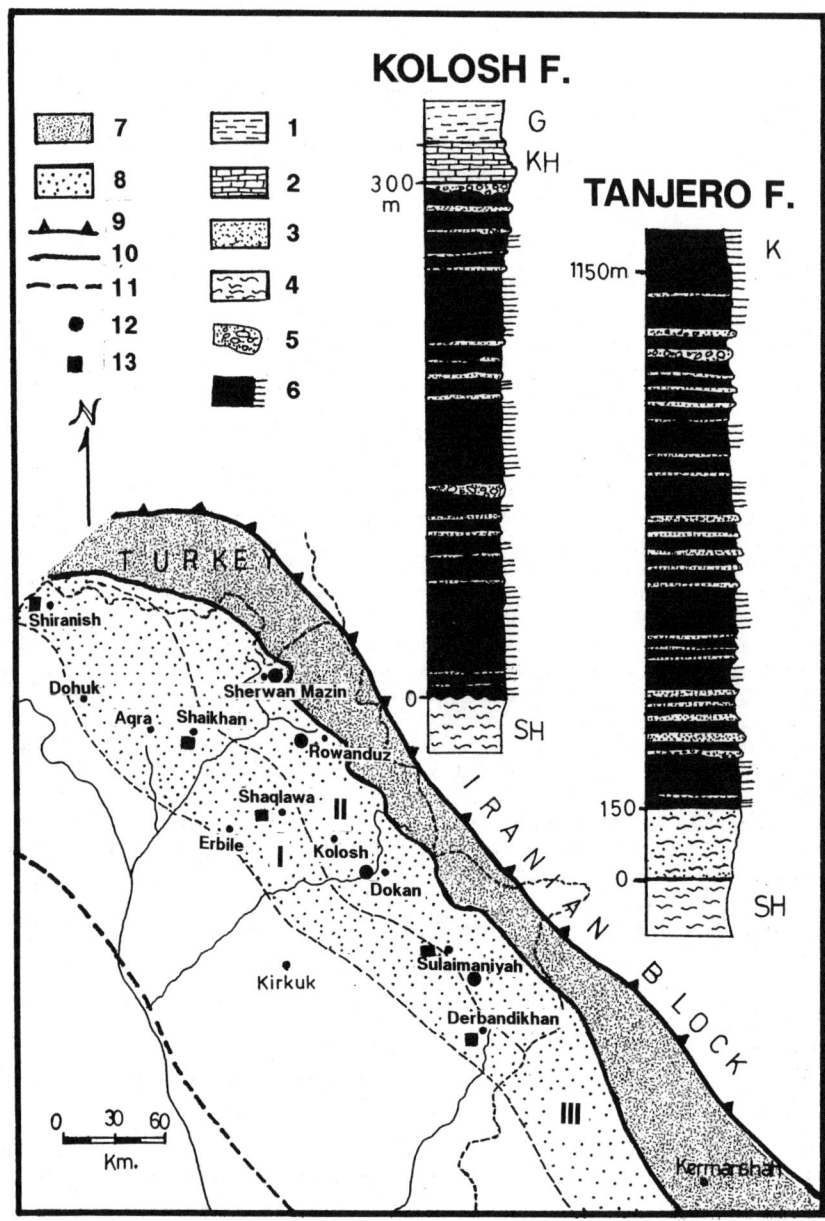

Figure 1. Index map of northeast Iraq, showing the major tectonic zones, distribution of flysch sediments and studied localities, as well as the representative lithologic sections of the Kolosh and Tanjero Formations.
1: shale, 2: limestone, 3: sandstone, 4: marlstone, 5: channel conglomerate, 6: thin-bedded turbidite, 7: thrust zone (Zagros crush zone), 8: flysch sediments (I; Kolosh Formation, II; Tanjero Formation, III; Ameran Formation), 9: Main Zagros thrust, 10: front of Zagros crush zone, 11: limit of Zagros folded belt, 12: localities for Tanjero Formation, 13: localities for Kolosh Formation, G: Gercus Formation, KH: Khurmalah Formation, K: Kolosh Formation, SH: Shiranish Formation.

in thickness. The thrust zone on the other hand constitutes a part of the Zagros crush zone which is represented by the eugeosynclinal rocks sandwiched between the Arabian plate and the Sanandaj-Sirjan block of the Iranian plate [6]. This zone generally consists of a complex association of ophiolites, radiolarites, volcanics, plutonics, as well as sedimentary and metamorphic rocks. All are intensively folded and faulted into mega-thrust-sheets overriding southwestwardly the adjacent imbricated zone. The Zagros folded belt is continued southwestward with decreasing folding intensity (Figure 1).

The flysch sediments under investigation are believed to have been deposited in a deep and linear trough [7]. It extends southeastward into Iran, and recognized as the undifferentiated (Maastrichtian to Paleocene) Ameran Formation [8]. The Ameran flysch is believed to have been deposited in a similar basin along the northern part of the Zagros basin [6]. Northwestward in the southeastern part of Turkey, the flysch belt is continued, and known as the undifferentiated (Maastrichtian to Paleocene) Germave Formation [9]. In Iraq the flysch sequence is differentiated into two units, the older Tanjero Formation and the younger Kolosh Formation. The former is more thickly and better developed, and distributed immediately adjacent to the thrust zone. The latter has a smaller thickness, less developed flysch sequence, and distributed southwestward (Figure 1). The Kolosh Formation overlays the Tanjero Formation at many localities, i.e., Kolosh village, and thus furnishes the complete section of the flysch sequence.

Four localities for the Tanjero Formation and five localities for the Kolosh Formation were chosen for detailed examination and sampling. The localities for the Tanjero Formation include: (1) Sherwan Mazin about 15km north of Mergasur nearby the Iraqi-Turkish border, (2) Rowanduz locality: along Rowanduz river at outskirts of Rowanduz town, (3) Dokan locality: outsides Dokan, about 1 km west of the road to Dokan, (4) Sulaimaniah locality 5 km south of Sulaimaniah town along the road to Derbandikan. The Kolosh Formation localities are (1) Shiranish locality: about 0.5 km south of Shiranish Islam village, (2) Shaikhan locality: along the road to Ain Sifni, about 5 km southwest of town, (3) Shaqlawa locality: along the road between Salahuddin and Shaqlawa, (4) Sulamaniah locality: along the road to Krikuk, about 15 km west of Sulaimaniah town, and (5) Derbandikhan locality: nearby the Derbandikhan Dam site (Figure 1).

Seventy sandstone samples from the Tanjero and seventythree samples from the Kolosh were collected, and thin sectioned for petrographic analysis.

STRATIGRAPHY AND SEDIMENTOLOGY

The stratigraphic status of the two units is originally reviewed by Bellen et al.[10]. A distinction between the two units is mainly based on micropaleontological criteria. The Tanjero Formation is named from the type locality near Halabja, about 80 km southeast of Sulaimanyiah city. The thickness of the formation exceeds 2,000 m. It consists of two lithologic parts. The lower part is 484 m thick, and consists of silty greenish marlstone of basinal type sediments. The upper part is 1,532 m thick, and consists of dark olive-green shale and mudstone which alternate with medium to thick bedded sandstone and thin bedded siltstone, conglomerate, and lenticular fossiliferous limestone. The upper part bears the flysch-proper features. At the type locality, the Tanjero Formation is topped by 30 m thick conglomerate layer which separates it from the overlying Kolosh Formation. This conglomerate and the faunal break are taken to mark a disconformable relation between the two units [10]. A similar relationship is noticed in the Dokan [11] and Rowanduz areas [10].

A representative stratigraphic section is measured and investigated in detail at the Rowanduz locality by the author. It is about 1,150 m in thickness as shown in Figure 1. The lower part, 150 m thick, consists of grayish-green silty marlstone and shale, and represents a transitional interval from the underlaying Shiranish Formation.

The upper part is 1000 m thick and characterized by olive green to grayish green shale and silty marlstone which rhythmically alternate with 10 to 60 cm, medium- to coarse-grained, graded sandstone beds (Figure 2a). Sandstone beds reflect typical characters of turbidite sequence. At certain parts of the sequence sandstone beds become thicker, and more frequent, and even amalgamated (Figure 2c). At Sherwan Mazin the Tanjero Formation is characterized by thick sandstone (exceed 2 m) with ball and pillow structures. The thickness of the formation at this locality is relatively thin, 560 m thick. It is probable that the thickness is reduced due to the subsequent erosion. At Dokan locality, the formation is about 1,350 m thick, and characterized by the occurrence of fossiliferous limestone bed near the top of the formation [11]. At Sulaimaniah locality, the formation shows no significant variation from the type section except its thickness which reaches 1,400 m. The sediments of the Tanjero Formation represent the miogeosynclinal facies of Northeast Iraq basin [7], and received materials from a source area located to the east and northeast of the basin [12, 13].

The Kolosh Formation on the other hand is described for the first time from the type locality at Kolosh village near Koisanjaq (Figure 1). It consists of two distinctive lithologic parts. The upper part is 294 m thick, and characterized by alternation of shelf carbonates, gray shale, marlstone and sandstone. The lower part is 483 m in thickness, and shows features of flysch type sediments. It shows alternation of thin sandstone bed with thick dark gray silty shale interlayers. At this locality, the Tanjero Formation disconformably underlies the Kolosh Formation. The upper contact is transitional to the overlying Gercus Formation (Figure 1). The upper part of the formation which embraces shelf carbonates is often considered as a tongue from the lagoonal (Khurmala Formation) or shoal (Sinjar Formation) equivalents of the Kolosh [7].

A detailed section measured at Shaqlawa shows domination of turbidite sequence with distal facies [14]. The thickness of this sequence is 300 m. The dominating lithologic type is silty shale which alternate rhythmically with thin-bedded sandstone 10 to 30 cm thick (Figure 2b). Sandstones are brownish in color and calcareous with variable tool mark structures. Sometimes the sandstone beds become thicker up to 5 m, lenticular in shape and pebbly. Thin, 10 to 25 cm, marlstone or limestone beds are frequently encountered. Lenticular conglomerate horizons are characteristic, but infrequent (Figure 1). At the Shaikhan locality the formation is noticeably thin, 45 m, and is characterized by conglomeratic and pebbly sandstone with lenticular bedding structure (Figure 2d). At this locality the Kolosh Formation is unconformably underlain by the Aqra-Bekhme Limestone Formation. The small thickness of the formation in this locality and the pebbly nature of the formation suggest the uplifting and subsequent erosion of the area. A similar situation is reported from the Aqra area [15]. To the northwest in the Shiranish area the formation regains its normal thickness (200 m) and flysch type characters. It, however, lacks conglomerate and shelf carbonate beds suggesting a relatively distal location in the basin [16]. Toward the southeast, at Sulaimaniah and Derbandikhan, the formation becomes thicker (more than 500 m) with a typical flysch-type feature. Sandstone beds become thicker and more frequent, and coarser in grain size, and conglomerate horizons become more frequent. All these features suggest that the formation is approaching to the source land southeastward.

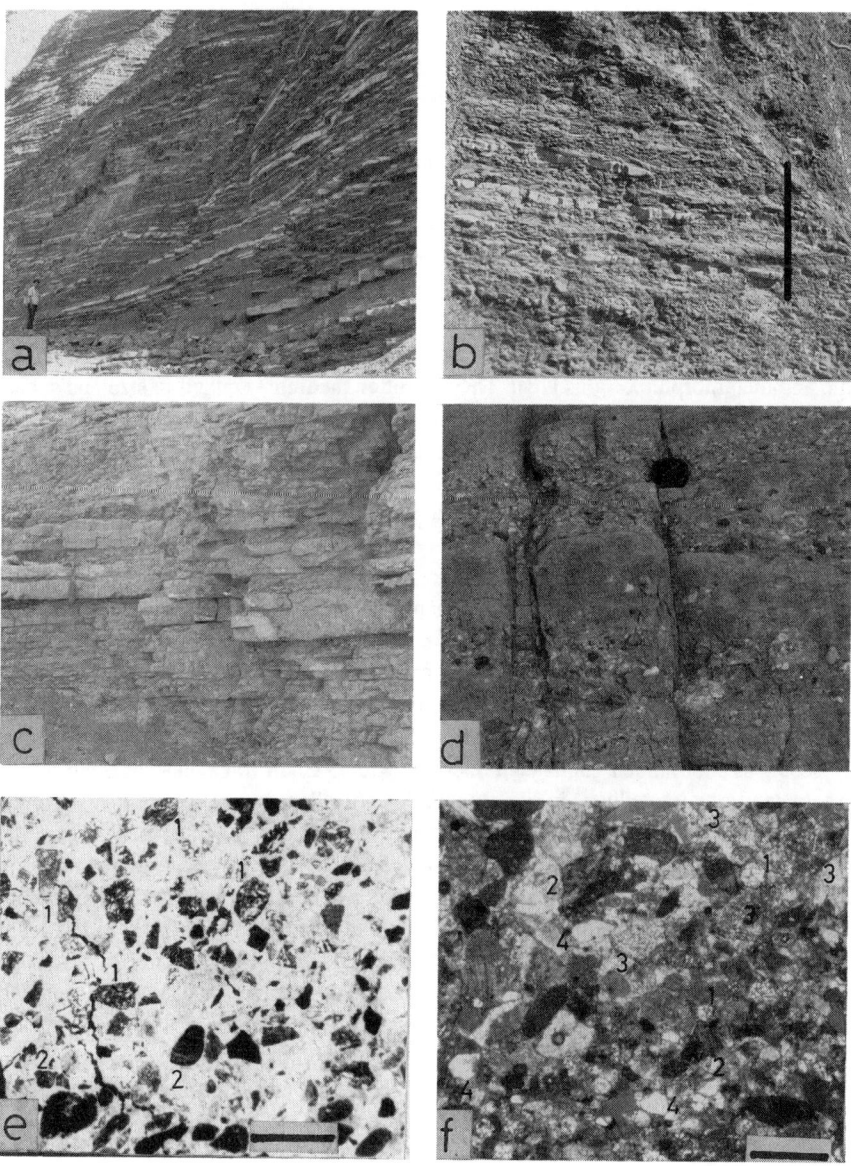

Figure 2. (a) Proximal turbidite of the Tanjero Formation at Rowanduz locality. (b) Distal turbidite of the Kolosh Formation at Shaqlawa area. Bar is 1 meter long. (c) Thick sandstone beds of the Tanjero Formation at Rowanduz. (d) Intercalations of pebbly sandstone and conglomerate are the characteristic features of the Kolosh Formation at Shaikhan. (e) Photomicrograph of Kolosh sandstone showing domination of angular chert fragments(1), metamorphic rock fragments(2) and calcareous cement, Shaqlawa Loc. Cross Nicols. Bar is 1 mm long. (f) Photomicrograph of Tanjero sandstone showing derived radiolaria (1), metamorphic rock fragments (2), chert fragments (3) and subrounded quartz grains (4). Sulamaniah Loc. Cross Nicols. Bar is 1 mm long.

SANDSTONE PETROGRAPHY

More than 140 thin sections were made for sandstones from both units. Sandstone samples were collected from various parts of each section to attain optimum representation. Emphasis were given to relatively thick beds with medium to coarse grain size and considerable lateral continuity. A point-count technique was applied to evaluate the percentage of most important petrographic constituents in each thin section. At least 500 points were counted per one thin section.

The sandstones of the Tanjero Formation are rich in rock fragments and are of lithic arenite type [17]. However, it shows relatively high contents of quartz and feldspar as compared to the sandstone of the Kolosh Formation. Quartz grains are of various types including volcanic, plutonic and metamorphic origins [13]. They are often medium-grained in size and subangular in shape (Figure 2f). Feldspar grains are usually coarse-grained, and of orthoclase type. Lithic detrital grains include abundant coarse and subangular radiolarian chert. Derived detrital radiolarians are commonly noticed (Figure 2f). Subrounded carbonate fragments are common. Argillaceous fragments are fairly common and finer in grain size. Igneous rock fragments exist both of plutonic and volcanic types. Metamorphic rock fragments are fairly common with a characteristic occurrence of green schist, and amphibolite. Clayey matrix is variably recognized, and sometimes it becomes partly calcareous and silty. Carbonate cement is uncommon and often developed by diagenetic replacement of carbonate grains.

The sandstones of the Kolosh Formation are generally of lithic arenite type. Dominant detrital grains include carbonate, chert and metamorphic rock fragments in decreasing order. The carbonate fragments are most abundant with an average exceed 50 % in some cases. It usually occurs as subrounded grains with micritic texture. Chert fragments are often subangular in shape and fine to medium in size (Figure 2e). Argillaceous grains include subrounded, medium sand-size shale, mudstone and marlstone fragments. Igneous rock fragments are common coarse to medium in size, and subrounded shape. Most of these grains are of basic and ultrabasic types. Metamorphic rock fragments are characteristic and fairly common. They are dominated by amphibolite schist. Quartz grains occur in a fairly small amount, usually of fine sand size and monocrystalline type. Feldspars are generally of low content. They occur mainly as K-feldspar type with an occasional occurrence of plagioclase. Matrix is clayey micrite and usually replaced by calcareous cement.

In general, sandstones from the Tanjero Formation are comparatively coarser in grain size, less clayey than usual, and contain higher amount of quartz and feldspars as well as igneous rock fragments. The sandstones from the Kolosh Formation, on the other hand, are generally finer in grain size, and contain more calcareous cement and higher amount of chert, carbonate and metamorphic rock fragments.

PETROFACIES ANALYSIS

The sandstone detrital modal data were recalculated into new categories to fit the triple poles of the QFL and QmFLt diagrams of Dickinson and Suczek [2]. The mean values of these modal data at each locality and for each formation are shown in Table 1.

The QFL and QmFLt plotting of modal data of both formations shows that the two units are closely associated with the field of the "recycled orogen"(Figure 3). Most of the examined localities tend to be associated with the lower end (L and Lt poles) of the recycled orogen field. Such a tendency which reflects certain variance within the studied assemblages is interpreted to represent a provenance characterized by either foreland uplift and/or subduction complex (QFL diagram), or subduction complex (QmFLt diagram) [2]. Moreover, it imply a high oceanic/continental component ratio [18].

The characteristic abundance of chert debris of the examined flysch sediments which exceed the combined amount of quartz and feldspar is considered as a key signal which indicate derivation of such sand from an uplifted subduction complex piled up into the adjacent trench [2]. The relatively low content of quartz and feldspar in these flysch sediments (Table 1), further indicates a derivation from such a source land [18].

The QFL and QmFLt plots of the Tanjero Formation, however, show tendency toward magmatic arc association (Figure 3). Such admixture of debris from arc terrain with detritus from subduction complex is conceivable [2]. The implication of an uplifted foreland belt in the source area, especially for the Kolosh Formation could be acceptable. The high content of carbonate rock fragments is often considered as an important signal to such an implication. Evidently, a redeposited upper Cretaceous fauna reported from the lower part of the Kolosh Formation at many localities [10, 19] suggests an incorporation of upper Cretaceous sediments into the uplifted complex.

The general composition of the assigned subduction complex as a major source for the sand of the studied flysch is mirrored in the type and relative abundance of these detrital grains. It seems obvious that the significant component of this subduction complex is radiolarite, carbonate,and igneous rocks of ophiolite assemblages.

The green schist and other type of metamorphic rock debris, though less common, but conspicuous, reflects regional metamorphism of the frontal parts of thrust sheets [20, p. 244].

TECTONIC ENVIRONMENT AND EVOLUTION

The petrofacies analysis of the studied sandstones points towards a subduction complex-trench setting for the evolution of the flysch sediments of northeast Iraq. The thrust zone of NE Iraq which extends by the northeast edge of the flysch belt, seems to represent the remains of that subduction complex. Remnant patches of ophiolite complexes are reported from three different areas within the thrust zone [21]. The Qulqula Formation of NE Iraq (Aptian to Albian) characterized by radiolarian chert and limestone is exposed within the thrust zone and is believed to have significantly contributed to the abundant chert debris of the flysch basin [7]. Furthermore, the Qulqula Formation is associated with syndepositional breccia and conglomerate which are believed to be equivalent to the colored melange unit of the Iranian crush zone [8, 20]. This distinguished association represents an indicative feature of a subduction complex [22]. Paleocurrent studies of the Kolosh and Tanjero sediment from many localities support derivation from source area located to the east-northeast of the basin. An area is occupied now by thrust zone [12, 13]. Thrusting, uplifting, and subsequent erosion of the subduction complex rocks during the Late Cretaceous time furnish most of the detrital sediments of Tanjero and Kolosh flysch. The deposition seems to be rapid and carried out into a trench-type basin developed adjacent to the subduction complex. Such a basin is characterized by an elongated, narrow and rapidly subsiding trough [23]. The elongated and narrow belt of the flysch suite, their relatively thick sequence, and continuous deep marine

Table 1. Average of modal point–count data of sandstones from the flysch sediments at the studied localities, northeast Iraq.

Fm.	LOCALITIES	SAMPLES	Q	F	L	Qm	F	Lt
K O L O S H	SHIRANISH	11	46	10	44	8	10	82
	SHAIKHAN	12	50	2	48	10	2	88
	SHAQLAWA	10	44	6	50	6	6	88
	SULAIMANIYAH	10	56	4	40	3	3	94
	DERBANDIKHAN	27	53	1	46	5	1	94
T A N J E R O	SHERWAN MAZIN	15	42	15	43	17	15	68
	ROWANDUZ	28	50	13	37	22	13	65
	DOKAN	18	58	10	32	15	10	75
	SULAIMANIYAH	12	50	3	47	3	3	94

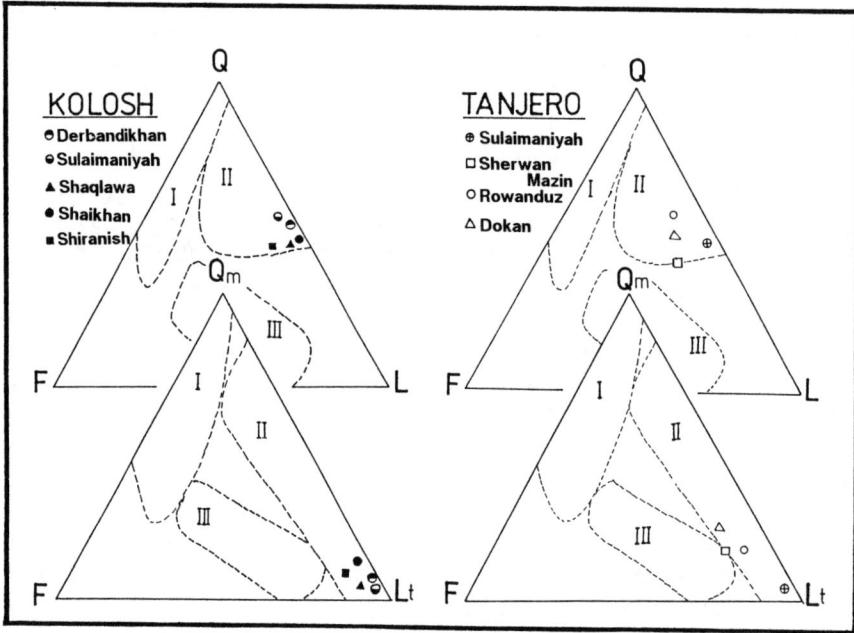

Figure 3. Averaged QFL and QmFLt plots of sandstones from the flysch sequence of northeast Iraq, showing tectonic environment of the studied units. Modal data from Table 1.
(I) Continental Block, (II) Recycled Orogen, (III) Magmatic Arc.

and turbidite facies reflected the basic characters of such a basin.

The flysch sediments of west Iran (Lorestan) which represent a continuation to the flysch belt of northeast Iraq are believed to be deposited in a trench or remnant oceanic basin of similar setting. Their depositional basin is designated as the "Kirmanshah basin" [24].

Since those flysch sediments represent an integral part of northeast Iraq flysch, therefore both should be assigned to one continuous basin. This basin is called the Rowanduz-Kermanshah basin (Figure. 4).

The deposition of flysch sediments in this basin began by the Early Maastrichtian time [10, 8]. Prior to this time the basin was receiving pelagic sediments of the Shiranish Formation (in Iraq) and Gurpi Formation (in Iran). The flysch deposition continued throughout the Maastrichtian time. By the end of the Cretaceous period, deposition of the flysch was interrupted in many areas in Iraq, and marked by the erosional and/or disconformable transition from the Tanjero to the Kolosh Formation [10, 25, 11]. Such an interruption implies an uplifting and erosion of (at least) certain parts of the already deposited Tanjero Formation before the resumption of the flysch deposition of the Kolosh Formation (Paleocene). However, no angular unconformity between the two units is reported in Iraq or elsewhere, which indicates mild tectonic interruption of the basin. Incidentally, no such interruption is reported from the flysch sequence of Iran [8]. Areas of pervasive tectonic interruption in Iran or Iraq could have been exist, but may be hidden underneath the overthrust sheets of the thrust zone after subsequent compression and suturing. At any rate, based on the available data, it seems fairly certain that a part of the Tanjero flysch was uplifted, eroded and incorporated into the sediments of the Kolosh Formation.

The Paleocene flysch sediments are characterized by relatively small thickness, distal facies and southwestward shifting of basin axis. Such changes indicate shallowing of the Tertiary flysch basin as well as migration away from the orogenic belt. The tectonic interruption of the basin and its shallowing and migration southwestward off the orogenic belt might suggest a transition from a remnant oceanic basin onto a peripheral foreland basin similar to the one described by Dickinson [26]. The shelf carbonates which characterize the upper part of Kolosh Formation at many localities mark the ultimate shoaling of the basin and a period of calming tectonism [27].

CONCLUSIONS

The Tanjero and Kolosh Formations of northeast Iraq constitute an integral part of the Maastrichtian to Paleocene flysch belt which extends along the southwest side of the Zagros orogenic belt. Petrofacies analysis of sandstones from both units reveal their "recycled orogen" provenance. The petrographic variances within these sandstones suggest a source area of an uplifted subduction complex. This subduction complex is now constituting a part of the thrust zone of northeast Iraq and the northeastern wing of the so-called Zagros crush zone. These flysch sediments were accumulated in a closing remnant oceanic basin (trench) developed to the southwest of the subduction complex, and to the northeast of Arabian plate margin.

The early stage of flysch filling (Maastrichtian) is characterized by a thick sequence of proximal turbidite facies. The late stage of basin filling was commenced during the Paleocene time after a local uplift of the basin and a partial erosion of the precursor Maastrichtian flysch sediments (Tanjero Formation). Flysch deposition was resumed as the basin continued its

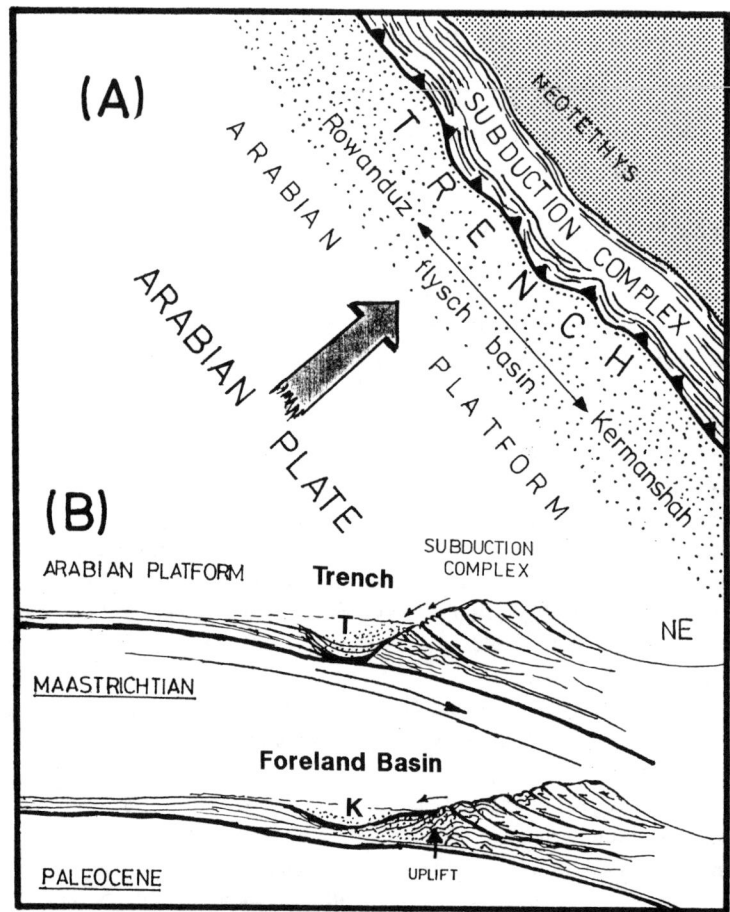

Figure 4. Schematic presentation of tectonic setting and evolution of flysch sediments of northeast Iraq. (A) Map view. (B) Cross-sectional presentation of basin evolution. T: Tanjero Formation, K: Kolosh Formation.

subsidence. Flysch sediments of this stage is of distal facies, and smaller in thickness, and migrating further southwestward suggesting a shallowing and migration of the basin axis away from the orogenic belt. The basin during this stage could have been transformed into peripheral foreland basin evolved southwestward from the precursor trench. An ultimate shoaling of the basin around the early Eocene time is marked by a local development of shelf carbonates on the top of the flysch sequence.

REFERENCES

1. F. Schwab. Framework mineralogy and chemical composition of continental margin-type sandstone. *Geology*, **3**, 487–490 (1975)

2. W. Dickinson and C. Suczek. Plate tectonics and sandstone compositions, *AAPG Bull.*, **63**, 2164-2182 (1979).
3. W. Dickinson, L. Beard, G. Brakenridge, J. Erjavec, R. Fergunson, K. Inman, R. Knepp, F. Lindberg and P. Ryberg. Provenances of North American Phanerozoic sandstones in relation to tectonic setting, *GSA Bull.*, **94**, 222-235 (1983).
4. B. Al-Qayim. Petrofacies analysis of Paleocene flysch-like sediments form north Iraq, *8th Geol. Cong. Baghdad, Abstract with program*, 39, (1988)
5. B. Al-Qayim. Petrogenic association and evolution of a first order foreland basin, northern Iraq, *13th ISC, Nottingham, Abstracts*, 412 (1990).
6. M. Berberian and, G. King . Towards a paleogeography and tectonic evolution of Iran, *Canad. Jour. Earth Sci.*, **18**, 210-265 (1981).
7. T. Buday. Regional geology of Iraq, Vol.I: Stratigraphy and Paleogeography, *Geol. Surv. Iraq Pub. Baghdad* (1980).
8. G. James and J. Wynd. Stratigraphic nomenclature of Iranian oil consortium agreement area, *AAPG Bull.*, **49**, 2182-2245 (1965).
9. I. Altinli. Geology of eastern and southeastern Anatollia. *Turkey Min. Rep. Expl. Inst. Bull.*, 35-76, (1966).
10. V. Bellen, H. Dunnington, R. Wetzel and D. Mirton. Lexique Stratigraphique International, *Asia, fasc. 10a, Iraq. Paris*, 333 (1959).
11. S. Al-Shaibani, B. Al-Qayim and L. Salman. Stratigraphic analysis of the Tertiary-Cretaceous contact, Dokan area, north Iraq. *Jour. Geol. Soc. Iraq*, **19**, 101-110 (1986) .
12. I. Al-Rawi. Sedimentology and petrology of the Tanjero clastic formation in north and northeast Iraq, *Ph.D. Thesis, Univ. of Baghdad* (1986).
13. A. Saadallah and A. Hassan. Sedimentological study of selected sections of the Tanjero Formation (Iraq), *Iraqi Jour. Sci.*, **28**, 483-518 (1987).
14. B. Al-Qayim and l. Salman. Lithofacies analysis of Kolosh Formation, Shaqlawa area, north Iraq, *Jour. Geol. Soc. Iraq*, **19**, 107-121 (1986).
15. H. Dunnington. Generation, migration, accumulation and dissipation of oil in Iraq, In: *Habitat of Oil.* W. Weeks (ed.). Symp., Tulsa, 1194-1251 (1958).
16. B. Al-Qayim, B. Nisan and M. Al-Mutuali. Petrology and mineralogy of the Paleogene clastic strata of northwest Iraq. *Jour. Sci. and Nature, Univ. Salauaddin, Iraq*. (IN PRESS)
17. B. McBride, Classification of common sandstone, *Jour. Sed. Petrology*, **33**, 664-669 (1963).
18. F. Schwab. Sedimentary signatures of foreland basin assemblages: real or counterfeit? In: *Foreland basins*. P. Allen and P. Howmewood (eds). IAS Spec. Pub., 8, 395-410 (1984).
19. I. Kassab. Planktonic foraminifeidal ranges in the type Kolosh Formation (Middle-Upper Paleocene) of NE Iraq, *Jour. Geol. Soc. Iraq*, **9**, 54-99 (1976).
20. T. Buday and S. Jassim. The regional geology of Iraq, Vol. II. Tectonism, Magmatism. *Geol. Surv. Iraq Pub. Baghdad* (1980).
21. S. Jassim, J. Walahausrova and M. Suk. Evolution of magmatic activity in Iraq Zagros complexes, *Krystalin-ikum*, **16**, 87-108 (1982).
22. W. Dickinson and D. Seely. Structure and stratigrapy of forearc regions, *AAPG Bull.*, **63**, 2-31 (1979).
23. M. Underwood and S. Bachman. Sedimentary facies associations within subduction complexes. In: *Trench-forearc Geology: Sedimentation and Tectonics on Modern and Ancient Active Plate Margins*. J. Leggett (ed.). 537-550, Blackwell Scientific Publications, Oxford (1982).
24. V. Cherven. Tethys-marginal sedimentary basins in western Iran, *GSA Bull.*, **97**, 516-522.
25. I. Kassab, F. Al-Omari and N. Safawee. The Cretaceous-Tertiary boundary in Iraq (represented by the subsurface section of Sasan well No.1, NW Iraq), *Jour. Geol. Soc. Iraq*, **19**, 129-167 (1986).
26. W. Dickinson. Plate tectonics and sedimentation, In: *Tectonics and Sedimentation*. W. Dickinson (ed.). SEPM Sep. Pub., **22**, 1-27, Tulsa (1974).
27. B. Al-Qayim and B. Nisan. Sedimentary facies analysis of a Paleogene mixed carbonate-clastic sequence, Haibat-Sultan ridge, northeast Iraq, *Iraqi Jour. Sci.*, **30**, 525-558 (1989).

(Manuscript received 28 November 1992; accepted 8 February 1993.)

Permian to mid-Triassic evolution of sandstone composition in a complex back-arc extensional to foreland basin: the Bowen Basin, eastern Queensland, Australia

CHRIS R. FIELDING[1], PATRICE DE CARITAT[2], JULIAN C. BAKER[3]
and MELVILLE M. WILKINSON[4]

[1] *Department of Earth Sciences, The University of Queensland, 4072, Queensland, Australia*
[2] *Department of Geology and Geophysics, The University of Calgary, Calgary, Alberta T2N 1N4, Canada*
[3] *Centre for Microscopy & Microanalysis, The University of Queensland, 4072, Queensland, Australia*
[4] *AGL Petroleum Pty. Ltd., P.O. Box 1010, Brisbane, 4001, Queensland, Australia*

Abstract: The Permo-Triassic Bowen Basin is a back-arc extensional to foreland basin which developed landward of an intermittently-active continental volcanic arc associated with the eastern Australian convergent plate margin. The basin has a complex, polyphase structural history which began during the Early Permian with limited back-arc crustal extension. This created a series of north-trending grabens and half grabens which, in the west, were infilled by quartz-rich sediment derived locally from surrounding uplifted continental basement. In the east, coeval calc-alkaline volcanolithic-rich and volcaniclastic sediment was derived from the active volcanic arc.

This early extensional episode was followed by a phase of passive thermal subsidence and episodic compression during the late Early Permian to early Late Permian, with little contemporaneous volcanism. In the west, quartzose sediment was shed from stable, polymictic, continental basement lying immediately to the west and south of the basin, whereas volcanolithic-rich sediment which entered the eastern side of the basin during this time was presumably derived from the inactive and possibly partly submerged volcanic arc.

During the late Late Permian, flexural loading and increased compression occurred, and renewed volcanism took place in the arc system to the east of the Bowen Basin. The reactivation of this arc led to the westward and southward spread of volcanolithic-rich sediment over the entire Bowen Basin. The Bowen Basin was from this stage onward a true retroarc foreland basin.

Arc-derived sediment continued to accumulate in the Bowen Basin until about the latest Middle Triassic. Compression climaxed during the earliest Late Triassic, finally leading to major uplift, deformation and erosion of the basin fill.

Key words: Sandstone petrology, Provenance, Bowen Basin

INTRODUCTION

The Bowen Basin is a large, elongate Permo-Triassic sedimentary basin extending in a north-south direction for about 900 km in eastern Queensland and northeastern New South Wales, Australia (Figure 1). It contains up to 10 km of variably deformed, marine and non-marine siliciclastic sedimentary rocks which, in the south, are unconformably overlain by relatively flat-lying Jurassic and Cretaceous sedimentary rocks of the Surat Basin. Hosting economic accumulations of coal, natural gas and precious metals, the Bowen Basin is one of the most intensely explored sedimentary basins in Australia.

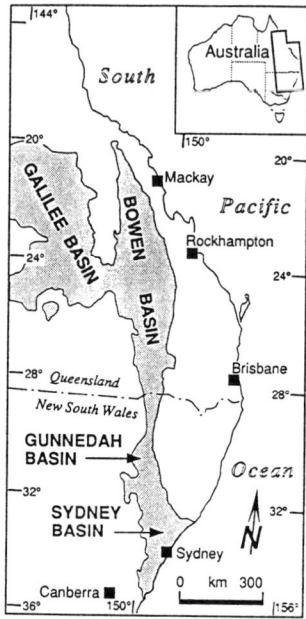

Figure 1. Location of the Bowen Basin.

STRATIGRAPHY

Figure 2 shows the Permian and part of the Triassic stratigraphy of the Bowen Basin. Local stratigraphic terminologies for the Permian sequence reflect the difficulty in correlating the Permian units across the entire basin, a problem compounded by the occurrence of major depositional breaks within the Permian sequence. Only the sections in the southwestern, southeastern and central parts of the basin will be discussed in this paper.

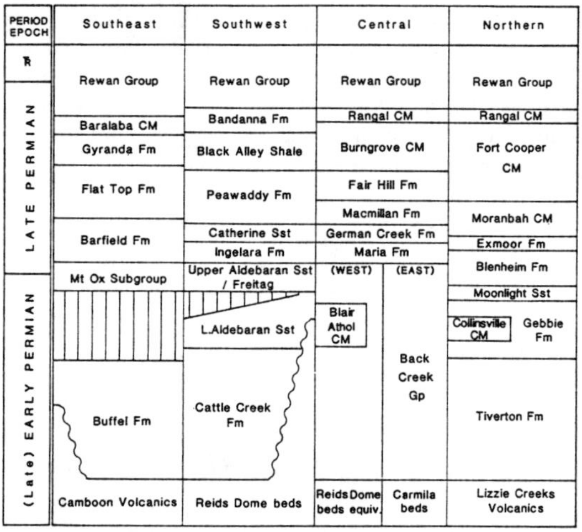

PERIOD EPOCH	Southeast	Southwest	Central		Northern
⃪R	Rewan Group	Rewan Group	Rewan Group		Rewan Group
LATE PERMIAN	Baralaba CM	Bandanna Fm	Rangal CM		Rangal CM
	Gyranda Fm	Black Alley Shale	Burngrove CM		Fort Cooper CM
	Flat Top Fm	Peawaddy Fm	Fair Hill Fm		
			Macmillan Fm		Moranbah CM
	Barfield Fm	Catherine Sst	German Creek Fm		Exmoor Fm
		Ingelara Fm	Maria Fm		Blenheim Fm
	Mt Ox Subgroup	Upper Aldebaran Sst / Freitag	(WEST)	(EAST)	Moonlight Sst
(Late) EARLY PERMIAN		L.Aldebaran Sst	Blair Athol CM	Back Creek Gp	Collinsville CM / Gebbie Fm
	Buffel Fm	Cattle Creek Fm			Tiverton Fm
	Camboon Volcanics	Reids Dome beds	Reids Dome beds equiv.	Carmila beds	Lizzie Creeks Volcanics

Figure 2. Lithostratigraphic correlations for the Bowen Basin. Modified from [3].

TECTONIC SETTING OVERVIEW

The Bowen Basin is interpreted to be a back-arc extensional to foreland basin which developed landward of an intermittently-active continental volcanic arc associated with the eastern Australian convergent margin [3]. To the east, the basin was bounded by the intermittently-active volcanic chain, whereas to the west it was bounded by cratonised Precambrian and Palaeozoic rocks. Geologic evidence for a back-arc extensional to foreland basin origin for the Bowen Basin includes 1) east-west basinal asymmetry, 2) an abundance of westward-propagating thrust faults and a concentration of compressive deformation along the eastern basin margin, 3) parallelism of the basin with a chain of contemporaneously-active volcanic mountains, and 4) sediment infilling patterns. As will be discussed in this paper, the temporal evolution of sandstone composition in the Bowen Basin from craton-derived to orogen-derived is also consistent with a back-arc extensional to foreland basin origin.

SEDIMENTATION PATTERNS

Permo-Triassic sedimentation patterns in the Bowen Basin have been deduced based on a variety of field evidence, including facies relationships, palaeocurrent directions, isopach trends and gross lithology [3], and serve as an important framework for interpreting petrological data for the Permo-Triassic sandstones in a tectonic context.

Sediment accumulation in the Bowen Basin commenced with the opening of a series of grabens and half grabens during the Early Permian. While a variety of volcaniclastic sediments accumulated in the east, coal-bearing continental sediments infilled the grabens and half grabens which developed to the west (Figure 3A).

A marine transgression, which probably entered the Bowen Basin from the east, followed infilling of the grabens and half grabens in the west and waning of volcanism in the east. This transgression led to accumulation of marine mudrocks over most of the western part of the basin. Areally restricted fluvial-wave moulded deltas also prograded eastward from the western margin of the basin during this time (Figure 3B).

During the early Late Permian, fluvial-wave moulded deltas continued to prograde eastwards from the western margin of the basin. To the east, while the largely inactive volcanic chain continued to subside, coastal mixed carbonate/siliciclastic deposits were succeeded by a thick shelf mud sequence (Figure 3C).

Re-establishment of an emergent volcanic arc along the eastern Bowen Basin margin during the middle Late Permian caused vast quantities of sediment to be shed westward into the eastern part of the basin. In the north, arc-derived drainage systems joined to feed a major axially-flowing fluvio-deltaic system which quickly advanced southward and eventually westward (Figure 3D).

The final major phase of Permian sediment accumulation in the Bowen Basin was marked by prograding deltaic and alluvial systems directed southwards and westwards. Hence, during the late Late Permian, the entire Bowen Basin was covered by peat-forming wetland environments and associated drainage systems, which operated during a period of subdued volcanic activity in the east. Alluvial fan deposits locally preserved on the eastern margin of the Bowen Basin are probably associated with uplift and onset of thrusting of the arc (Figure 3E).

Arc-derived sediment continued to accumulate in the Bowen Basin until about the latest Middle Triassic (Figure 3F). Compression climaxed during the earliest Late Triassic, finally leading to major uplift, deformation and erosion of the basin fill.

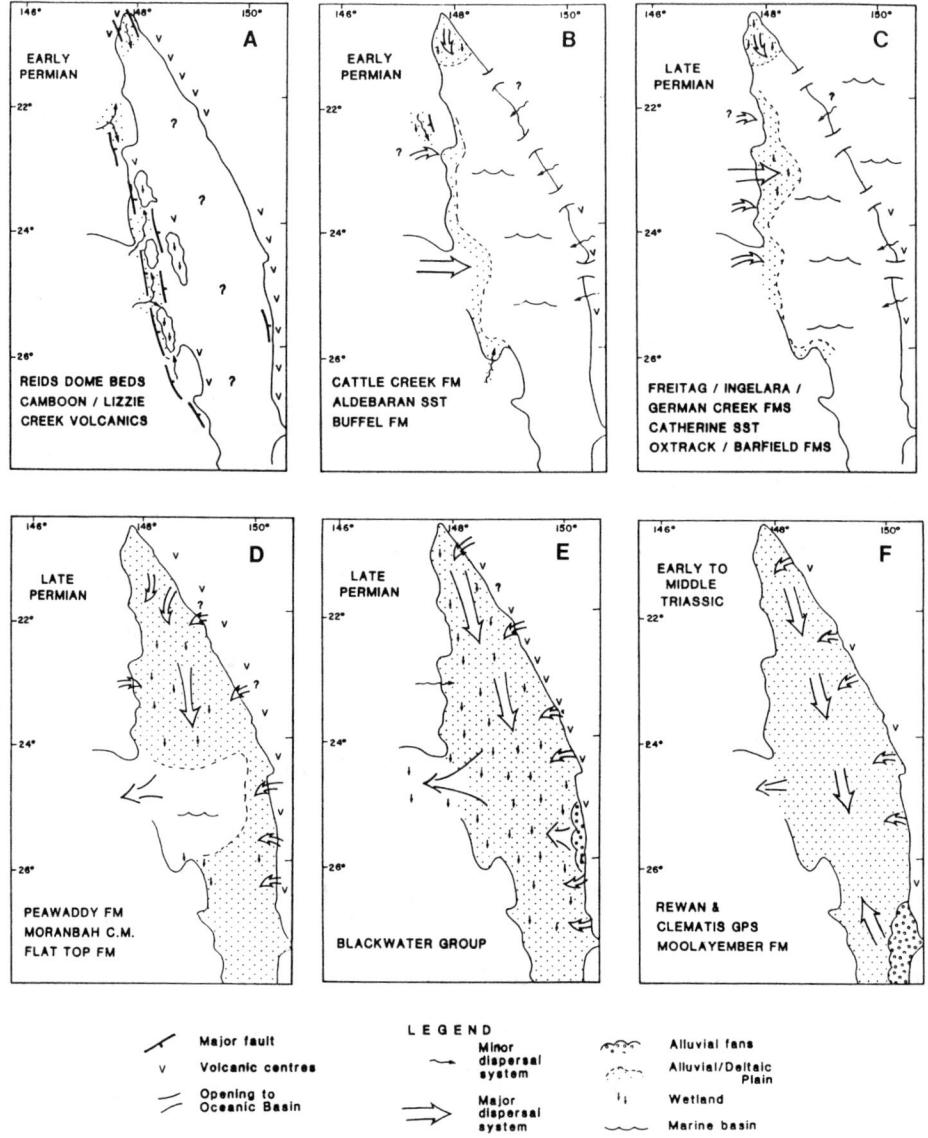

Figure 3. Palaeogeographic maps for the Bowen Basin showing major changes in sediment dispersal patterns during the Permian and Triassic. Modified from [3]. See text for discussion.

SANDSTONE PETROLOGY

Modal analyses were carried out on 285 sandstone samples collected from the subsurface and outcrop in the three regions shown in Figure 4. By far, most samples are medium-grained, moderately well-sorted arenites.

QFR compositions of sandstones from the three regions are shown in Figure 5. The majority of samples from the southwestern region are quartz rich, with most plotting as sublitharenites and quartz-rich feldspathic litharenites and litharenites [4]. Significantly, the three stratigraphically-highest Permian units (Peawaddy Formation, Black Alley Shale and Bandanna Formation) are seen to be generally more rock-fragment rich than the underlying units.

In contrast to samples from the southwestern region, those from the southeastern region are all quartz-poor litharenites, feldspathic litharenites and lithic arkoses.

Samples from the central region vary widely in their QFR compositions, ranging from quartz-rich sublitharenites to quartz-poor litharenites. Like in the southwestern region, the higher Permian units, particularly the Burngrove Formation and Rangal Coal Measures, are seen to be considerably more enriched in rock fragments than the underlying section.

SEDIMENT SOURCE ROCK TYPES

A variety of igneous, metamorphic and sedimentary rock types contributed sediment to the Bowen Basin during the Permian and Triassic Periods. In the quartz-rich sandstones, granitic source rocks are indicated by the presence of granitic rock fragments and locally-abundant first-cycle microperthite. Sedimentary sources are indicated by the presence of fine-grained siliciclastic fragments. In the quartz-rich sandstones, the abundance of monocrystalline quartz with straight extinction further indicates granitic and sedimentary sources. Also in the quartz-rich sandstones, the presence of minor acid to intermediate volcanic rock fragments, rare volcanic quartz and minor metamorphic rock fragments indicates subordinate volcanic and metamorphic sources.

In the case of the volcanolithic sandstones, the lack of quartz and the abundance of intermediate volcanic rock fragments and volcanic plagioclase indicate the dominance of intermediate volcanic rocks as sediment sources for these sandstones.

By integrating the petrological evidence for source rock types with the sediment dispersal maps shown in Figure 3, specific sediment source rocks for the Permo-Triassic sandstones can be identified. All identified rock units occur within 300 km of sediment accumulation sites in the basin.

RELATION BETWEEN SANDSTONE PETROLOGY AND TECTONIC SETTING

As noted previously, a large body of field evidence indicates that the Bowen Basin is a back-arc extensional to foreland basin which developed landward of an intermittently-active continental volcanic arc associated with the eastern Australian convergent margin. The basin is seen to have a polyphase structural history which can be divided into three main phases; 1) an initial extensional phase during the Early Permian, 2) a passive thermal sag phase with episodic compression during the late Early Permian to early Late Permian, and 3) a compressive phase with flexural loading from the late Late Permian to Mid Triassic. It was at the commencement of the third phase, when there was a resurgence in arc volcanism along the eastern basin margin, that the Bowen Basin became a true retroarc foreland basin.

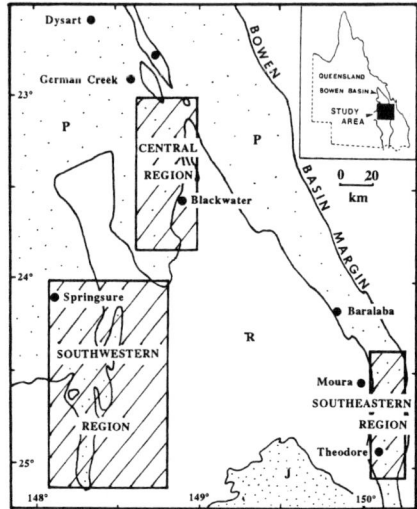

Figure 4. Location of sampled regions referred to in text.

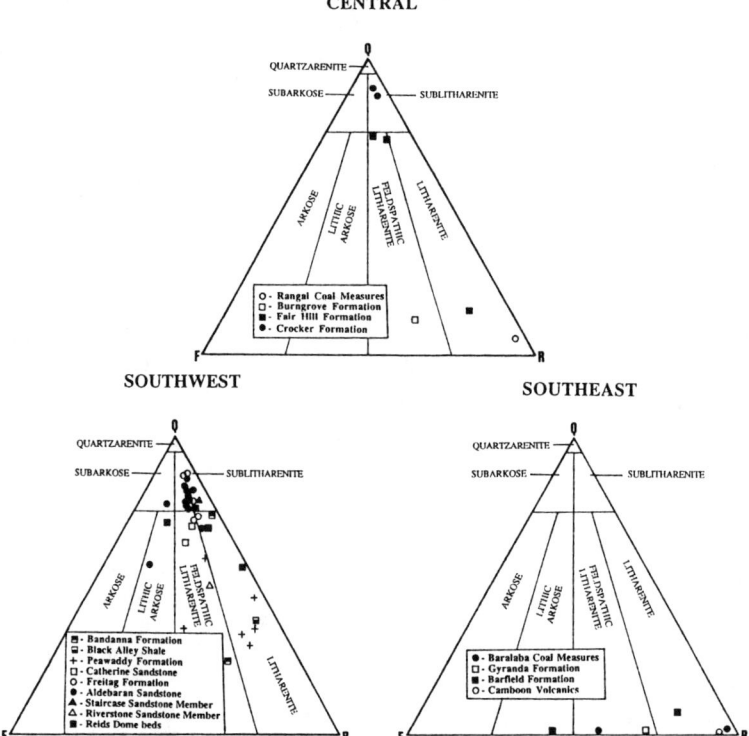

Figure 5. QFR compositions [4] of Bowen Basin sandstones from regions shown in Figure 4 (285 analyses grouped and averaged by geographic location).

Extensional phase

During the initial extensional phase, quartz-rich sediment shed from fault-bounded uplifted continental basement lying immediately to the west and south of the basin infilled the developing grabens and half grabens. Despite being derived from fault-bounded uplifted continental basement, these sediments, containing a significant proportion of rock fragments, produced sandstones with recycled orogen provenance compositions (cf., [2]), reflecting the fact that granitic basement in the source areas was largely covered by sedimentary, low-grade metasedimentary and volcanic rocks of several older sedimentary basins which had been earlier uplifted.

While the grabens were being filled by quartz-rich sediment in the west, to the east, coeval calc-alkaline volcanolithic and plagioclase-rich sediment was shed westward from the active volcanic arc to produce sandstones with undissected arc provenance compositions (cf., [2]).

Hence, during this initial extensional phase, a western cratonic source supplied quartz-rich sediment to the western side of the basin while an active volcanic arc supplied volcanolithic-rich sediment to the eastern side (Figure 6A).

Passive thermal sag phase with episodic compression

During the subsequent phase of passive thermal subsidence, when only subdued volcanic activity was occurring to the east, quartz-rich sediment continued to be shed eastwards from the uplifted continental basement into the western side of the basin. These sediments produced sandstones that, like the underlying graben-filling sandstones, incorrectly indicate a recycled orogen provenance. In the nearshore marine environments, quartz contents of the craton-derived sediment were increased by up to 15% at the expense of feldspars and rock fragments as a result of sediment reworking prior to accumulation (e.g., [1]). In the east, volcanolithic-rich sediment which entered the basin was derived from the inactive and possibly submerged volcanic arc (Figure 6B).

Compressive phase

The resurgence of the volcanic arc to the east during the Late Permian led to the westward and southward spread of volcanolithic-rich sediment over the entire Bowen Basin. Hence, areas in the west that were receiving quartzose, craton-derived sediment from the west and south were now inundated by volcanolithic-rich, arc-derived sediment from the east and north (Figure 6C). This explains the upward increase in rock fragments evident in the QFR plots for the southwestern and central regions (Figure 5).

Arc-derived sediment continued to accumulate in the Bowen Basin until about the latest Middle Triassic, producing sandstones of widespread volcanolithic composition. Compression climaxed during the earliest Late Triassic, finally leading to major uplift, deformation and erosion of the basin fill.

CONCLUSIONS

The Bowen Basin is a Permo-Triassic back-arc extensional to foreland basin with a structural history which can be divided into three main successive phases, these being extension, passive subsidence, and compression during which the basin became a true retroarc foreland basin. During the extensional and passive subsidence stages, the basin exhibited bimodality in sediment provenance, with a stable craton to the west supplying quartz-rich sediment to the western side of the basin, and a volcanic arc to the east supplying volcanolithic-rich detritus

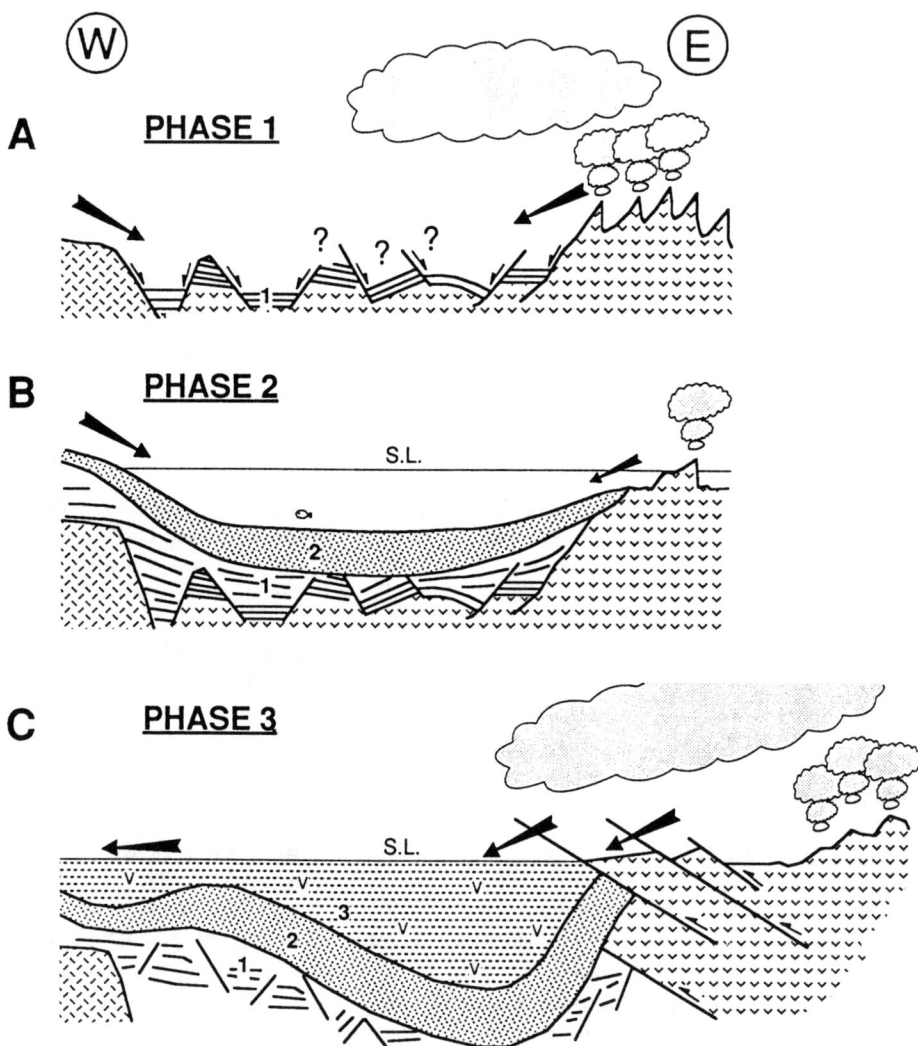

Figure 6. Conceptual model for evolution of sandstone composition in the Bowen Basin during the Permian. Phase 1 (extension) – large amounts of sediment shed from both the craton in the west and active volcanic arc in the east; Phase 2 (thermal sag) – modest amounts of sediment shed from the craton, very little coarse sediment shed from the inactive and subsiding arc; Phase 3 (compression) – large amounts of sediment shed westward from reactivated arc, overwhelming any cratonic sediment input.

to the eastern side of the basin. The cratonic sediments produced quartz–rich sublitharenites, litharenites and feldspathic litharenites, whereas the arc–derived sediments produced quartz-poor litharenites and feldspathic litharenites. During the compressive phase, volcanolithic-rich detritus from the reactivated arc system spread from the east over the entire Bowen Basin and produced quartz–poor volcanic litharenites and feldspathic litharenites. This overwhelming of early quartz–rich craton–derived sediments by later volcanolithic–rich sediments derived from a newly evolved arc orogen is consistent with the interpreted back–arc extensional to foreland basin origin for the Bowen Basin.

REFERENCES

1. J.C. Baker. Diagenesis and reservoir quality of the Aldebaran Sandstone, east–central Queensland, Australia, *Sedimentology.* **38**, 819–838 (1991).
2. W.R. Dickinson and C.A. Suczek. Plate tectonics and sandstone compositions, *Bull. Am. Ass. Petrol. Geol.* **63**, 2164–2182 (1979).
3. C.R. Fielding, A.J. Falkner, J. Kassan and J.J. Draper. Permian and Triassic depositional systems in the Bowen Basin, In: *Proc. Bowen Basin Symposium.* pp. 21–25 (1990).
4. R.L. Folk, P.B. Andrews and D.W. Lewis. Detrital sedimentary rock classification and nomenclature for use in New Zealand, *N. Z. J. Geol. Geophys.* **13**, 947–968 (1970).

(Manuscript received and accepted 18 November 1992.)

Compositional changes of the sandstone in relation to the evolution of a mobile belt −an example of the Paleozoic to Mesozoic sandstones of North Japan−

KAZUYOSHI OKAMI[1], MASAYUKI EHIRO[2] and SHIN KOSHIYA[1]

[1]*Department of Engineering Geology, Faculty of Engineering, Iwate University, Morioka 020, Japan.*
[2]*Department of Earth Sciences, Tohoku University, Sendai 980, Japan.*

Abstract: The sandstones of the Silurian to Upper Cretaceous distributed in the Kitakami and Abukuma Mountains in North Japan were studied to clarify the changes of the mineralogical and chemical compositions. The compositional changes of the sandstones of North Japan displayed in the Q-F-R ternary diagram well represent the tectonic evolution of North Japan which was sited in an island arc and/or continental margin from Middle Paleozoic to Mesozoic.
The sandstones plotted along the Q-R or Qm-Lt line in the ternary diagrams of Q-F-R or Qm-F-Lt respectively, suggest the existence of intense tectonic movements before the deposition of these sandstones. The tectonic discrimination diagrams previously proposed, however, can not be always applicable to the sandstones deposited in such a mobile belt as the Japanese Islands. Detailed studies on the Late Jurassic arkosic and quartzose sandstones provide the information on the different tectonic environments among the three belts, namely South Kitakami Belt, Kuzumaki-Kamaishi Belt and Akka-Tanohata Belt.

Key words: sandstone composition, mobile belt, Paleozoic, Mesozoic, North Japan, Kitakami Mountains, Abukuma Mountains

INTRODUCTION

The Paleozoic and Mesozoic strata distributed in the Kitakami and Abukuma Mountains yield the characteristic sandstones. The peculiar compositions of the sandstones reflect the geology of the provenance areas and tectonic movements during the deposition. There are many ideas to classify the sandstones and to interpret the depositional environments by using the ternary diagram. Recently, tectonic discrimination frameworks based on the study of the North American Phanerozoic sandstones were proposed by Dickinson et al. [1].

In the Kitakami and Abukuma Mountains of North Japan, sandstones of various types ranging from Silurian to Paleogene in age are distributed. In particular, arkosic and quartzose sandstones deposited in the different sedimentary environments, such as the trench fill sediments and the shallow marine to terrestrial sediments, are widely distributed.

The writers studied the mineral and chemical compositions of the sandstones in the Kitakami and Abukuma Mountains to clarify the tectonic movements of each period in North Japan, and to examine the validities of the above mentioned discrimination ternary diagrams to the sandstones of such a mobile belt as the Japanese Islands.

GEOLOGIC OUTLINE

The pre-Miyakoan (pre-Late Neocomian in European standard) strata in the Kitakami and Abukuma Mountains are developed in the five belts, namely, Abukuma Belt, South Kitakami

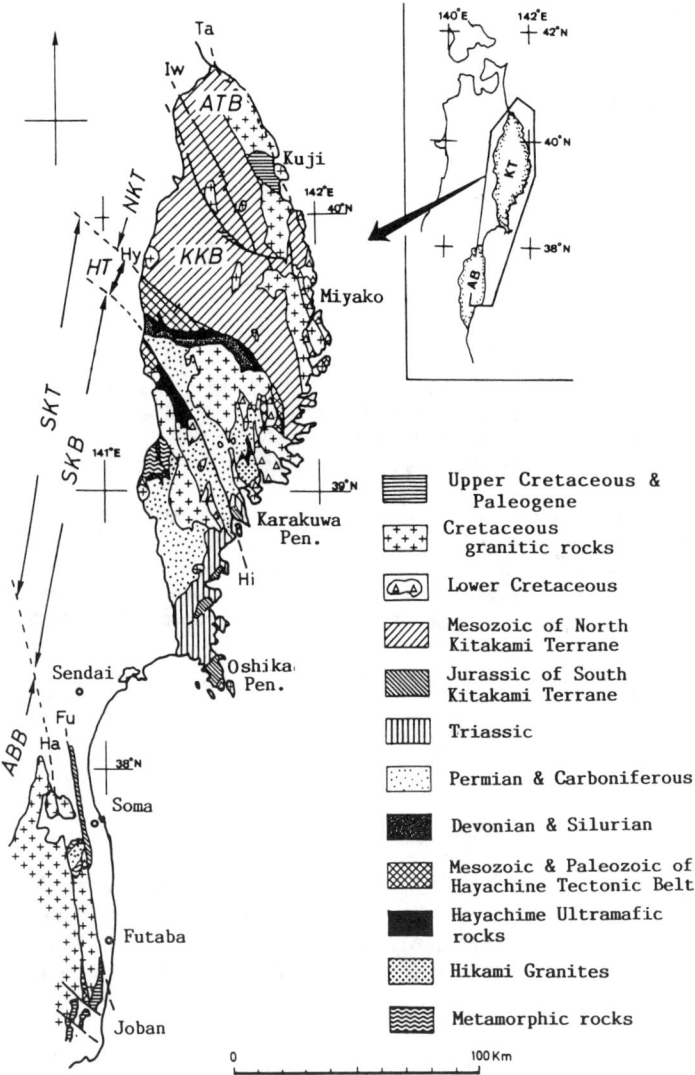

Figure 1. Simplified geologic map of the Kitakami and Abukuma Mountains.
KT: Kitakami Mountains, AB: Abukuma Nountains, SKT: South Kitakami Terrane, NKT: North Kitakami Terrane, ABB: Abukuma Belt, SKB: South Kitakami Belt, HT: "Hayachine Tectonic Belt", KKB: Kuzumaki-Kamaishi Belt, ATB: Akka-Tanohata Belt, Ta: Taro Fault, Iw: Iwaizumi Tectonic Line, Fu: Futaba Fault, Ha: Hatakawa Shear Zone.

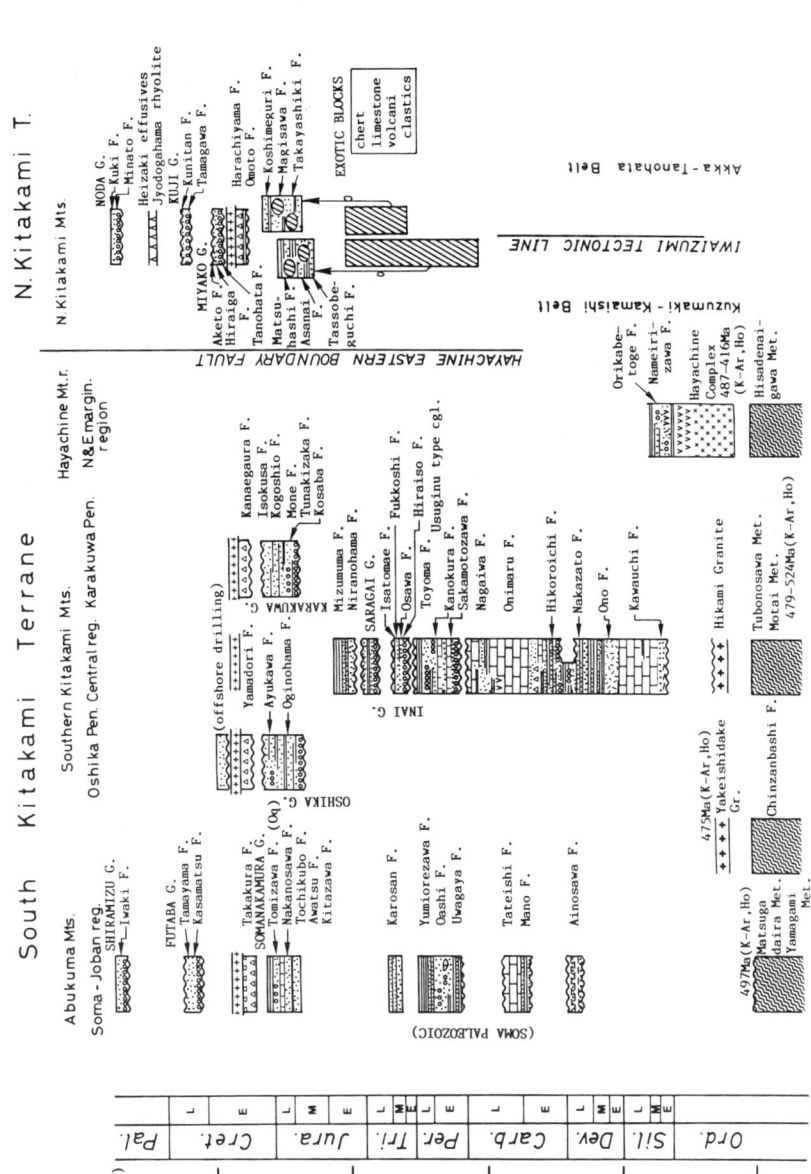

Figure 2. Simplified pre-Neogene sequences of the Kitakami and Abukuma Mountains. Arrow shows the horizon of the sandstones treated in this study.

Belt, "Hayachine Tectonic Belt", Kuzumaki-Kamaishi Belt and Akka-Tanohata Belt from southwest to northeast (Figure 1). The last two are collectively called the North Kitakami Terrane. The pre-Miyakoan sediments in the Kitakami Mountains had been considered to be divided by the "Hayachine Tectonic Belt" into the South Kitakami Terrane of shallow marine sediments and the North Kitakami Terrane of pelagic sediments [2]. However, the basic to ultrabasic rocks, which are main constituents of the "Hayachine Tectonic Belt" and are named Hayachine Complex, were recently clarified to be the basement rocks of the Paleozoic sediments of the South Kitakami Belt [3,4,5]. So, the significance of the "Hayachine Tectonic Belt" as a tectonic belt became obscure. Now, the "Hayachine Tectonic Belt" is included into the South Kitakami Terrane, and the Hayachine Eastern Boundary Fault [4] is referred as the boundary between the South Kitakami Terrane and North Kitakami Terrane (Kuzumaki-Kamaishi Belt).

In the Southern Kitakami Mountains (main part of the South Kitakami Belt), the sequences from Silurian to Jurassic which consist of shallow marine coarse clastics and thick limestone piles, are extensively distributed overlying unconformably the basement rocks of Hikami Granites, Motai Metamorphic Rocks and Hayachine Complex. The oldest sediments are considered to be the Late Ordovician [3]. Similar sequences are also distributed in the eastern marginal area of the Abukuma Mountains, and they are included in the South Kitakami Terrane. On the contrary, the pre-Miyakoan accretional complexes which include the exotic blocks of pelagic sediments such as cherts and limestones of the Carboniferous to Early Jurassic ages are distributed in the North Kitakami Terrane. The age of the clastic sediments are assigned to Late Jurassic to Early Cretaceous, based on the radiolarian fossils which are found from the matrix. Until the 1970s, the pre-Miyakoan strata were subdivided into the three belts, that is, Northern Kitakami Belt, Iwaizumi Belt and Taro Belt from west to east, whose ages were supposed to be Carboniferous to Permian, Triassic to Jurassic and Jurassic to Early Cretaceous, respectively [6]. However, the discoveries of such microfossils as conodonts and radiolarians from cherts, siliceous shales and limestones during 1980s enable to revise the division of the North Kitakami Terrane. The pre-Miyakoan strata in the North Kitakami Terrane are divided into the Kuzumaki-Kamaishi Belt in the west and the Akka-Tanohata Belt in the east, on the basis of the ages of the exotic blocks and sandstone compositions [7]. The two belts are bordered by the Iwaizumi Tectonic Line. The former corresponds to the previous Northern Kitakami Belt and the western half of the Iwaizumi Belt, and the latter coincides with the eastern half of the Iwaizumi Belt and the Taro Belt. The former belt includes Paleozoic and Mesozic blocks, and the latter includes only Mesozoic ones [7].

These geological framework of the pre-Miyakoan strata has been formed by the crustal disturbance called the Oshima Orogeny which occurred between the deposition of the Harachiyama Formation (Late Neocomian) and the Miyako Group (Aptian to Albian) [8]. The orogeny was accompanied by a large volume of granitic rocks (115-125Ma, [9]) intruded to the pre-Miyakoan sediments. The pre-Miyakoan sediments are extensively folded and faulted by the orogeny. After the orogeny, Early Cretaceous to Paleogene shallow marine sediments werw deposited in the eastern marginal areas of the both mountains (Figure 2).

GEOLOGY AND SANDSTONE COMPOSITION OF THE PALEOZOIC

Silurian

In the South Kitakami Belt, the Silurian beds are distributed in the central district (Hikoroichi, Yokamachi and Okuhinotsuchi areas), and in the northern district (Ohasama area). In the Hikoroichi and Yokamachi areas, the Silurian Kawauchi Formation overlies unconformably the Hikami Granites [10, 11, 12], and in the Okuhinotsuchi area the formation unconformably

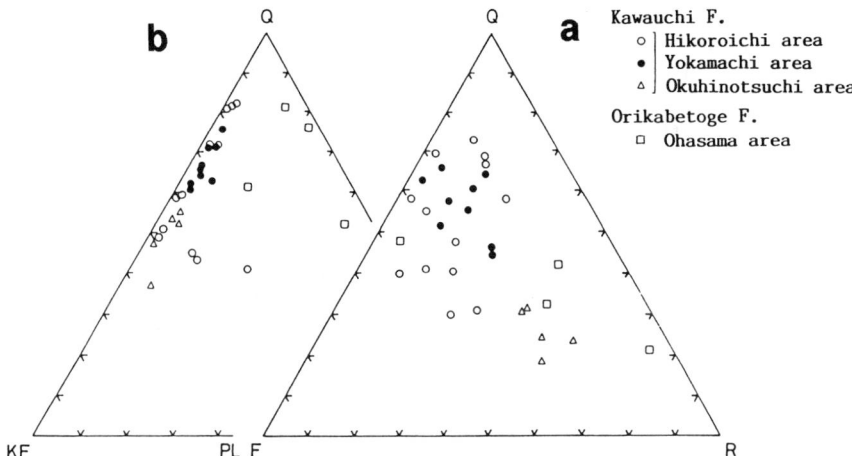

Figure 3. Q-F-R and Q-KF-PL plots of the Silurian sandstones in the South Kitakami Terrane [3,11,13]. Q: quartz, F: feldspars, R: rock fragments, KF: potash feldspars, PL: plagioclase.

covers the Okuhinotsuchi Pyroclastics which are composed of welded tuff [11, 12]. The Okuhinotuchi Pyroclastics overlies unconformably the Hikami Granites. The Kawauchi Formation consists mainly of limestone accompanied with clastic sediments in it's basal part. In the Ohasama area, the Silurian named the Orikabetoge Formation consists of coarse clastic sediments, contrasting with those in the central district. Thin acidic tuffs are intercalated in the formation. The formation is in contact with the Early Silurian or Late Ordovician Nameirizawa Formation by a fault. The sandstones of the Kawauchi Formation in the Yokamachi and Kawauchi areas are arkosic sandstones which are composed mainly of monocrystalline quartz and feldspars, associated with a subordinate amount of lithic fragments of granitic rocks and granoporphyry. Sandstones in the Okuhinotsuchi area are dominant in lithic fragments of welded tuff which must have been been derived from the underlying Okuhinotsuchi Pyroclastics [3]. If the welded tuff fragments are excluded, however, the relative proportions of quartz and feldspars are similar to those of other central district (Figure 3a).

In the northern district, the sandstones of the Orikabetoge Formation are rich in various kinds of lithic fragments such as granitic rocks, dacite, welded tuff, basalt, metamorphic rocks and so on, and the total amount of lithic fragments exceeds 50 %. The sandstones of the central district are rich in K-feldspars compared with those of the northern district (Figure 3b).

Devonian

The Early to Middle Devonian Ono and Nakazato Formations are composed of shallow marine clastic sediments, intercalating a large amount of acidic volcaniclastics (Figure 2). The Devonian sandstones are rich in plagioclase and lithic fragments (Figure 4). The lithic fragments are composed of granoporphyry, basalt, dacite, andesite and dolerite [14, 15]. The sandstones of the Ono Formation (Early Devonian) include granitic rock fragments referred to the Hikami Granites, and a little amount of quartz. These facts suggest the existence of supply also from the Hikami Granites in Early Devonian.

Carboniferous

The Carboniferous sequence in the central district of the South Kitakami Belt consists mainly of tuffaceous sandstone and pyroclastics in the lower horizon, and limestone in the upper horizon (Figure 2). The sandstones of the Hikoroichi Formation (Early Carboniferous) are poor in quartz, and rich in lithic fragments of volcanic rocks and feldspars [15]. The sandstone compositions of the Hikoroichi Formation resemble those of the Devonian sandstones (Figure 4), and well represent the depositional environments under the volcanic activities.

Permian

Both the Early Permian Sakamotozawa and Middle to earliest Late Permian Kanokura Formation are rich in coarse clastic sediments in the lower half, and rich in limestone in the upper half (Figure 2). The sandstones of the Sakamotozawa Formation are of two kinds in composition; one is poor in quartz and the other is poor in feldspars as shown in Figure 4 [16]. The sandstones of the Kanokura Formation are rich in lithic fragments. Lithic fragments of the both formations are mostly shale, siliceous shale and volcaniclastics.

GEOLOGY AND COMPOSITIONS OF THE MESOZOIC SANDSTONES

Triassic to Jurassic sandstones of the South Kitakami Terrane

The Triassic to Jurassic sediments are distributed in the southern part of the South Kitakami Mountains and in the eastern marginal part of the Abukuma Mountains. They are mainly composed of coarse- to fine-grained clastic sediments deposited in shallow marine to terrestrial environments.

Early to Middle Triassic

The Early to Middle Triassic Inai Group is subdivided into the Hiraiso, Osawa, Fukkoshi and Isatomae Formations in ascending order (Figure 2), and consist mainly of sandstone and

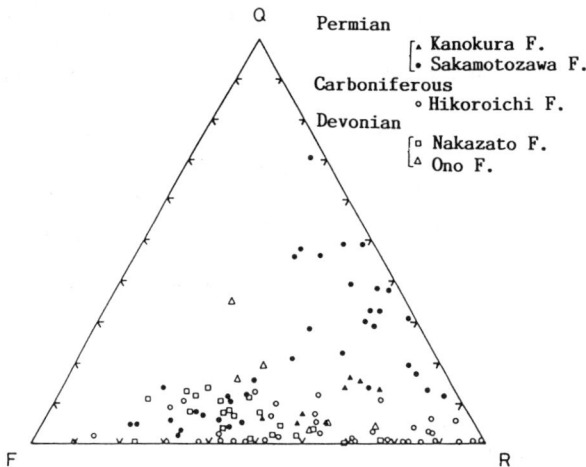

Figure 4. Q-F-R plots of the Devonian, Carboniferous and Permian sandstones from the South Kitakami Terrane [14,15,16]. Alphabetical keys are same as Figure 3.

shale. The sandstones of each formation possess similar composition which is rich in quartz and feldspars. Quartz grains are mostly monocrystalline, which contain a small amount of undulatory extinction quartz. Feldspars comprise mainly plagioclase, though K-feldspars are contained commonly in a little amount [17]. The sandstones are characterized by the abundance of various kinds of lithic fragments, such as granitic rocks, andesite, porphyrite, basalt, hornfels and a small amount of other crystalline rocks. This fact is harmonized with the compositions of the conglomerate pebbles of the Early Triassic Osawa Formation. The pebbles consist of various kinds of rocks, such as granodiorite, quartz porphyry, porphyrite, sandstone, limestone and other crystalline rocks.

Jurassic

The Oshika Group distributed in the Oshika Peninsula, is divided into the Tsukinoura, Oginohama and Ayukawa Formations from the lower upwards (Figure 2). It is composed of coarse- and fine-grained, deltaic to shallow marine clastic sediments [18, 19]. The sandstones of the Oginohama Formation are arkosic, and feldspars are rich in plagioclase (Figure 5). The sandstones of the Ayukawa Formation correlative to the Tomizawa Formation of the Abukuma Mountains, are rich in quartz and feldspars. They are plotted along the Q-F line of the Q-F-R ternary diagram, and feldspars are rich in K-feldspars (Figure 5). In those sandstones, the quartzose sandstones plotted in the orthoquartzite clan of Folk's classification [20] are included [19]. The Karakuwa Group distributed in the Karakuwa Peninsula, is divided into the Kosaba, Tsunakizaka, Ishiwaritoge, Mone, Kogoshio and Isokusa Formations in ascending order. The sandstones of the Kosaba Formation are rich in quartz and feldspars, and include the granitic rock fragments in a considerable amount. The sandstones of the Kogoshio Formation are also arkosic and are rich in K-feldspars [18] (Figure 5).

The Somanakamura Group distributed in the eastern marginal part of the Abukuma Mountains is divided into the Kitazawa, Awatsu, Yamagami, Nakanosawa, Tomizawa and Oyamada Formations in ascending order [21]. The Tomizawa Formation deposited in the delta and shallow marine conditions, is composed mainly of sandstone and shale, intercalating coal

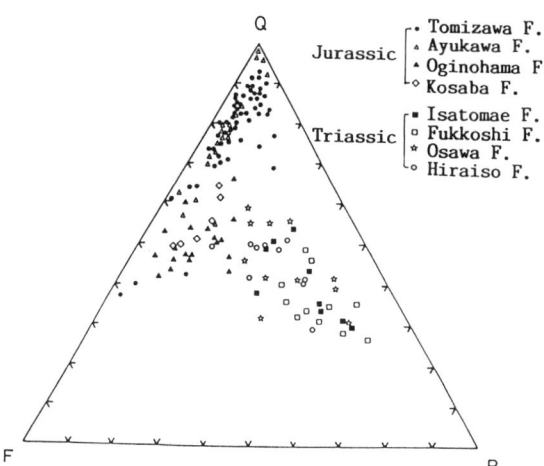

Figure 5. Q-F-R plots of the Triassic and Jurassic sandstones from the South Kitakami Terranes [17,18,19,22]. Alphabetical keys are same as Figure 3.

beds in the lower horizons. The sandstones of the Tomizawa Formation consist of quartz and a small amount of K–feldspars, accompanied a very little amount of plagioclase and lithic fragments of granitic and felsic volcanic rocks [22]. They are plotted along the Q–F line in the Q–F–R ternary diagram (Figure 5), and include the quartzose sandstones classified into the orthoquartzite clan of Folk's classification [22]. The sandstone compositions of the Nakanosawa and Tochikubo Formations are similar to those of the Tomizawa Formation, though they are rich in plagioclase [19, 23].

Thus, the compositions the Jurassic sandstones distributed in the South Kitakami and Abukuma Mountains are similar to each other, that is, in the abundance of qaurtz and scarcity of rock fragments.

Accretional sediments (Jurassic to Early Cretaceous) of the North Kitakami Terrane

Pre–Miyakoan strata in the North Kitakami Terrane are divided into the Kuzumaki–Kamaishi Belt and Akka–Tanohata Belt bordered by the Iwaizumi Tectonic Line [7], and the latter is also subdivided into the western and eastern areas by the Taro Fault. Because the detailed revision of the stratigraphic units have not been done enough, the formation names of previous works [6, 24, 25, 26] are used in this study. The Tassobeguchi, Matsuhashi, Asanai and Kassenba Formations belong to the Kuzumaki–Kamaishi Belt. The Magidai and Takayashiki Formations belong to the western area of the Akka–Tanohata Belt, and Makisawa and Koshimeguri Formations belong to the eastern area of the Akka–Tanohata Belt. The sandstones of the both belts consist of monocrystalline quartz, feldspars and a small amount of lithic fragments of chert, shale and felsic volcanics. They are classified into arkose to quartzose sandstone (Figure 6a). Although there are no apparent differences between the both belts in the Q–F–R ternary diagram, the sandstones of the Kuzumaki–Kamaishi Belt ("Northern Kitakami Belt": old division name) are rich in plagioclase and fragments of volcanic clasts, and include commonly epidote and chlorite. On the other hand, those of the Akka–Tanohata Belt ("Iwaizumi" and "Taro Belts": old division name) are rich in K–feldspars and granitic rock fragments [25, 27]. Especially, the sandstones of the Tassobeguchi Formation distributed in the Kawai area are almost lacking in K–feldspars [28]. The difference is shown in Q–KF–PL ternary diagram [29] (Figure 6b).

Chemical compositions of the Late Jurassic sandstones [30, 31] are examined to clarify the tectonic setting of the North Kitakami and South Kitakami Terranes by using the K_2O/Na_2O–SiO_2 diagram [32]. The chemical compositions of the sandstones from the Tomizawa Formation, the Kuzumaki–Kamaishi Belt and the Akka–Tanohata Belt display the passive margin, oceanic island arc and active continental margin settings, respectively (Figure 7). The tectonic environment of the Tomizawa Formation corresponds with its lithofacies. The sandstones of the North Kitakami Terrane are clearly subdivided into two groups also by the chemical composition, coinciding with the subdivision based on the mineralogical composition.

Early Cretaceous (Miyakoan)

The Miyako Group distributed in the eastern marginal part of the Kitakami Mountains is divided into the Raga, Hiraiga, Tanohata and Aketo Formations in ascending order. They are deposited in the shallow marine environment having appeared after the Oshima Orogeny [8]. The Miyako Group overlies the pre–Miyakoan strata of the Akka–Tanohata Belt and Early Cretaceous granites with a distinct unconformity. The compositions of the sandstones vary considerably according to their basement rocks. Most sandstones are poor in feldspars, and are rich in lithic fragments such as chert, shale, sandstone, andesite, acidic volcanic rocks, hornfels and granitic rocks [33]. They are scattered along the Q–R line in the Q–F–R diagram, and the amount of quartz and rock frgamets are very variable. The sandstones of the

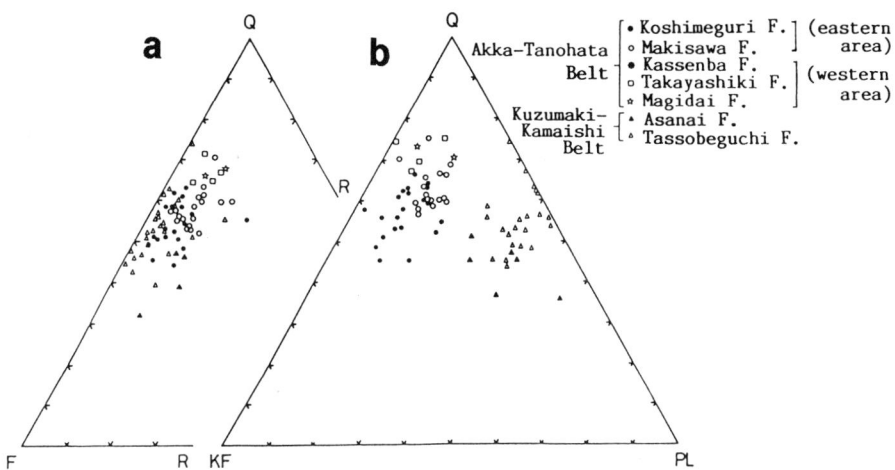

Figure 6. Q–F–R and Q–KF–PL diagrams of the sandstones from the accretional sediments in the North Kitakami Terrane [24,25,28,29]. Alphabetical keys are same as Figure 3.

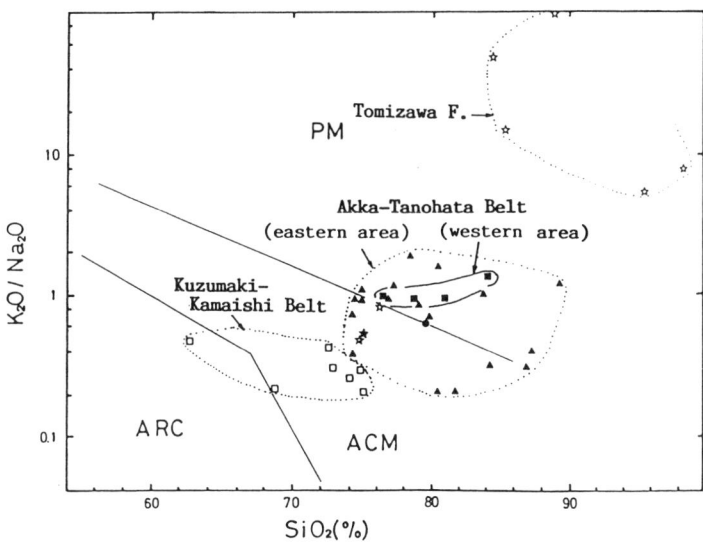

Figure 7. Tectonic discrimination diagram for the Jurassic sandstones of the Nortern Kitakami and Abukuma Mountains. Diagram after Roser and Korsh [32].
ACM: active continental margin, ARC: island arc, PM: passive continental margin

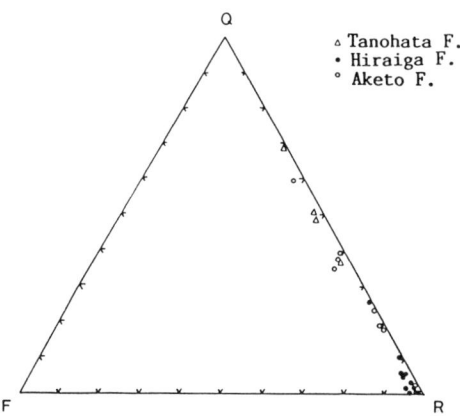

Figure 8. Q-F-R plots of the sandstones from the late Early Cretaceous Miyako Group [33]. Alphabetical keys are same as Figure 3.

Tanohata Formation are typical lithic sandstone, and often include lithic fragments more than 90% in amount (Figure 8).

Late Cretaceous and Paleogene

The Late Cretaceous sediments are distributed in the Kuji area (Kuji Group), the northeastern marginal part of the Kitakami Mountains, and in the Futaba area (Futaba Group), the eastern marginal part of the Abukuma Mountains. The Kuji Group is divided into the Tamagawa, Kunitan and Sawayamakawa Formations in ascending order, and overlies unconformably pre-Miyakoan strata of Akka-Tanohata Belt and Early Cretaceous granites. They were deposited in terrestrial and shallow marine environments, and consist of coarse-grained clastics, intercalating coal beds and acidic to andesitic tuffs. Similar to those of the Early Cretaceous, the sandstones are poor in feldspars, and rich in lithic fragments such as chert, shale, sandstone, andesite, dacite, tuff, hornfels, and others (Figure 9) [34].

The Futaba Group unconformably overlies the Early Cretaceous granitic rocks, and is divided into the Ashizawa, Kasamatsu and Tamayama Formations in ascending order. The sediments consist mainly of terrestrial to shallow marine sandstones and shales, intercalating a thin beds of conglomerates and coal. The sandstones of the Kasamatsu and Tamayama Formations are arkosic, and contain little amount of lithic fragments such as granitic rocks and pre-Miyakoan sedimentary rocks (Figure 9). The sediments were mostly supplied from the underling granitic rocks [29].

The Paleogene sediments also distributed in the Kuji area (Noda Group) and in the Futaba and Joban areas (Shiramizu Group) have a close relation with the Upper Cretaceous strata. The Noda Group deposited in a terrestrial environment, is divided into the Minato and Kuki Formations in ascending order (Figure 2). The group consists of four cyclothematic sequences which are composed of conglomerate, sandstone, shale and coal in ascending order, intercalating andesitic tuffs. The sandstones are poor in feldspars and rich in lithic fragments of volcaniclastics. The sandstones resemble those of the underlying Upper Cretaceous (Figure 9).

The Shiramizu Group is divided into the Iwaki, Asagai and Shirasaka Formations in ascending order, and has a transgressive lithofacies as a whole [35]. The Iwaki Formation consists

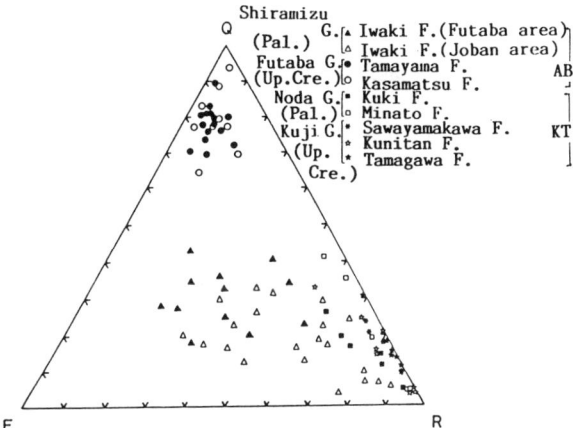

Figure 9. Q–F–R diagram of the Late Cretaceous and Paleogene sandstones in the Kitakami and Abukuma Mountains [29,34,35]. Alphabetical keys are same as Figure 3.

of coal-bearing cyclothematic sediments which resemble those of the Noda Group. The sandstones of the Iwaki Formation are lithic in nature, though a considerable amount of quartz and feldspars are included (Figure 9). The lithic fragments are rich in basaltic andesite [35]. If the basaltic andesite fragments are excluded, the relative proportion of quartz and feldspars of the sandstones in the Iwaki Formation resembles that of the Upper Cretaceous sandstones.

DISCUSSION ON THE RELATIONSHIP BETWEEN TECTONIC DEVELOPMENT OF NORTHERN JAPAN AND SANDSTONE COMPOSITIONS

The basement rocks such as the older metamorphic and granitic rocks of North Japan were formed in the subduction zone in the Ordovician period [3]. The metamorphic rocks named the Matsugadaira (Abukuma Mountains), Motai, Tsubonosawa and Hisadenaigawa (Southern Kitakami Mountains) were formed under a high pressure and low temperature metamorphism in the subduction zone [36, 37]. The Hikami Granites which intruded into the metamorphic rocks, are considered to be an island arc type granite [3]. The Hayachine Complex were formed in a extensional back-arc region in the Late Ordovician time [3, 5, 38]. These facts suggest that an intense tectonic movement took place in the Late Ordovician (Figure 10). After the orogeny, North Japan began to separate as a matured island arc from a continent [39].

The tectonic evolution of North Japan from the Silurian to Paleogene is briefly illustrated as follows;

Silurian: matured island arc [39],
Devonian: island arc with acidic volcanism,
Early Carboniferous: island arc with the extensional stress field accompanied by the bimodal volcanism in the central area [39, 40, 41],
Middle to Late Carboniferous: non-active island [41],
Permian to Triassic: dissecting stage of island accompanied by the basement uplift and block tilting [16,29],
Jurassic: the South Kitakami Terrane was the dissected and matured island arc, and the North Kitakami Terrane was situated in a convergent margin, and the accretional sediments were

formed under a oblique subduction of the Izanagi plate loaded with the South Kitakami and Abukuma Mountains [43, 44],
Late Jurassic to early Early Cretaceous: collision of the South Kitakami Terrane to a continent, the Oshima Orogeny,
Early Cretaceous: active continental margin,
Late Cretaceous and Paleogene: active continental margin, volcanism of 50 to 60 Ma [9] are referred to the ridge subduction origin of the Kula-Pacific plate, as shown in the Southwest Japan [45].

The stratigraphical change of sandstone compositions corresponds to the evolutional change of the northern Japanese Islands. The change from the Silurian arkosic sandstones to the Devonian to Permian quartz-poor lithic ones coincides with the transition from a matured and dissected island arc to a volcanic arc with violent volcanism. The compositions of the Permian and Triassic sandstones represent the absence of volcanism, and the decreasing of rock fragments means a gentle uplift and/or block tilting of the basements and the dissecting process of the previous island arc, but these interpretations need the other geologic information. In Jurassic time, arkosic and quartzose sandstones indicate that the islands were deeply dissected

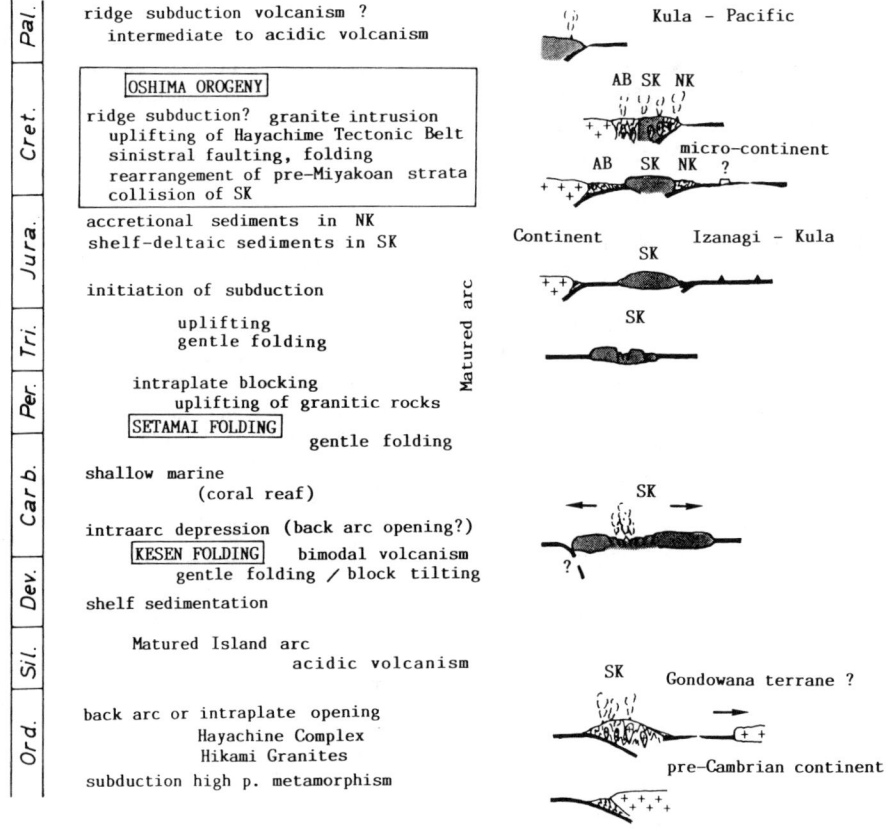

Figure 10. Tectonic developments of the Kitakami and Abukuma Mountains[3,8,34,43,45]. NK:North Kitakami Terrane, SK:South Kitakami Terrane, AB:Abukuma Belt.

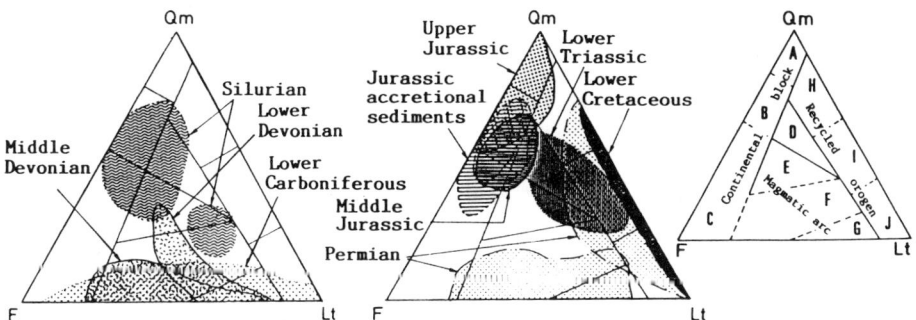

Figure 11. Qm-F-Lt diagrams for the Paleozoic to Mesozoic sandstones of the Kitakami and Abukuma Mountains on the discrimiantion diagram after [1].
Qm: monocrystalline quartz, F: feldspars, Lt: lithic fragments including quartz varieties, A: craton interior, B: transitional continental, C: basement uplift, D: mixed, E: dissected arc, F:transitional arc, G: undissected arc, H: quartzose recycle, I: transitional recycle, J: lithic recycle.

down to the basement of the matured islands arc. Post-orogenic sandstones of the Lower Cretaceous are very poor in feldspars and variable in rock fragments, and may represent the uplifting of older sedimentary rock basement. In the Abukuma Mountains, the arkosic sandstones of the Late Cretaceous may indicate the intrusion and dissection of acidic pulton. The compositions of the Paleogene correspond to the initiation of a new acidic volcanism. In the North Kitakami Mountains, the sandstone compositions of Late Cretaceous and Paleogene represent the uplift of basement accompanied with volcanism. Thus it is possible to interpret the provenance in relation to the tectonic setting based on the sandstone compositions by using the ternary diagrams of Q-F-R or Q-PL-KF.

The Paleozoic to Mesozoic sandstones of North Japan are classified into five types; 1. quartzose sandstone (Late Jurassic, Late Cretaceous of the Futaba area), 2. arkosic sandstone (Kuzumaki-Kamaishi Belt, Akka-Tanohata Belt, Middle Jurassic), 3. quarztose-lithic sandstone (a part of Permian, Early Cretaceous, Late Cretaceous and Paleogene of Kuji area), 4. lithic-feldspathic sandstone (Devonian, Early Carboniferous, a part of Permian), 5. intermediate sandstone (a part of Silurian, Triassic, Paleogene in the Abukuma Mountains). The compositional areas are shown on Qm-F-Lt diagram (Figure 11).

The discrimination diagrams were proposed by Dickinson et al. [1], based on the Phanerozoic sandstone compositions from North America. One of the diagrams shown in Figure 11 can not exactly be applied to interpret the tectonic provenance of the sandstones from the mobile belt such as the Japanese Islands (Figure 11). Especially, the Jurassic arkosic to quartzose sandstones of the South Kitakami Terrane are plotted in the continental block realm, but they were actually deposited in an arc-trench setting. The detritals derived from a matured and dissected island arc, because the Kuzumaki-Kamaishi and Akka-Tanohata Belts was being formed as a accretional deposits during the same period. They never belonged to a craton continent. Additionally, more than a half of the Jurassic arkosic sandstones are classified into the active continental margin based on the chemical compositions (Figure 7).

In conclusion, the studies of the Paleozoic to Mesozoic sandstones of North Japan indicate thatthe compositions of the sandstones change in clockwise trend in Qm-F-Lt ternary diagram, corresponding to the evolution of the island arc. The sandstones plotted along the Q-R or Qm-Lt line in the ternary diagrams of Q-F-R and Qm-F-Lt respectively, coincide with the recycled orogen provenance by Dickinson et al. [1], and may suggest the existence of an intense tectonic movements before the deposition of the sandstones. Although the discrimination diagrams proposed by Dickinson et al. [1] are effective for the provenance study, it is necessary to examine the geological data synthetically from many view points for interpreting the tectonic setting of the basin where the sandstones deposited.

ACKNOWLEDGEMENTS

This work was partly supported by the Grant-in-Aid from Japanese Ministry of Education, Science and Culture (Nos. 63302019 and 02452063).

REFERENCES

1. W. R. Dickinson, L. S. Beard, G. R. Brakenridge, J. L. Erjavec, R. C. Fergunson, K. F. Imman, R. A. Knepp, F. A. Lindberg and P. T. Ryberg. Provenance of North American Phanerozoic sandstones in relation to tectonic setting, *Geol. Soc. Am. Bull.* **94**, 222-235 (1983).
2. T. Yoshida and M. Katada. *Geological sheet map 1:50,000 "Otsuchi and Karodake" and its explanatory text*, Geol. Surv. Japan (1964).*
3. K. Okami, M. Ehiro and M. Oishi. Geology of the Lower-Middle Paleozoic around the northern marginal part of the southern Kitakami Massif, -with reference to the geologic development of the "Hayachine Tectonic Belt", In: *Prof. N. Kitamura Commemorative Vol.* H. Nakagawa et al.(Eds), 313-330, Toko Print, Sendai (1986).*
4. M. Ehiro, K. Okami and S. Kanisawa. Recent progress and further subjects in the studies on the "Hayachine Tectonic Belt" in the Kitakami Massif, Northeast Japan, *Chikyu Kagaku* **42**, 317-335 (1988).*
5. K. Ozawa, K. Shibata and S. Uchiumi. K-Ar ages of hornblendes in gabbroic rocks from the Miyamori ultra mafic complex of the Kitakami Mountains, *Jour. Japan Assoc. Min. Petr. Econ. Geol.* **83**, 150-159 (1988).*
6. M. Sugimoto. Stratigraphical study in the outer belt of the Kitakami Massif, Northeast Japan, *Contr. Inst. Geol. Paleont. Tohoku Univ.* no.74, 1-48 (1974).*
7. K. Okami and M. Ehiro. Review and recent progress of studies on the pre-Miyakoan sedimentary rocks of the Northern Kitakami Massif, Northeast Japan, *Chikyu Kagaku.* **42**, 87-201 (1988).*
8. T. Kobayashi. The Sakawa orogenic cycle and its bearing on the origin of the Japanese Islands, *Jour. Fac. Sci. Imp. Univ. Tokyo* **5**, 219-578 (1941).
9. K. Shibata, T. Matsumoto, T. Yanagi and R. Hamamoto. Isotopic ages and stratigraphic control of Mesozoic igneous rocks in Japan, In: *Contr. Geol. Time Scale*, G.V. Cohee et al.(Eds), 143-164, *Assoc. Petr. Geol.* (1978).
10. M. Murata, S. Kanisawa, Y. Ueda and N. Takeda. Base of the Silurian system and the pre-Silurian granites in Kitakami Massif, Northeast Japan, *Jour. Geol. Soc. Japan.* **80**, 475-486 (1974).*
11. K. Okami and M. Murata. Basal sandstone of the Silurian Kawauchi Formation in the Kitakami Massif, Northeast Japan, *Jour. Geol. Soc. Japan.* **81**, 339-348 (1975).
12. M. Murata, K. Okami, S. Kanisawa and M. Ehiro. Additional evidence for the pre-Silurian basement in the Kitakami Massif, Northeast Honshu Japan, *Mem. Geol. Soc. Japan.* no.21, 245-259 (1982).
13. K.Okami, M. Murata, S. Kanisawa and M. Ehiro. Recent progress study on the Hikami Granites, In: *Geology and petrological studies on the Lower Paleozoic-Upper Proterozoic Erathums in the Japanese Islands and its vicinity.* H. Kano(Ed.). 69-72, Toko Print, Sendai (1981).*
14. T. Mikami. Preliminary notes on the Paleozoic sandstones from the Hikoroichi area, South Kitakami Mountains, *Mem. Geol. Soc. Japan.* no.6, 33-37 (1971).*
15. T. Kawamura. The Lower Carboniferous formations in the Hikoroichi region, Southern Kitakami Mountains -part 2, sedimentological study of the sandstone and limestone-, *Jour. Geol. Soc. Japan.* **90**, 831-847 (1984).*
16. T. Mikami. A sedimentological study of the Lower Permian Sakamotozawa Formation, *Mem. Fac. Sci. Kyushu Univ.* **19**, 331-372 (1969).
17. K. Kamada. The Triassic Inai Group in the Karakuwa area, Southern Kitakami Mountains, Japan -Part 1- Stratigraphy and paleogeography -, *Jour.Geol.Soc.Japan.* **85**, 12-31 (1979).*

18. F. Takizawa. Lower Cretaceous sedimentation in the Oshika Peninsula, Miyagi Prefecture, Northeast Japan, *Bull. Geol.Surv.Japan* **26**, 267-305 (1975).*
19. F. Takizawa. Jurassic sedimentation of the South Kitakami Belt, Northeast Japan, *Bull. Geol. Surv. Japan.* **36**, 203-320 (1985).
20. R. L. Folk. *Petrology of sedimentary rocks*. Hemphill's Book Store, Austin (1968).
21. K. Mori. Geology and paleontology of the Somanakamura Group, Fukushima Prefecture, Japan, *Tohoku Univ. Sci. Rep. 2nd.ser.* **35**, 33-65 (1963).
22. K. Okami. Sedimentary petrographic study of the quartzose sandstone of the Tomizawa Formation, *Tohoku Univ. Sci. Rep. 2nd. ser.* **41**, 95-108 (1969).
23. H. Kubo, Y. Yanagisawa, T. Yoshioka, T. Yamamoto and F. Takizawa. Geology of "Haramachi and Omika district". Quadrangle series, scale 1:50,000, Geol. Surv. Japan (1990).*
24. Y. Yamaguchi, H. Tsushima and N. Kitamura. Geologic development of the southern part of "Taro Belt" and "Iwaizumi Belt" in the Kitakami massif, Northeast Japan, *Contr. Inst. Geol. Paleont. Tohoku Univ.* no.80, 99-117 (1979).*
25. Y. Yamaguchi. Geological structure of the eastern part of the north Kitakami Mountains, Japan –with special reference to the structural subdivision–, *Contr. Inst. Geol. Paleont. Tohoku Univ.* no.83, 1-19 (1981).*
26. H. Osawa. Geological study on the "Hayachine Tectonic Belt", *Contr. Inst. Geol. Paleont. Tohoku Univ.* **85**, 1-34 (1983).*
27. M. Katada. Volcanic activities during the deposition of the Harachiyama Formation and the Paleozoic and Mesozoic sandstones of the Northern Kitakami Mountains, *Mem. Geol. Soc. Japan* no.10, 41-45 (1974).**
28. K. Minoura. The pre-Cretaceous geology and tectonics of northern Kitakami region, In: *Pre-Cretaceous Terranes of Japan.* K. Ichikawa et al. (Eds). IGCP Proj. no.224, 267-279 Nippon Insatsu Shuppan, Osaka (1990).
29. K. Okami, S. Koshiya and M. Ehiro. Compositions of Paleozoic to Mesozoic sandstones distributed in the Kitakami and Abukuma Mountains, Northeast Japan, *Mem .Geol. Soc. Japan* no.38, 43-57 (1992).*
30. M. Katada, H. Isomi and E. Omori. A preliminary report on Paleozoic sandstones and slates of the Northern Kitakami zone, *Jour. Japan. Assoc. Min. Petr. Econ. Geol.* **65**, 129-143 (1971).*
31. M. Katada and Y. Teraoka. Chemical composition of sandstones in Japan, *Annual Rep. Fac. Educ. Iwate Univ.* **40**, 55-66 (1981).*
32. B. P. Roser and R. J. Korsch. Determination of tectonic setting of sandstone mudstone suites using SiO_2 content and K_2O/Na_2O ratio, *Jour. Geol.* **94**, 635-650 (1986).
33. S. Murai, K. Okami and M. Oishi. Geology around the discovered area of "Moshi-ryu" (Dinosaurs), *Iwaizumi Town Board of Education*, Toryo Shuppan, Morioka (1983).**
34. K. Terui and H. Nagahama. Source and sedimentation of the Upper Cretaceous and Paleogene clastics in the Kuji area, northern Kitakami Mountains, In: *Prof. N. Kitamura Commemorative Vol.*, H. Nakagawa et al. (Eds). 545-570, Toko Print, Sendai (1986).*
35. K. Okami. Sedimentological study of the Iwaki Formation of the Joban Coal Field, *Tohoku Univ. Sci. Rep. 2nd. ser.* **44**, 1-53 (1973).
36. S. Kanisawa. Metamorphic rocks of the southwestern part of the Kitakami Mountains, Japan, *Tohoku Univ. Sci. Rep. 3rd. ser.* **9**, 155-198 (1964).
37. H. Maekawa. Geology of the Motai Group in the Southwestern part of the Kitakami Mountains, *Jour. Geol. Soc. Japan* **87**, 543-554 (1981).*
38. T. Yoshida, S. Kanisawa and M. Ehiro. Trace elemental abundances of the Hayachine Complex, *Jour. Japan. Assoc. Min. Petr. Econ. Geol.* **85**, 183 (1981).**
39. Y. Saito and M. Hashimoto. South Kitakami Region: An allochthonous terrane in Japan, *Jour. Geophys. Res.* **87**, 3691-3699 (1982).
40. S. Kanisawa. Basic and intermediate volcanic rocks from the Paleozoic formations in the southern Kitakami Mountains, Northeastern Japan, *Jour. Japan. Assoc. Min. Petr. Econ. Geol.* **81**, 12-31 (1971).
41. R. Sugisaki. Paleozoic volcanicity in the Kitakami Massif, *Mem. Geol. Soc. Japan.* no.10, 21-24 (1974).*
42. T. Kawamura and M. Kawamura. The Carboniferous system of the South Kitakami terrane, northeast Japan (part 2)–sedimentary and tectonic movement–, *Chikyu Kagaku.* **43**,157-167 (1989).*
43. S. Maruyama and T. Seno. Relative plate motions and orogeny around the Japanese Islands, *Kagaku.* **55**, 32-41 (1985).**
44. S. Maruyama and T. Seno. Orogeny and relative plate motions –an example of the Japanese Islands, *Tectonophysics* **127**, 1-25 (1986).

45. K. Kiminami, S. Miyashita and K. Tabata. Pass of the Kula-Pacific ridge along the Japanese Islands: its geologic significance, *Chikyu* **12**, 507–515 (1990).**

* in Japanese with English abstract.
** in Japanese.

(Manuscript received 28 September 1992; accepted 28 January 1993)

Modal and chemical compositions of the representative sandstones from the Japanese Islands and their tectonic implications

FUJIO KUMON[1] and KAZUO KIMINAMI[2]

[1] *Department of Geology, Fac. Sci., Shinshu Univ., Asahi 3-1-1, Matsumoto, 390 Japan.*
[2] *Department of Geology and Mineralogy, Fac. Sci., Yamaguchi Univ., Yoshida 1677-1, Yamaguchi 753 Japan.*

Abstract: The Japanese Islands have been situated in an active continental margin or island arc with intense magmatism since late Paleozoic time. Therefore, the sandstones in the islands are good records for tectonic movement and magmatism in a mobile belt on plate convergence. We selected the sandstones from four large turbidite basins as the representatives of the Japanese Islands which reflect the different provenances and tectonic settings.

We used the traditional classification for discriminating the provenance of sandstone, because the most data of the modal compositions of sandstones from the Japanese Islands have been point-counted by the traditional method. Sandstones from the Japanese Islands in different tectonic situations are plotted in the fields separated each other on QFR diagram. On QFR diagram, primitive volcanic arc, evolved and mature magmatic arc, dissected magmatic arc, and renewed magmatic arc provenances are distinguishable in magmatic arc provenance.

Based on the chemical analysis, we propose a new scattered diagram of Al_2O_3/SiO_2 vs. $(FeO+MgO)/(SiO_2+K_2O+Na_2O)$ which enables to discriminate the immature island arc, evolved island arc and mature magmatic arc provenances. The first two correspond to the primitive volcanic arc provenance based on modal composition. This diagram is especially effective for distinguishing the early stages of magmatic arc evolution, but the distinction of the evolved and mature magmatic arc and the dissected magmatic arc is impossible.

These diagrams seem to be useful tools for the comparison among the various basins and tectonic settings of magmatic arc.

Key words: sandstone, provenance, petrography, chemical composition, magmatic arc, Japanese Islands, Yubetsu Group, Izumi Group, Shimanto Belt

INTRODUCTION

Dickinson and Suczek [1] proposed firstly the discrimination diagrams for tectonic provenance based on a modal composition counted by the Gazzi-Dickinson method [2]. Then, Dickinson et al. [3] followed the former proposal by the new data collected from various geologic settings in and around North American Continent of Phanerozoic time, and refined their proposal on the basis of statistic treatment. On the other hand, Bhatia [4], Roser and Korsch [5], and Bhatia and Crook [6] proposed discrimination diagrams for tectonic settings based on chemical compositions. Although these proposals are useful keys to clarify the ancient tectonic settings of sedimentary basins, we try to get more detailed information on provenance and tectonic setting especially for a magmatic arc, using modal and chemical compositions of sandstones.

The Japanese Islands with a thick semi-continental crust are situated in a mobile zone where oceanic crust is subsiding into mantle now. In the geologic past, these islands formed an eastern marginal part of Asian Continent where violent igneous activity and tectonic movement took place intermittently from the Late Paleozoic to the Recent. The sandstones deposited in or near the Japanese Islands may provide key infromation to clarify the provenances of sandstones derived from magmatic arc source land.

METHODOLOGY

We selected four large turbidite basins for the representatives or standards of various settings of the Japanese Islands, that is, the Yubetsu, Tamba, Izumi, and Shimanto basins (Figure 1). The turbidite sandstones are generally immature in texture and mineralogy, and expected to reflect their provenance natures directly. Additionally, those basins are supposed to represent different situations of magmatic arc evolution.

We had also chosen six sandstone specimens from the basins to check some methodological problems. One problem is the difference between the traditional method and the Gazzi-Dickinson method [2]. As pointed by Ingersoll et al.[2] and Zuffa [8], we also had the result that the Gazzi-Dickinson method is more excellent way to get the information on provenance factor, because it decreases the grain-size effect and operator error [9]. The traditional way, however, have also advantageous points that it has more information about rock fragments, and is effective to clarify the relationship between grain size and modal composition which reflects sedimentary environments. Then, we concluded that it was better to use a new data sheet which enables to get point-counting data adaptable for the both methods [9]. The difference of the both ways is mainly in the recognition of rocks fragments as pointed by Zuffa [10]. It is not difficult to count the points on coarse-grained rock fragments according to both procedures at the same time. You can use either the traditional modal composition or new one corresponding to the purpose.

In this work, we used the data of modal composition counted by the traditional method to characterize the sandstones in the Japanese Islands, because a lot of compositional data of sandstones have been accumulated on the basis of traditional point-counting in japan.

Major chemical elements of sandstones were analyzed by ICP-AES method, and FeO was determinded by titration method of Chemex Labs Ltd., Canada. We use a scatter diagram, Al_2O_3/SiO_2 vs $(FeO+MgO)/(SiO_2+K_2O+Na_2O)$=Basicity Index(B.I.) to describe the chemical characters. This diagram enables to discriminate the provenance types of magmatic arc as will be discussed later.

SANDSTONE PETROGRAY AND THE SUPPOSED TECTONIC SETTINGS

Yubetsu Group in the Tokoro-Nemuro Belt, Hokkaido

The Yubetsu Group is a very thick clastic pile which consists mainly of sandstone, shale, and flysch-type alternations of sandstone and shale. The basin was formed in a trench-forearc region where a oceanic plate (Kulla-Pacific Plate?) was subducted eastward under the Okhotsk paleoland during the late Cretaceous to early Paleogene [11, 12]. Detrital materials were supplied from an immature volcanic island arc situated to the east of the basin [13]. The arc was composed mainly of basic to intermediate volcanic rocks accompanied with a small amount of older sedimentary rocks and granitic rocks [14, 15].

The sandstones of the Yubetsu Group are lithic to feldspathic greywackes, and very poor in both quartz and K-feldspar [14]. They are mostly plotted along the F-R line on QFR diagram (Figure 2). Rock fragments are abundant in intermediate volcanic rocks, but also contain fine-grained sedimentary rocks, and rarely semischist and granitic rocks. Feldspars are

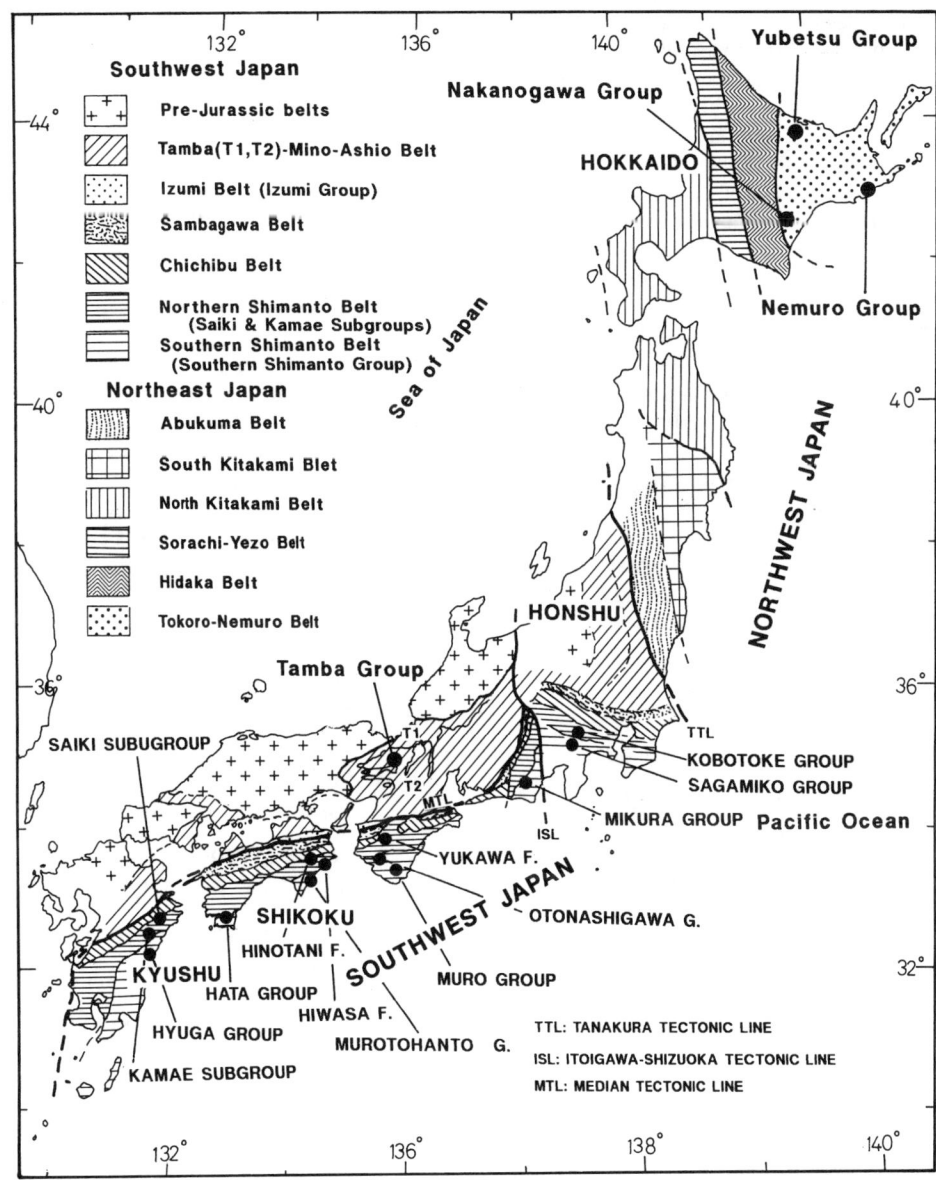

Figure 1. Geologic outline of the pre–Tertiary strata and the location of the studied formations in the main Japanese Islands. Modified from [7].

mostly plagioclase, and K-feldspar such as orthoclase and microcline are rarely found. Accessory heavy minerals such as clinopyroxene and hornblend are commonly included in 1 to 5 %.

Sandstones of the Saroma and Nakanogawa Groups in the Tokoro Belt are similar to those of the Yubetsu Group in petrography as shown in Figure 2 [16, 17]. The both groups distributed in the separated areas can be correlated mostly to the Yubetsu Group on the basis of radiolarian biostratigraphy and lithology [15]. Sandstones of the Nakanogawa Group have slightly abundant quartz, and those of the Saroma Group are rich in plagioclase.

Sandstones of the lower Nemuro Group also have similar characters to those of the Yubetsu Group (Figure 2). They are abundant in feldspars, especially in plagioclase, and very poor in quartz [18]. They were derived mainly from an island arc named Paleo-Kurile arc to the north of the basin. Some authors regarded the Nakanogawa and Yubetsu basins as the western extension of the Nemuro basin along the Paleo-Kuril islands which had been bent in N-S direction by the Tertiary tectonic movement [19, 11], although Nanayama [20] recently considered that quatrz-rich sandstones of the Nakanogwa Group were derived from a more mature island arc located to the west of the basin, probably Paleo-Japan arc.

We believe that these sandstones were derived from the same immature or semi-mature volcanic island arc, probably Paleo-Kurile island arc.

Izumi Group in the Izumi Belt, Shikoku

The Izumi Group distributed along the Median Tectonic Line in Shikoku and western Kinki district consists of sandstone, shale, conglomerate and acidic tuff, and forms a gentle synclinal structure. Turbidite facies composed of interbedded sandstone and shale is predominant. It rests unconformably on the Ryoke Granites, Ryoke metamorphic rocks and middle Cretaceous Sennan "Rhyolites". The thickness is more than 8,000 m in the axial part of the basin,

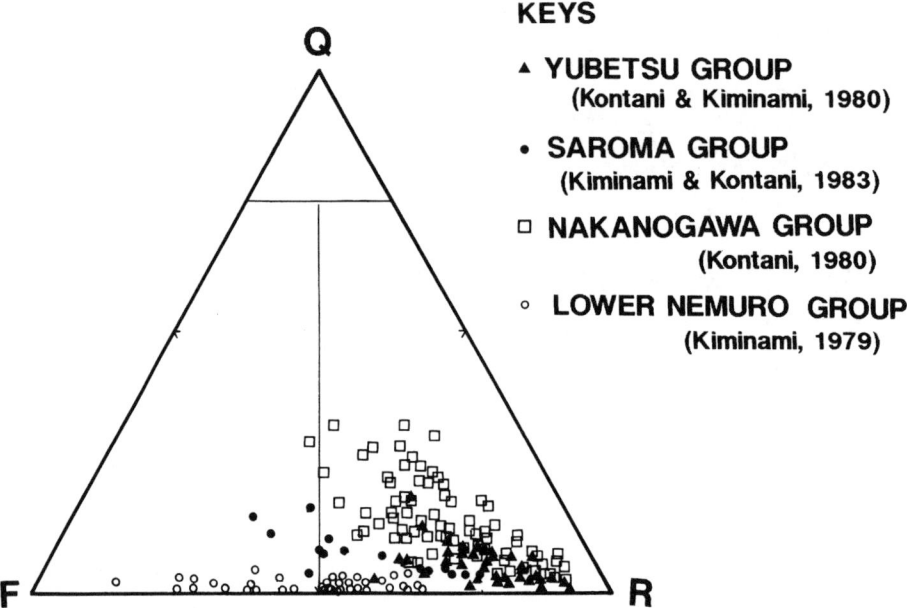

Figure 2. QFR plots of the Late Cretaceous sandstones from the Tokoro-Nemuro Belt in Hokkaido.

and the depocenter moved eastward stage by stage. It yields abundant ammonites and inoceramids, and the age is confirmed to be Campanian to Maastrichtian of Cretaceous.

Sandstones of the Izumi Group are abundant in volcanic rock fragments of acidic to intermediate nature, and classified mostly into lithic wacke (Figure 3)[21, 22]. Matrix is mostly clay minerals such as illite and chlorite, and cement of calcite and slicates are commonly observable. Their total amount ranges from 14 to 36 %. Rock fragments are most abundant constituent, ranging mostly from 38 to 73 %. Then, the sandstones are classified into lithic graywacke with a considerable amount of quatz and feldspars. The rock fragments are mainly acidic volcanic rock fragments such as glassy rhyolite and rhyolitic tuff some of which have welded textures. There are also contained commonly chert, shale, andesitic rocks, and rarely pelitic hornfels and granitic rocks. One sample exceptionally poor in rock fragments in Figure 7 belongs to the marginal facies of the basin, and its composition must reflect the sedimentary environments different from that of the other sandstones. Conglomerates are mostly composed of the clasts of acidic volcanic rocks.

Paleocurrents deduced from current marks on sole plane of turbidite sandstone indicate western axial and northern lateral supplies of clastic sediments [21, 22, 23].

In the late Cretaceous time, there was violent acidic to intermediate volcanic activity in the Inner Zone of Southwest Japan. Volumetric pyroclastics such as the Takada "Ryolites", Nohi "Ryolites", etc. are still widely distributed there. The Japanese Islands were situated in the eastern margin of Asian Continent. On the other hand, a oceanic crust, probably Kulla-Pacific plate, was subducted under Asian Continent. The subduction formed the Cretaceous Shimanto accretional terrane (Northern Shimanto Blet), and seems to be the major cause of the volcanic activity in the Inner Zone. The Izumi Group was located between the volcanism area and the subduction site. Therefore, the Izumi basin is regarded as a forearc basin on continental crust or intra-massif basin by Dickison and Seely [24].

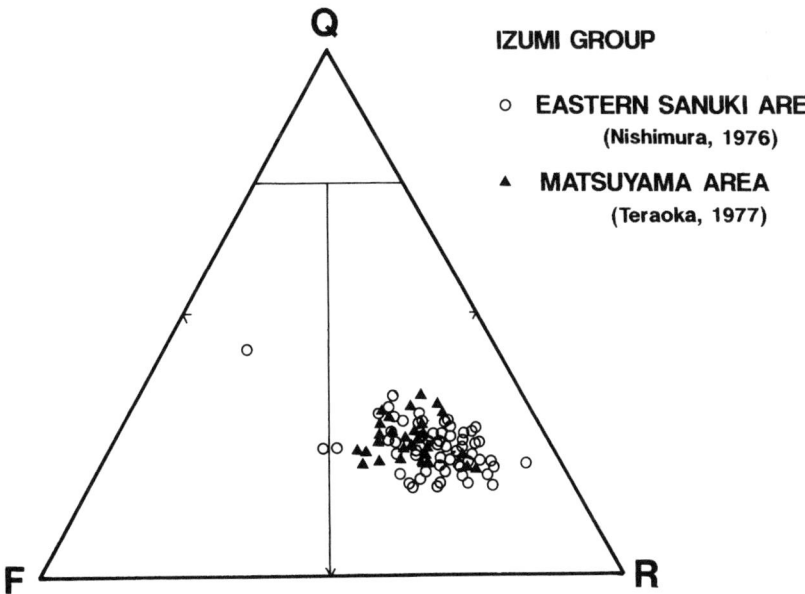

Figure 3. QFR plots of the latest Cretacous sandstones from the Izumi Group in Shikoku, Japan.

Shimanto Supergroup in the Shimanto Blet, Southwest Japan

The Shimanto Supergroup is distributed widely in the outermost zone of Southwest Japan, and regarded as a typical accretion complex in Japan. The Shimanto terrane is divided into the Cretaceous Northern Shimanto Belt and the Paleogene to early Neogene Southern Shimanto Belt. The Northern Shimanto Group in the Northern Belt is composed mainly of sandstone, alternating beds of sandstone and shale, and shale associated with chert, greenstones and acidic tuff. It has so-called eugeosynclinal facies. Radiolarian fossils indicate that chert and associated greenstones are generally much older than surrounding muddy sediments. They form exotic blocks of various size. On the other hand, the Southern Shimanto Group in the Southern Belt consists mainly of sandstone, flysch-type alternating beds of sandstone and mudstone, mudstone, conglomerate and small amount of greenstones without chert. It is supposed to have been deposited in a trench-forearc region.

Kumon [25, 26] once subdivided the Northern Shimanto Belt into three zones, namely northern, middle and southern zones. Recently, Teraoka and Okumura [27] divided the northern belt into the Saiki and Kamae subbelts on the basis of stratigraphy and sandstone composition. The northern zone by Kumon corresponds to the Saiki subbelt, and the middle and southern zones to the Kamae subblet. The strata in the Saiki subbelt range from Barremian to Turonian, and the strata in the Kamae subbelt from Turonian to Maastrchtian in age.

Modal composition of sandstone and conglomerate from the supergroup was extensively investigated [22, 25, 26, 28, 29, 30, 31,32, 33,34, 35, 36, 37, 38, 39, 40, 41]. The representative data are listed in Table 1, and shown in Figures 4 and 5.

Sandstones from the Northern Shimanto Belt are mostly feldspathic to lithic wackes of which matrix exceeds 15 %, and have distinct difference between the Saeki and Kamae subbelts in composition (Figure 4). The framework grains are angular to subangular, and matrix

Table 1. Modal composition of the sandstones from the formations of the Shimanto Supergroup.

Formation	n	ratio among framework grains						%	%	Ref. No.
		Qm	Qp	Qt	Pl	Kf	RF	Ot	Mtx	
Saiki subbelt										
Yukawa Formation	35	30.8	4.3	35.1	32.0	10.4	22.4	3.8	21.4	[25]
Hinotani F.	72	29.2	3.2	32.5	35.7	11.2	20.6	4.5	21.2	[25]
Saiki Subgroup	146	-	-	26.4	31.2	11.8	30.6	-	18.5	[41]
Kamae subbelt										
Kobotoke Group	28	-	-	26.2	22.2	9.0	42.6	1.2	17.3	[40]
Hiwasa Formation	83	-	-	34.9	19.8	7.9	37.5	3.5	19.1	[25]
Kamae Subgroup	129	-	-	22.3	24.6	7.6	45.6	-	19.8	[41]
Southern Belt										
Sagamiko Group	21	-	-	42.4	17.0	10.4	29.1	1.1	15.7	[40]
Mikura Group	13	33.5	7.4	40.9	22.9	7.2	29.1	3.7	16.8	[38]
Otonashigawa G.	51	42.2	6.0	48.2	24.3	13.4	14.0	3.8	9.0	[25]
Muro Group	214	44.6	9.1	53.7	20.7	13.4	12.2	3.3	10.9	[25]
Murotohanto G.	29	-	-	54.0	18.0	13.1	14.9	4.5	15.0	[30]
Hata Group	42	-	-	42.6	24.5	6.5	26.4	-	-	[28]
Hyuga Group	194	-	-	42.0	22.8	5.7	29.4	-	-	[28]

n: number of specimens, Qm: monocrystalline quartz, Qp: polycrystalline quartz, Qt: total quartz, Pl: plagioclase Kf: K-feldspar, RF: total rock fragments including chert, Ot: others, Mtx: matrix and cement

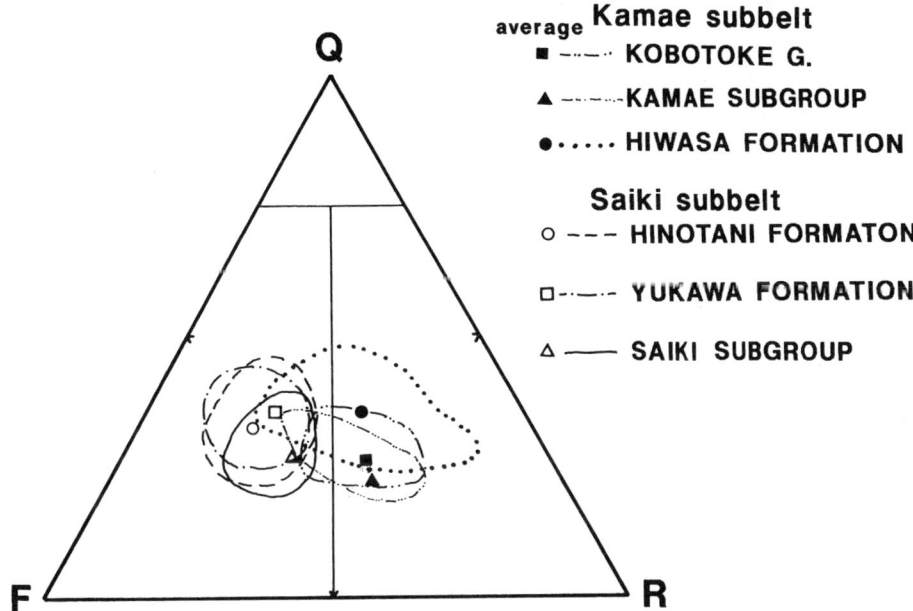

Figure 4. Modal compsosition of the Cretaceous sandstones from the Northern Shimanto Belt, Southwest Japan.

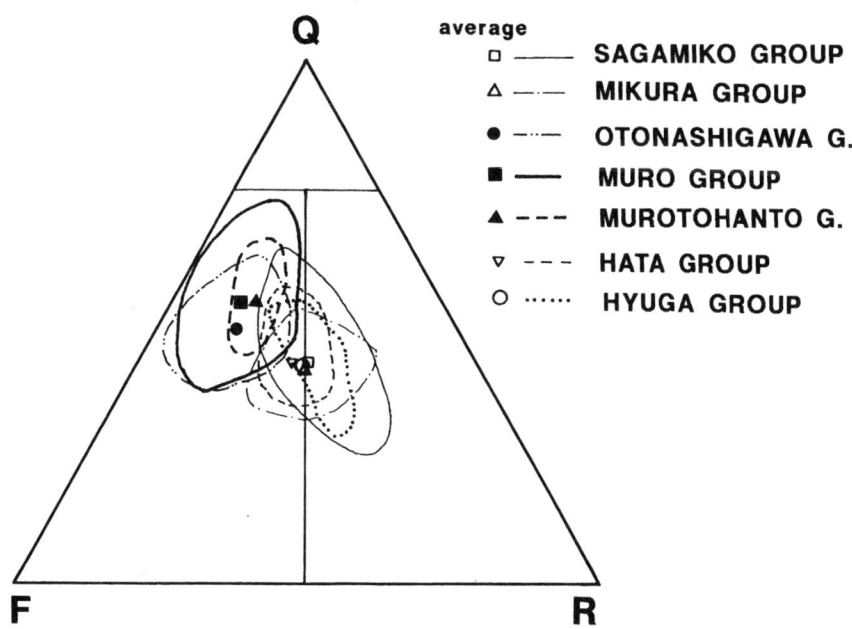

Figure 5. Modal composition of the Paleogene sandstones from the SouthernShimanto Belt, Southwest Japan.

is mostly clay and silica minerals. Sandstones of the Saiki subbelt are rich in feldspars, and plagioclase is abundant as twice or triple as K-feldspar. Rock fragments are mostly intermediate to acidic volcanic rocks associated with shale, chert, quartz schist, etc. On the other hand, sandstones of the Kamae subbelt are abundant in rock fragments, and poor in feldspars. Rock fragments are mostly acidic volcanic rocks, accompanied with shale and granitic rocks. There exists a progressive change of modal composition among the sandstones from the Northern Shimanto Belt.

Kumon[25, 26] pointed out that the stratigraphical change of modal composition reflects the evolution of igneous activity which took place in the Inner Zone of Southwest Japan under the continental arc-trench system. This process named "roofing" means the initiation of intermediate volcanic activity on continental crust and successive violent acidic volcanism represented by the Nohi "Rhyolites" which cover extensively the older rocks as a roof.

The Paleogene sandstones from the Southern Shimanto Belt in eastern Shikoku and the Kii Peninsula are mostly quartz-rich arkosic arenite with a small amount of rock fragments. The Paleogene sandstones of the whole Shimanto terrane have similar properties to each other, but a slight areal difference can be recognized (Figure 5). Sandstones from the Kanto, Chubu, western Shikoku and Kyushu districts are slightly abundant in rock fragments mainly of acidic volcanics. The difference regarded as petroprovince is considered to be due to areal variation of the dissection degree of the magmatic source land once covered by acidic volcanic rocks [42].

Tamba Group in the Inner Zone, Southwest Japan

The Tamba Group in Kinki and Chugoku districts is composed of sandstone, shale, flysch-type alternations of sandstone and shale, siliceous shale, limestone, chert and greenstones. There occur a lot of radiolarian and fusulinid microfossils. Limestone and the associating

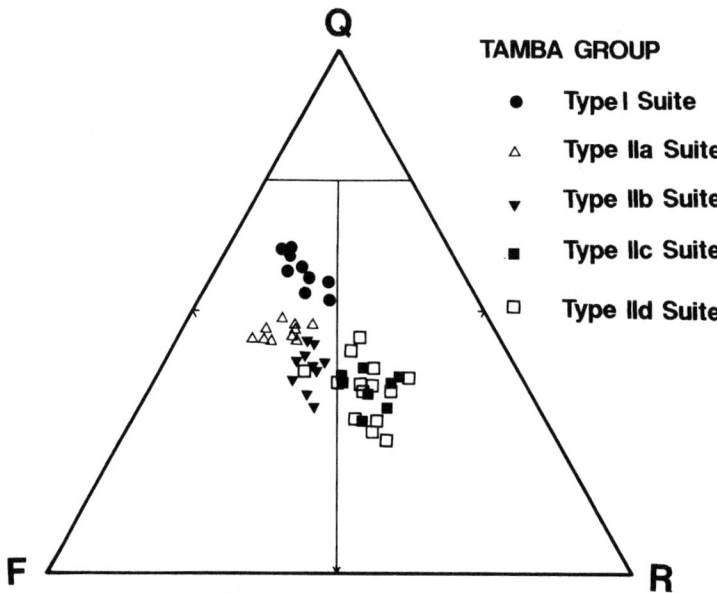

Figure 6. QFR plots of the Jurassic sandstones from the Tamba Group in Kinki district, Japan. After [44, 45].

greenstones are mostly late Paleozoic, and chert are late Paleozoic to early Jurassic in age. They occur as older blocks in muddy clastic sediments of Jurassic to earliest Cretaceous age, and are considered to be exotic in origin. Therefore, the sedimentation age of the group is the middle Jurassic to earliest Cretaceous. The Tamba Group consists of two distinct suites designated as Type I and Type II [43]. Type I suite is composed mainly of Triassic-Jurassic chert, middle Jurassic siliceous shale, and middle to late Jurassic shale and sandstone. Type II suite consists mostly of Permian chert and greenstones, Triassic chert and early-middle Jurassic mudstone and sandstone. Type II suite has been recently subdivided into Type IIa to IId subsuites [44, 45]. The age of the subsuite become old form Type IIa to IId, judging from radiolarian fossils. The Tamba Group and the equivalent Mino Group in the neighbouring Chubu district are an accretion complex of late Jurassic to earliest Cretaceous time.

Sandstones of the Tamba Group are mostly arkosic wacke (Figure 6). They are abundant in quartz and feldspars, and poor in rock fragments [44, 45]. They are mainly fine-grained to medium-grained, and are subangular to subrounded, rarely well-rounded in texture. Muscovite is relatively abundant in accessory minerals. There are differences in composition between those of Type I and II suites, and further, among Type IIa to Type IId suite sandstones.

Sandstones of Type I are rich in quartz and poor in rock fragments compared with those of Type II suite. They have following characters. Matrix is mostly clay minerals with calcite cement, and its amount ranges form 10 to 33 % (20 % on an average). Quartz is most abundant component, and the ratio of monocrystalline quartz to polycrystalline quartz which is an aggregate of a few crystals of quartz is 0.5 on an average. Plagiocase is a slightly more abundant than K-feldspar, or nearly same. Rock fragments are acidic to intermediate volcanic rocks, quartz-schist, slate, chert, etc.

Sandstones of Type II suite are slightly poor in quartz and rich in rock fragments. Furthermore, the amount of rock fragments increases from Type IIa to Type IId subsuites [44, 45]. Sandstones of Type IIa subsuite contain a little smaller amount of quatrz, but nearly same amount of rock fragments, compared with those of Type I suite. Sandstones of Type IIb subsuite are distinctly poorer in quartz and slightly richer in rock fragments than those of Type I suite. Sandstones of Type IIc includes more rock fragments than those of Type IIb subsuite. Acidic volcanics are dominant in the rock fragments. Sandstones of Type IId subsuite are almost same as those of Type IIc subsuite in composition.

The sandstones of the Tamba Belt were considered to have been derived from a dissected magmatic arc [44, 46]. It should be mentioned that the decreasing trend of rock fragments from Type IId to Type I suite might correspond to the dissecting process of the magmatic arc source land.

PETROGRAPHIC CRITERIA FOR MAGMATIC ARC PROVENANCE

The following four tectonic provenances can be recongized among the magmatic arc source land based on the petrographic criteria as shown in Figure 7. The tectonic situations and provenance natures are also illustrated in Figure 8.

Primitive volcanic arc provenance

The sandstones which are abundant in rock fragments or plagioclase and very poor in quartz must represent primitive volcanic arc where basic to intermediate volcanism takes place violently. The crust of the islands is thin and immature like as that of the present Izu-Marina Islands arc. In the later and more evolved stage of this provenance, there may be some acidic volcanism and granitic intrusions which are indicated by the presence of quartz, microcline feldspars and acidic volcanic fragments. These sandstones are plotted along or near the F-R line on QFR diagram (Figure 7).

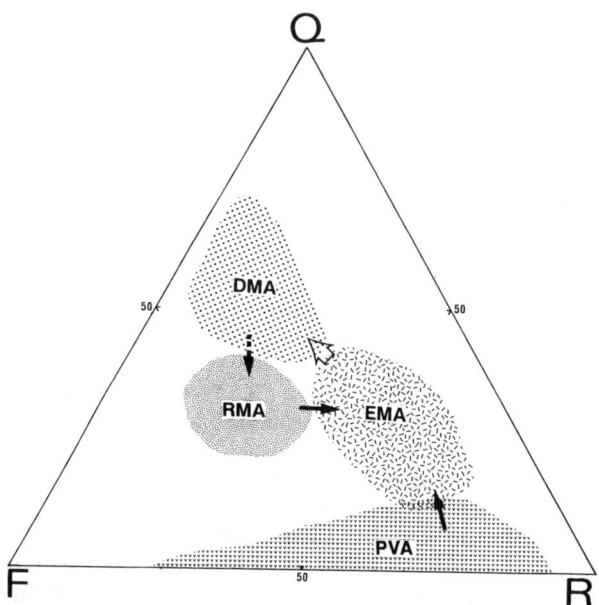

Figure 7. Tentative proposal for discriminating provenance type of magmatic arc, based on the traditional point-counting method. Solid arrows show evolving and maturing process, and open arrow indicates unroofing and dissecting process.
PVA: primitive volcanic arc, EMA: evolved and mature magmatic arc, DMA: dissected magmatic arc, RMA: renewed magmatic arc, Q: quartz, F: feldspars, R: rock fragments

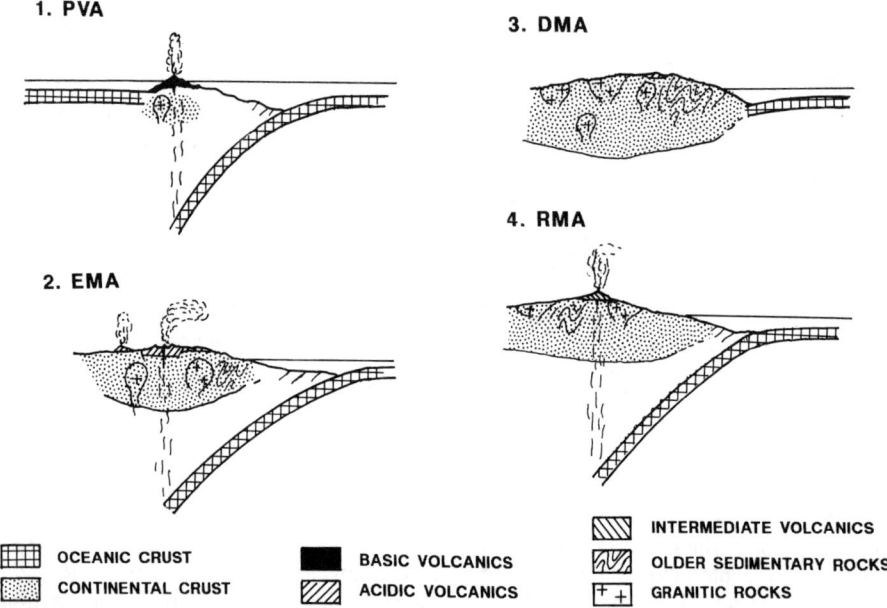

Figure 8. Schematic cross sections of magmatic arc provenances and their tectonic situations. Alphabetic symboles are same as those in Figure 7.

The representative or typical sediments of this provenance are the Yubetsu Group, Nakanogawa Group and Nemuro Group mentioned before. Similar sediments were also reported from the basins in a modern primitive volcanic arc-trench system, that is, the forearc and backarc basins in the Izu-Bonin arc (ODP Leg 126; [47]), Atsuka Basin in the Alutian arc [48] and Middle American trench off Guatemala [49]. Sandstones from the Western Facies of New Zealand [50] and the middle Permian strata in the Maizuru Belt of Southwest Japan [51] have similar properties of modal composition, and their provenance were supposed to be volcanic islands.

Evolved and mature magmatic arc

Sandstones plotted in the middle to lower right position on QFR diagram are indicative of an evolved and mature magmatic arc provenance. The sandstones are dominant in rock fragments, but contain also fairly abundant quartz and feldsparss. Rock fragments are mostly acidic volcanic rocks. This provenance may be characterized by thick continental crust and acidic to intermediate volcanic activity associated with acidic plutonism. It forms a mature island arc or continental arc.

Sandstones of the Izumi Group and from the Kamae subbelt of the Northern Shimant Belt are typical examples derived from this provenance. The Izumi Group was deposited in a intra-massif basin or forearc basin on continental crust next to violent volcanism site. It partly overlies the acidic pyroclastic rocks of middle Cretaceous time, Sennan "Rhyolites". It was elucidated that the coarse clastic sediments of the late Cretaceous Shimanto Supergroup corresponding to the Kamae Subgroup were derived also from the Inner Zone of Southwest Japan where violent volcanic activity took place extensively [25].

Sand compositions from the modern analogue of this provenance type are rarely reported. Sands from the late Pleistocene fluvial sediments north of Tokyo Bay has very similar composition [52]. There are distributed Mesozoic strata, older crystalline rocks and Quaternary volcanoes, some of which volcanoes are still active, to the north of Tokyo Bay. The sediments were derived surely from the surrounding areas.

Tertiary sandstones of the Bristol basin along the Alaska Peninsula and the Chehalis-Grays Harbor basins have a similar modal compositions [53]. These basins are interpreted as forearc or backarc basins, and volcanic activity took place near the basins.

Therefore, we can regard these lithic arkose sandstones to represent evolved and mature magmatic arc provenance. The provenance has continental crust where active volcanism of intermediate to acidic nature take place. The older sedimentary rocks and granitic intrusives form the basements of the volcanic rocks.

Dissected magmatic arc

Arkosic sandstones plotted in the area of low rock fragment amount, and moderate quartz and feldsparss content must represent dissected magmatic arc provenance (Figure 7). The stratigraphical or time-sequential changes from lithic sandstones to arkosic sandstones is recognized in a few basins [54, 44, 39]. There is extensive cropping out of granitic rocks which formed deeper facies of previously-existed volcanic rocks, and older sedimentary rocks. The characters of source land may resemble partially those of continental block provenance.

Renewed magmatic arc

What happen when new volcanism occurs on dissected magmatic arc? Kumon [25] considered that the sandstones from the Saiki subbelt of Northern Shimanto Belt were derived from the Inner Zone of Southwest Japan where the activity of intermediate volcanic rocks named the Kanmon Group initiated on thick continental crust constituting the eastern margin of

Asian Continent. The basement rocks are older sedimentary rocks, granitic rocks and metamorphic rocks.

The sandstones of the Saiki subbelt are feldspathic wackes abundant in plagioclase and slightly poor in quartz. The rock fragments are mostly intermediate to acidic volcanic rocks accompanying sedimentary rocks and granitic rocks. These characters are common throughout the Shimanto Belt from Kanto Mountains to Kyushu [22, 25]. Additionally, the Cenomanian sandstones of the Onogawa Group in the Inner Zone of Kyushu have the same composition as those of the Saiki subbelt [22].

It is difficult to find a modern analogue of this type sandstones and provenance, because of the scarcity of modern sand composition based on this viewpoint. Pliocene sands in the Middle American Trench off South Mexico may represent an example of this type of provenance [49]. Another reason is the difficulty to distinguish the renewed magmatic arc from the dissected magmatic arc. The both resemble each other in rock constitution. The key of discrimination is the stratigraphic change of the sandstone composition. The properties of rock fragments may be supporting means, that is, relative abundance of plagioclase and volcanic fragment of intermediate nature. We suppose that some of sands (sandstones) regarded as dissected magmatic arc by Dickinson and Suczek [1] and Dickinson et al. [3] might have been derived from a renewed magmatic arc (for example, Pliocene to Miocene Queen Charlotte basin in western Canada; [53]).

CHEMICAL PROPERTIES AND TECTONIC SETTING

After trial and error, we concluded that the diagram of $(Al_2O_3+SiO_2)$ vs. $(FeO+MgO)/(SiO_2+K_2O+Na_2O)$ can discriminate best the sandstones derived from magmatic arc [55]. Al_2O_3 amount in sandstone depends mainly on feldsparss and clay amount. SiO_2 amount is proportional to quartz amount. Then, Al_2O_3/SiO_2 roughly correspond to the ratio of feldspars to quartz. In general, FeO and MgO are contained much more in mafic volcanics. In contrast, Na_2O, K_2O and SiO_2 are abundant in felsic volcanics. Therefore, $(FeO+MgO)/(SiO_2+Na_2O+K_2O)$ means rough measure of basicity of the source volcanic rocks. Then, we called the ratio, $(FeO+MgO)/(SiO_2+Na_2O+K_2O)$, Basicity Index (B.I.) [55].

Figure 9 shows chemical characteristics of the sandstones from the Yubetsu Group, Nakanogawa Group, Izumi Group, Tamba Group, Saiki Subgroup and Kamae Subgroup on the diagram. The sandstones from the Yubetsu Group are highest in Al_2O_3/SiO_2 ratio and B.I. Those from the Nakanogawa Group are plotted in slightly lower positions in respect of B.I. Sandstones from the Izumi Group, Tamba Group and Kamae subgroup of the Shimanto Belt are distributed in the same field of the lowest B.I. and Al_2O_3/SiO_2 ratio. Sandstones of the Saiki Subgroup of the Shimanto Belt are located between the second and the third group. These plots show a definite trend of distribution as a whole. Miyashiro [56] clarified that the magmatism of arc changes progressively from basic magmatism to acidic one with the thickening of the crust of arc. The trend of decreasing of B. I., and Al_2O_3/SiO_2 ratio reflects the general evolution of magmatic arc. We can recognize several clusters which correspond to the provenance types defined based on modal composition, that is, immature island arc, evolved island arc, and mature magmatic arc provenances for the sandstones. The first two correspond to the immature volcanic arc provenance based on the modal composition. The last one contains large categories based on modal composition. Sandstones of the evolved and mature magmatic arc provenance and the dissected magmatic arc are plotted all together, and can not be distinguished from each other, because the both provenances are almost the same in magma composition. Sandstones of the renewed magmatic arc are located in the right margin of the mature magmatic arc province of the diagram. This diagram is very sensitive for discriminating the early stage of magmatic arc evolution. The name of provenance type has been partly revised here from that of Kiminami et al. [55] to avoid confusion with the

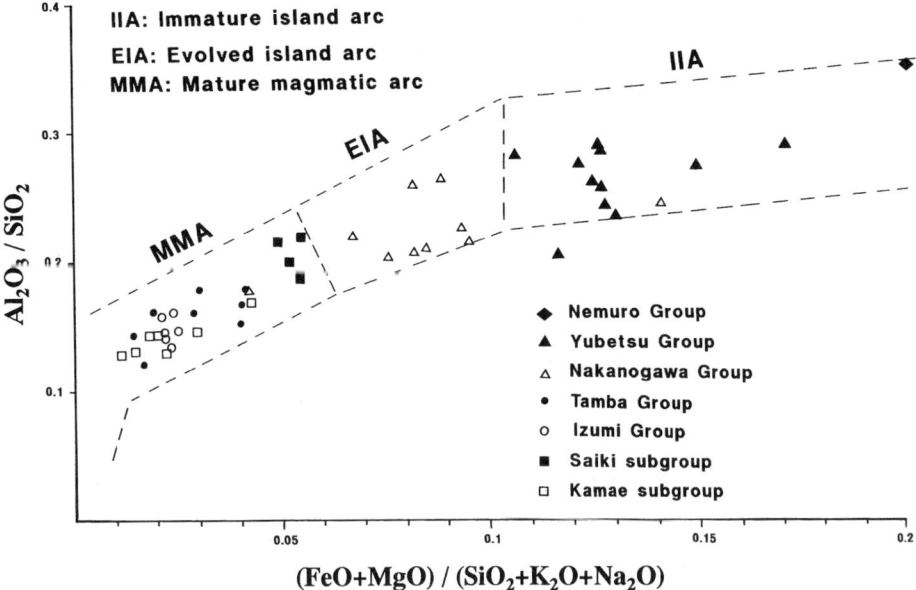

Figure 9. Tentative proposal of a discrimination diagram for provenance of sandstones, based on chemical composition.

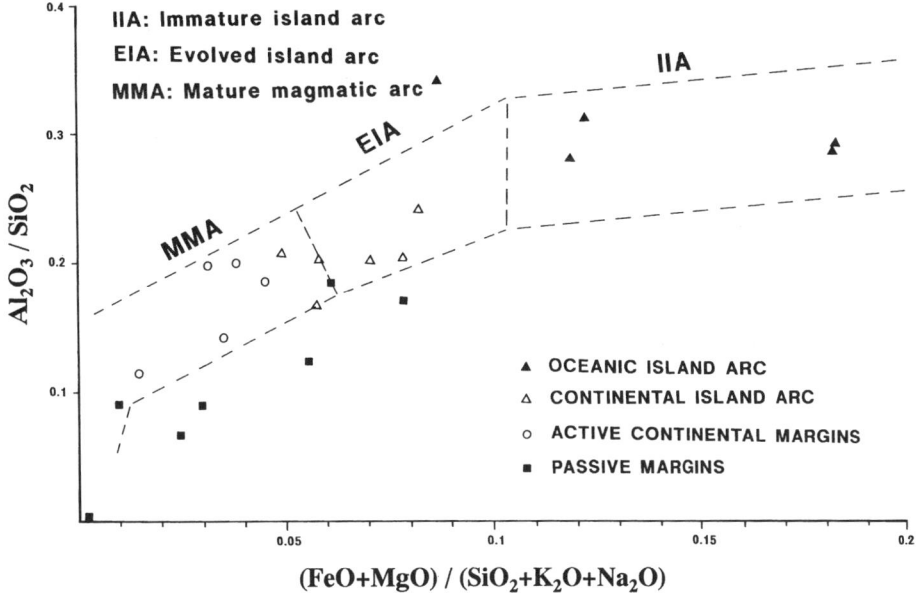

Figure 10. A test of the $(Al_2O_3+SiO_2)$–B.I. diagram to discriminate provenance, using the data of Bhatia [4].

provenance name based on modal composition.

Sandstone compositions reported by Bathia [4] were also plotted on this diagram (Figure 10), and tested the reliability of the diagram. Sandstones classified as oceanic island arc provenance correspond to the immature island arc provenance, and those of continental arc provenance are plotted in evolved island arc area. Sandstones of active continental margin are scattered in mauture magmatic arc, and those of passive continental margin are mostly out of the any fields. Then, we believe that the diagram proposed above is effective for analysing magmatic arc provenance.

DISCUSSION

As mentioned above, it is concluded that the sandstones derived from the magmatic arc provenance have peculiar properties of modal and chemical compositions depending on their provenance characters. The new point of this proposal is the recognition of tectonic provenances which correspond to the progressive stage of arc evolution. Magmatic arc is initiated as a volcanic island arc of basic to intermediate magma like the Izu-Bonin Islands arc, and grows up to a large island arc of intermediate to acidic volcanism like the Japanese Islands arc which have continental or semi-continental crust [56]. This process is called "maturing" in this paper. The mature magmatic arc starts to be eroded immediately after the volcanism, and the erosion results in the wide outcropping of the deeper facies of igneous rocks and the older sedimentary rocks after the decease of volcanism. This dissecting process was called "unroofing" by Mansfield [54]. The deeply dissected magmatic arc provenance may have similar provenance properties to those of continental block provenance. The revival of the volcanism in the dissected magmatic arc or continental margin may take place by the change of plate motion. The volcanic rocks begin to cover the granitic rocks and older sedimentary rocks again, and overlis them extensively at later stage. This covering process was called "roofing" by Kumon [25, 26]. The diagrams proposed here enable to discriminate the evolutional stage of magmatic arc provenance based on the modal and chemical composition of sandstones.

The diagram based on modal composition, however, still has some weak points. One is that the boundary between the provenance types on QFR diagram is somewhat arbitrary. The QFR mode by traditional point-counting method is affected largely by grain size. The data used for the diagram include all range of sand grain size, because most compositional data lack grain-size data. Another point is the scarcity of composition data of the modern sands derived from the provenances corresponding to the present proposal.

Mack [56] warned that the routine use of the plate-tectonic provenance diagrams lead to the recognition of error populations. We agree with him in principle, but we think that the careful treatments enable us to eliminate the unsuitable data from the consideration. One way is to use the data from large basins which reflect the large area of source land in the detrital compositions. This may avoid the influence of relict source rocks distributed in a local area.

Another way is to weight the data of immature sediments such as fluvial sandstone and turbidite sandstone, decreasing the influence of various depositional environments. As generally believed, sedimentary environment affects the composition and texture of sandstone. For example, Ito and Masuda [58] reported the systematic change of modal composition from fluvial to shallow marine in the same drainage basin. It shows a progressive change of mineralogic maturity. It is important to study on immature sandstones for recognizing the provenance, considering the textural maturity of sandstones which may be a guide of mineralogic maturity.

We examined only the sandstones supposed to be derived from magmatic arc provenance in this paper. There may be some difficulty to distinguish it from the other tectonic provenances. Most probable confusion may exists between the dissected magmatic arc provenance and a part of continental block provenance. Properties of rock fragments, that is, the ratio of volcanic rocks, sedimentary rocks, schistose rocks and plutonic rocks, may be useful to avoid misunderstanding of the provenance. The stratigraphical or time-sequential change which corresponds to the evolution of magmatic arc, should also provide an important key of correct provenance interpretation.

CONCLUSION

The provenance of sandstones derived from magmatic arc can be discriminated as immature volcanic arc, evolved and mature magmatic arc, dissected magmatic arc and renewed magmatic arc provenances on QFR diagram based on the traditional point-counting method. This proposal for provenance determination is based on the evolution of magmatic arc. It provide the more detailed information on source land and tectonic setting.

The discrimination diagram based on the major chemical compositions also enable us to distinguish the plate tectonic setting and the evolution stage of magmatic arc. This diagram is effective for discriminating the early stages of magmatic arc evolution.

Then, these diagrams provide powerful tools to interpret the tectonic setting and magmatic provenance in the geologic past.

ACKNOWLEDGEMENTS

We would like to thank Dr. Y. Kontani of Ritsumeikan High School for co-working in the Tokoro-Nemuro Belt in Hokkaido, and T. Nishimura of Hyogo University of Teacher Education, M. Musashino of Kyoto University of Education and T. Kawamura of Hokkaido University for offering the chemical data of sandstones. We also heartily thank Professor Emeritus K. Nakazawa of Kyoto University for helpful advice and reviewing. This study was supported by the Grant-in-Aid for Scientific Research funded by the Ministry of Education, Science and Culture, Japan (No.63302019).

REFERENCES

1. W.R. Dickinson and C.A. Suczek. Plate tectonics and sandstone compositions, *AAPG Bull.*, **63**, 2164-2182 (1979).
2. R.V. Ingersoll, T.F. Bullard, R.L. Ford, J.P. Grimm, J.D. Pickle and S.W. Sares. The effect of grain size on detrital modes: a test of the Gazzi-Dickinson point-counting method, *Jour. Sed. Petrology*, **54**, 103-116 (1984).
3. W.R. Dickinson, L.S. Beard, G.R. Brakenridge, J.L. Erjavec, R.C. Ferguson, K.F. Inman, R.A. Knepp, F.A. Lindberg and P.T. Ryberg. Provenance of North American Phanerozoic sandstones in relation to tectonic setting,*GSA, Bull.*, **94**, 222-235 (1983).
4. M.R. Bhatia. Plate tectonics and geochemical composition of sandstones., *J. Geol.*, **91**, 611-627 (1983).
5. B.P. Roser and R.J. Korsch. Determination of tectonic setting of sandstone-mudstone suites using SiO_2 content and K_2O/Na_2O ratio, *Jour. Geology*, **94**, 635-650.
6. M.R. Bhatia and K.A.W. Crook. Trace element characteristics of graywackes and tectonic setting discrimination of sedimentary basins, *Contrib. Mineral. Petrol.*, **92**, 181-193 (1986).
7. K. Ichikawa. Pre-Cretaceous terranes of Japan, In: *Pre-Cretaceous Terranes of Japan*. K. Ichikawa, S. Mizutani, I. Hara, S. Hada and A. Yao (Eds), 1-11, Pub. IGCP Project, no.224 (1990).
8. G.G. Zuffa. Optical analyses of arenites: influece of methodology on compositional results. *In: Provenance of Arenites*. G.G. Zuffa(ed.). 165-189, D. Reidel Publishing Company, Dordrecht (1985).
9. F. Kumon, K. Kiminami, M. Adachi, T. Bessho, K. Kawabawa, T. Kusunoki, T. Nishimura, H. Okada, K. Okami, S. Suzuki and Y. Teraoka. Modal compositions of representative sandstones from the Japanese Islands and their tectonic implications, *Mem. Geol. Soc. Japan*, no.38, 385-401 (1992).*
10. G.G. Zuffa. Hybrid arenites: their composition and classification, *Jour. Sed. Petrology*, **50**, 21-29 (1980).

11. M. Sakakibara, K. Niida, H. Toda, N. Kito, G. Kimura, J. Tajika, T. Katoh, A. Yoshida and Research Group of the Tokoro Belt. Nature and tectonic history of the Tokoro belt, *Monograph Assoc. Geol. Collab.. Japan*, no.31, 173–187 (1986).*
12. K. Kiminami, K. Shibata and S. Uchiumi. K-Ar age of a tuff from the Yubetsu Group in the Tokoro Belt, Hokkaido, Japan. *Jour. Geol. Soc. Japan*, **96**, 77–80 (1990).**
13. K. Kiminami and Y. Kontani. Pre-Cretaceous paleocurrents of the northeastern Hidaka Belt, Hokkaido, Japan. *Jour. Fac. Sci., Hokkaido Univ., Ser.IV*, **19**, 179–188 (1979).
14. Y. Kontani and K. Kiminami. Petrological study of the sandstones in the pre-Cretaceous Yubetsu Group, northeastern Hidaka Belt, Hokkaido, Japan, *Earth Sci.*, **34**, 307–319 (1980).
15. Y. Kontani, K. Kiminami, J. Tajika and K. Maniwa. Cretaceous sedimentary rocks in the Tokoro and Nemuro belts, Hokkaido, *Monograph Asoc. Geol. Collab.. Japan*, no.31, 157–171 (1986).*
16. K. Kiminami and Y. Kontani. Sedimentology of the Saroma Group, northern Tokoro Belt, Hokkaido, *Earth Sci.*, **37**, 38–47 (1983).*
17. Y. Kontani. Geological study of the Hidaka Supergroup distributed on the east of the Hidaka metamorphic belt (Part II): petrographical study of the sandstones in the Nakanogawa Group, *Jour. Geol. Soc. Japan*, **86**, 1–14 (1980).*
18. K. Kiminami. Sedimentary history of the late Cretaceous-Paleocene Nemuro Group, Hokkaido, Japan: a forearc basin of the Paleo-Kuril arc-trench system, *Jour. Geol. Soc. Japan*, **89**, 607–624 (1983).
19. G. Kumura. The style of subduction in Hokkaido of Cretaceous time. *Kagaku* (Iwanami Pub.), **55**, 24–31 (1985).**
20. F. Nanayama. Three petroprovinces identified in the Nakanogawa Group, Hidaka Belt, central Hokkaido, Japan, and their geotectonic significance. *Mem. Geol. Soc. Japan*, no.38, 27–42 (1992).*
21. T. Nishimura. Petrography of the Izumi sandstones in the east of the Sanuki Mountain Range, Shikoku, Japan, *Jour. Geol. Soc. Japan*, **82**, 231–240 (1976).
22. Y. Teraoka. Comparison of the Cretaceous sandstones between the Shimanto Terrane and the Median Zone of Southwest Japan, with reference to the provenance of the Shimanto geosynclinal sediments, *Jour. Geol. Soc. Japan*, **83**, 795–810 (1977).*
23. T. Nishimura. Basin analysis of the Upper Cretaceous Izumi Group in western Shikoku, Japan, *Jour. Geol. Soc. Japan*, **90**, 157–174 (1984).*
24. W.R. Dickinson and D.R. Seely. Sturucture and stratigraphy of forearc regions, *AAPG Bull.*, **63**, 2–31 (1979).
25. F. Kumon. Coarse clastic rocks of the Shimanto Supergroup in eastern Shikoku and Kii Peninsula, Southwest Japan, *Mem. Fac. Sci., Kyoto Univ., Ser. Geol.& Miner.*, **49**, 63–109. (1983).
26. F. Kumon. Stratigraphic change of the coarse clastic rocks of the Shimanto Supergroup in eastern Shikoku, Southwest Japan, In: *Formation of active ocean margins*. Nasu et al.(eds). 819–833, TERRAUB, Tokyo. (1985).
27. Y. Teraoka and K. Okumura. Tectonic devision and Cretaceous sandstone compositions of the Northern Belt of the Shimanto Terrane, Southwest Japan, *Mem. Geol. Soc. Japan*, no.38, 261–270 (1992).*
28. Y. Teraoka. Provenance of the Shimanto geosynclinal sediments inferred from sandstone compositions, *Jour. Geol. Soc. Japan*, **85**, 753–769 (1979).*
29. F. Kumon. Shimanto Supergroup in the southern part of Tokushima Prefecture, Southwest Japan, *Jour. Geol. Soc. Japan*, **87**, 277–295 (1981).*
30. F. Kumon and Y. Inouchi. Stratigraphical and sedimentological studies of the Paleogene system of the Shimanto complex in the Shishikui-cho area in Tokushima Prefecture, the northeastern part of the Muroto Peninsula, *Jour. Geol. Soc. Japan*, **82**, 383–394 (1976).*
31. T. Tokuoka. The Shimanto Terrain in the Kii Peninsula, Southwest Japan -with special refference to its geologic development viewed from coarser clastic sediments-, *Mem. Fac. Sci. Kyoto Univ., Ser. Geol. & Mineral.*, **34**, 35–74 (1967).
32. H. Okada. Preliminary study of sandstones of the Shimanto Supergroup in Kyushu, with special reference to "Petrographic Zone", *Sci. Rep. Dept. Geology, Kyushu Univ.*, **12**, 203–214 (1977).*
33. T. Miyamoto. Comparison of the Cretaceous sandstones from the Chichibu and Shimanto terrains in the Odochi area, Kochi Prefecture, Shikoku, *Jour. Geol. Soc. Japan*, **82**, 449–462 (1976).*
34. Kishu Shimanto Research Group. The Hidakagawa Group in the southern part of the Ryujin village, Wakayama Prefecture -the study of the Shimanto Terrain in the Kii Peninsula, Southwest Japan (Part 8)-, *Earth Science*, **31**, 250–262 (1977).*
35. Kishu Shimanto Research Group. Terasoma and Shirama Formations of the Hidakagawa Group in the Shimanto Belt, Southwest Japan - the study of the Shimanto Terrain in the Kii Peninsula, Southwest Japan (Part 10) -, *Earth Sci.*, **37**, 235–249 (1983).*

36. Kishu Shimanto Research Group. Miyma Formation of the Hidakagawa Group around Nakatsu-mura in the western part of the Kii Peninsula -the study of the Shimanto Terrain in the Kii Peninsula, Southwest Japan (Part 11)-, *Earth Sci.*, **40**, 274-293 (1986).*
37. Kishu Shimanto Research Group. Yukawa and Miyama Formations of the Hidakagawa Group in eastern-middle part of Wakayama Prefecture -the study of the Shimanto Terrain in the Kii Peninsula, Southwest Japan (part 12)-, *Earth Sci.*, **45**, 19-38 (1991).*
38. T. Tokuoka and F. Kumon. The Shimanto Terrain in the Akaishi Mountainland and the Kii Peninsula -a consideration on mineral composition of sandstones -, *Monograph Japanese National Sci. Museum*, no.12, 41-54 (1979).*
39. I. Imai, Y. Teraoka, K. Okumura and K. Ono. *Geology of the Mikado district*, Quadrangle Ser., scale 1:50,000, 44p. Geol. Surv. Japan (1979).*
40. A. Sakai. *Geology of the Itsukaichi district*, With geological sheet map at 1:50,000, 75p. Geol. Surv. Japan (1987).*
41. Y. Teraoka, K. Okumura, A. Murata and H. Hoshizumi. *Geology of the Saiki district*, with geological sheet map at 1:50,000, 78p. Geol. Surv. Japan (1990).*
42. K. Kimura. Petroprovinces of the Eocene-early Oligocene Shimanto Supergroup, *Mem. Geol. Soc. Japan*, no.38, 299-309 (1992).*
43. H. Ishiga. Two suites of stratigraphic succession within the Tamba Group in the western part of Tamba Belt, southwest Japan, *Jour. Geol. Soc. Japan*, **89**, 443-454 (1985).*
44. T. Kusunoki and M. Musashino. The characteristics of sandstones in the Tamba Belt, Southwest Japan, *Earth Sci.*, **43**, 75-83 (1989).*
45. T. Kusunoki and M. Musashino. Permo-Triassic sandstones from the Ultra-Tamba Zone, the Tamba Belt and the Maizuru Belt -Modal compositions and their comparison-, *Earth Sci.*, **41**, 1-11 (1990).*
46. T. Kusunoki. The upper Paleozoic to Mesozoic sandstones in the Inner Zone of Southwest Japan -modal analysis and its regional comparison-, *Earth Sci.*, **46**, 309-324 (1992).*
47. K.M. Marsaglia. Petrography and provenance of volcaniclastic sands recovered from the Izu-Bonin arc, Leg 126, *Proceedings ODP*, **126**, 139-153 (1992).
48. R.J. Stewart. Neogene volcaniclastic sediments from Atka basin, Aleutian ridge, *AAPG Bull.*, **62**, 87-97 (1978).
49. K.J. McMillen, R.H. Enkeboll, J.C. Moore, T.H. Shipley and J.W. Ladd. Sedimentation in different tectonic environments of the Middle Ameria Trench, southern Mexico and Guatemala, In: *Trench-Forearc Geology: Sedimentation and Tectonics on Modern and Ancient Active Plate Margins*, J.K. Leggett (ed.), 107-119, Blackwell Sci. Publications, Oxford (1982).
50. W.R. Dickinson. Detrital modes of New Zealand graywackes, *Sed. Geology*, **5**, 37-56.
51. J.Y. Choi. Middle Permian to upper Triassic sandstones in the Maizuru Terrane, Southwest Japan, *News, synthetic research on the sandstones in mobile belt*, no.1, 42-44 (1989).
52. F. Masuda and M. Ito. Evolution of sand composition in Kanto District, central Japan, *Ann. Rep., Inst. Geosci., Univ. Tsukuba*, no.14, 39-41 (1988).
53. W.E. Galloway. Deposition and diagenetic alteration of sandstone in northeast Pacific arc-related basins: Implications for graywacke genesis, *GSA Bull.*, **85**, 379-390 (1974).
54. C.F. Mansfield. Upper Mesozoic subsea fan deposits in the southern Diablo Range, California: Record of the Sierra Nevada magmatic arc, *GSA Bull., Part I*, **90**, 1025-1046 (1979).
55. K. Kiminami, F. Kumon, T. Nishimura and T. Shiki. Chemical composition of sandstones derived from magmatic arcs, *Mem. Geol. Soc. Japan*, no. 38, 361-372 (1992).*
56. A. Miyashiro. Volcanic rock series in island arcs and active continental margins. *Amer. Jour. Sci.*, **274**, 321-355.
57. G.H. Mack. Exceptions to the relationship between plate tectonics and sandstone composition, *Jour. Sed. Petrology*, **54**, 212-220 (1982).
58. M. Ito and F. Masuda. Detrital mode and size-distribution of the Late Pleistocene Paleo-Tokyo Bay sands, Japan, *Ann. Rep., Inst. Geosci., Univ. Tsukuba*, no.13, 83-86 (1987).

* in Japanese with English abstract.
** in Japanese.

(Manuscript received 3 March 1993, accepted 30 April 1993.)

Serpentinite protruded into fore-arc region: implications of detrital chromian spinels in Cretaceous sandstones of the Kanto Mountains, Japan

KEN-ICHIRO[1] HISADA and SHOJI ARAI[2]

[1] Insititute of Geoscience, University of Tsukuba, Tsukuba 305, Japan
[2] Department of Earth Sciences, Kanazawa University, Kanazawa, Ishikawa 920, Japan

Abstract : Detrital chromian spinels in sedimentary rocks provide much information to comprehend the tectonics of the serpentinite protrusion. A Jurassic accretionary wedge (Chichibu Complex) and Lower Cretaceous fore-arc basin sediments (Sanchu Group) crop out in the Kanto Mountains of central Japan. We examined chromian spinels occurring in the Sanchu Group and the basalt-serpentinite complex to provide some constraints concerning the origin and history of the Sanchu fore-arc basin.
Detrital chromian spinels have been found in fine-grained sandstone and matrix of conglomerate intercalated in the basal part of the Sanchu Group. According to a microprobe analysis, two types of detrital chromian spinels are recognized: namely, low-Ti and high-Ti groups. High-Ti spinels occur as small euhedral crystals in serpentine or chlorite grains, and may be derived from unmetamorphosed volcanic rocks. Low-Ti spinels are considered to be derived from peridotitic rocks possibly originating from an arc, especially from a fore-arc upper mantle, and they are slightly different in chemistry from those in the presently exposed serpentinite. The spinel chemistry and bulk rock chemistry indicate that the presently exposed volcanic rocks (Sanchu green rocks) associated with serpentinite have the characteristics of intraplate basalt.
Transcurrent faulting may be responsible for the extrusion of basalt and the protrusion of serpentinite on the plate boundary in the fore-arc region. The extrusion and protrusion resulted in forming the basalt-serpentinite complex and damming up the terrigenous sediments as an oceanward border of the Sanchu fore-arc basin.

Key words : serpentinite, chromian spinel, transcurrent fault, fore-arc, pull-apart basin, Cretaceous

INTRODUCTION

Chromian spinel is a useful indicator of the physico-chemical conditions of formation and/or subsolidus re-equilibration of ultramafic-mafic rocks (e.g., [1]). Especially, chromian spinel in volcanic rocks can be a potential discriminant for magma chemistry [2]. If detrital chromian spinels can be found in clastic rocks, it becomes possible to determine their mafic-ultramafic rocks provenance [3].
 In the Kanto Mountains, central Japan (Figure 1), the Mesozoic accretionary wedge is widely exposed. This accretionary wedge is composed of the Sambagawa high P/T type metamorphic rocks and the Chichibu and Shimanto Complexes closely associated with serpentinite. The Chichibu and Shimanto Complexes consist of Jurassic to earliest Cretaceous and Late Cretaceous chaotic rocks which are referable to olistostrome and melange, respectively [4-6]. The southward (seaward) younging polarity is recognized in shaly matrices of these chaotic rocks and a large-scale imbricate structure is reported from these complexes [5,7]. During Early Cretaceous, the tectonically controlled

sedimentary basin was formed in the fore-arc region of the Chichibu accretionary wedge [8]. This basin fill sediments are called the Sanchu Group.

The basal part of the Sanchu Group is characterized by conglomerate-dominant facies and some pebble to cobble sized serpentinite clasts are contained in the conglomerate [9]. This evidence indicates that serpentinite was denudated during the time of an incipient stage of the Sanchu basin development history. In this paper, we pay attention to detrital chromian spinels contained in sandstones of the Sanchu Group instead of serpentinite clasts. Because detrital chromian spinels are more resistant against abrasion than serpentinite clasts during their transportation, it is expected that the former tends to be found in extensive areas far from its provenance. A part of this analytical result has already been published in two papers [10,11]. This paper deals not only with a description of sandstone petrography but also with the basin formative tectonics during an incipient stage based on the chemistry of detrital chromian spinels.

STRATIGRAPHY OF THE SANCHU GROUP

Many detrital chromian spinels are microscopically detected in some coarse-grained clastic rocks collected from the basal and lower parts of the Sanchu Group. On the other hand, we did not find any of them in the underlying Chichibu Complex. The distributional area of the Sanchu Group has been traditionally called the Sanchu graben. This "graben" does not mean a structural form, but it merely implies a

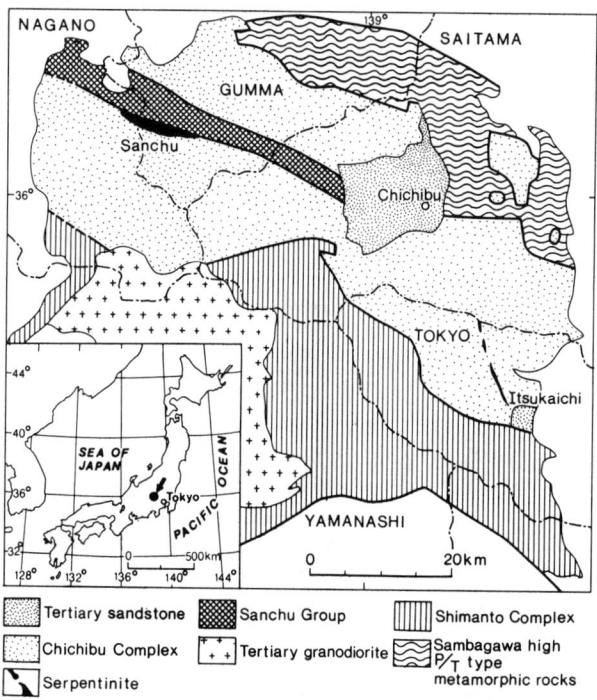

Figure 1. Geologic map of the Kanto Mountains.

geomorphological one caused by an erosional degree difference between the clastic (Sanchu Group) and siliceous-calcareous rocks (Chichibu Complex). The WNW-ESE trending extension of the Sanchu graben attains about 40 km in length and 4 km in width on average. The Sanchu Group consists of interbedded fine- to medium-grained sandstone, alternation of sandstone and mudstone, sandy shale, conglomerate or coarse-grained sandstone, white-gray tuffaceous siltstone and fossiliferous limestone. Total thickness of this group attains about 1700 m. Takei [12] showed that the Sanchu Group is composed of the Ishido, Sebayashi and Sanyama Formations in ascending order. In this paper, we revised Takei's division and adopted the tentatively proposed three-fold divisions; namely Units 1 to 3 in ascending order. The reason is that distributional patterns of Units 1 to 3 are conspicuously different from those of the Ishido, Sebayashi and Sanyama Formations.

Unit 1

Unit 1 is widely distributed in the western area of the Sanchu graben, while it crops out narrowly along the northern and southern margins in the eastern area. Unit 1 is composed of conglomerate, massive muddy fine-grained sandstone, and thin-bedded fine- or medium-grained sandstone. The muddy sandstone is locally bioturbated, and trails and burrows are commonly observed. Low-angled cross lamination is rarely developed in medium-grained sandstone underlain by basal conglomerate. Conglomerate is sandy matrix-supported, and clasts, commonly pebble or cobble sized, are mainly sedimentary rocks.

The occurrence of ammonites correlatable to the late Hauterivian to early Barremian fauna has been known [13,14]. Unit 1 yields abundant plant fossils such as Bennettitales, conifers, ferns and others [15] and prolific non-marine corbiculid bivalves [16]. Besides them, hexacorals, probably assignable to Barremian to early Aptian were recently discovered from limestone blocks, a few meters thick, intercalated with mudstone in the western area [17]. The age of Unit 1 can be regarded as late Hauterivian to early Aptian.

Unit 2

The distributional area of Unit 2 is limited in the western and central areas of the Sanchu graben. Unit 2 is characterized by repetition of upward-coarsening sequence, which begins with black siltstone and ends with conglomerate or conglomeratic sandstone. The contorted thin-bedded siltstone and the linguloid type rippled sandstone [18] are observed in the lower and upper portions in each upward-coarsening sequence. The lithology of clasts of conglomerate is markedly different from that in Unit 1; namely, conglomerate of Unit 2 contains numerous clasts of igneous rocks as well as sedimentary rocks [19].

This unit rarely yields reliable index fossils. The obtained fossils are plant fossils [15], non-marine corbiculid bivalves [16] and a piece of vertebra of dinosaur [20]. The age of this unit ranges probably from late Aptian to early Albian.

Unit 3

Unit 3 crops out extensively throughout the Sanchu graben, but it is prevalent in greater parts of the eastern area. Unit 3 is divided into three parts; namely the siltstone, thick-bedded sandstone, and thin-bedded sandstone in ascending order. The thin-bedded fine-grained sandstone has characteristics of distal turbidite. The thick-bedded sandstone is medium- to coarse-grained, 10 cm to 1 m in thickness of each bed. Saka and Koizumi [21] clarified a westward paleocurrent based on sedimentological data such as flute cast, prod cast, and others.

This unit yields some ammonites assignable to middle Albian to late Cenomanian [16]. Kamikawa et al. [22] reported that the occurrence of Albian, probably late Albian radiolarian assemblage from mudstone interbedded with sandstone. The age of Unit 3 ranges probably from late Albian to Cenomanian.

Sekiyama et al. [23] recognized the storm-generated offshore, slope to upper fan and lower fan environments for the Sanchu Group. The prolific bioturbated beds, some ripple marks, abundant plant fossils, brackish and non-marine molluscs, and some coralline limestone suggest that the shallow marine to non-marine environment is more appropriate for Unit 1 during late Hauterivian to early Aptian [11]. Unit 2 is characterized by repeated upward-coarsening sequence, which is marked at the lower and upper portions by intercalations of the contorted thin-bedded siltstone and the linguloid type rippled sandstone, respectively. This lithology of Unit 2 is suggestive of a deltaic environment. Afterward, the depth of the depositional site of Unit 3 seems to increase gradually, judging from frequent occurrence of distal turbidite.

PETROGRAPHY OF SANDSTONES

To elucidate petrographical characteristics of sandstone obtained from Unit 1 of the Sanchu Group, sandstone thin-sections are modally analyzed using an automatic point counter. Namely, 500 or more points including matrix per thin-section were counted to determine QmFLt percentages [8](Figure 2). The petrographical characteristics of sandstone of the Chichibu Complex were also similarly examined for the purpose of their comparison.

Sampling locations of sandstone are situated in the western part of the Kanto Mountains where both the Sanchu Group and Chichibu Complex occur. The Sanchu Group is less disturbed as compared with the Chichibu Complex. Sandstone samples collected from

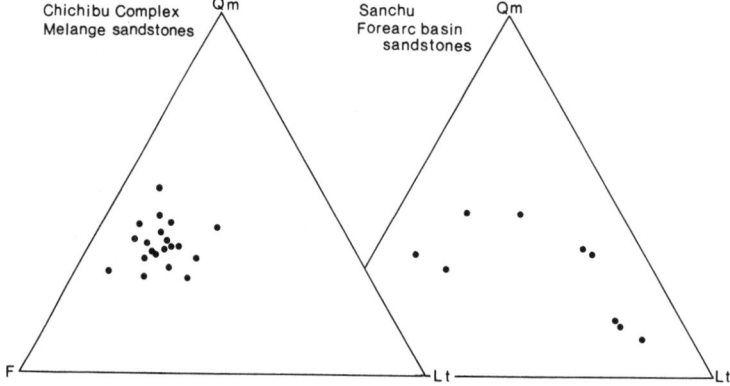

Figure 2. QmFLt plots of the sandstones from the Chichibu Complex and Sanchu Group. Qm = monocrystalline quartz, F = total feldspar grains, Lt = total lithic fragments.

Unit 1 (named here Sanchu fore-arc basin sandstones) are generally dark gray muddy sandstone. The Chichibu Complex is called the Hamadaira Group in this area [7]. This group is assigned to Jurassic to earliest Cretaceous based on radiolarian fossils, and is composed of disturbed and chaotic sedimentary rocks which can be classified into melange according to Cowan's classification [24]. Collected sandstone samples (named here Chichibu Complex melange sandstones) are contained in the early Jurassic melanges of I-1, I-2 and I-3 subzones proposed by Hisada and Kishida [7]. .

Quartzose grains are dominant framework grains in the fore-arc basin and melange sandstones. In fore-arc basin sandstones, chert and polycrystalline quartz make up 21 to 83 percent of quartzose grains, with the ratio of these grains to total quartzose grains (Qp/Qt) averaging 0.53. On the other hand, in melange sandstones, chert and polycrystalline quartz make up 7 to 47 percent of quartzose grains, with the ratio of these grains to total quartzose grains (Qp/Qt) averaging 0.20. Feldspar is largely twinned plagioclase. P/F averages 64 percent in fore-arc basin sandstones and 86 percent in melange sandstones.

The dominant lithic types in both sandstones are volcanic rocks and argillaceous rocks. Volcanic rock fragments present microlithic features [25], averaging 4 percent of total framework grains in fore-arc basin sandstones and 2 percent in melange sandstones. Argillaceous rocks average 4 percent of total framework grains in fore-arc basin sandstones and 6 percent in melange sandstones.

OCCURRENCE OF DETRITAL CHROMIAN SPINELS

Detrital chromian spinels are sparsely found in sandstones and conglomerates of the Sanchu Group. Sandstone thin-sections attaining 202 were microscopically examined concerning a relative amount of detrital chromian spinels. One thin-section corresponds approximately to one sampling locality. These samples are discriminated into four ranks according to a relative abundance of spinel grains; "none", "rare", "present" and "abundant". Namely, "rare", "present", and "abundant" indicate one or two grain(s), several grains, and more than ten grains per one thin-section, respectively. These discrimination criteria, however, are somewhat subjective. When counting through each sandstone thin-section (area of slide glass; 28 mm x 48 mm), spinel grain size and the occurrence such as a single grain or a phenocryst in groundmass are not taken into consideration.

Numerical data concerning the number of thin-sections are as follows:

	none	rare	present	abundant	total
Unit 1	79	22	11	6	118
Unit 2	11	1	0	0	12
Unit 3	60	11	1	0	72

202

(The number of thin-sections shown in Figure 3 is not included in the above numerical data.)

There may be a bias, because the sampling number depends on a frequency of sandstone exposures. These numerical data, however, seem to suggest that there is a distinct difference between Units 1 and 3.

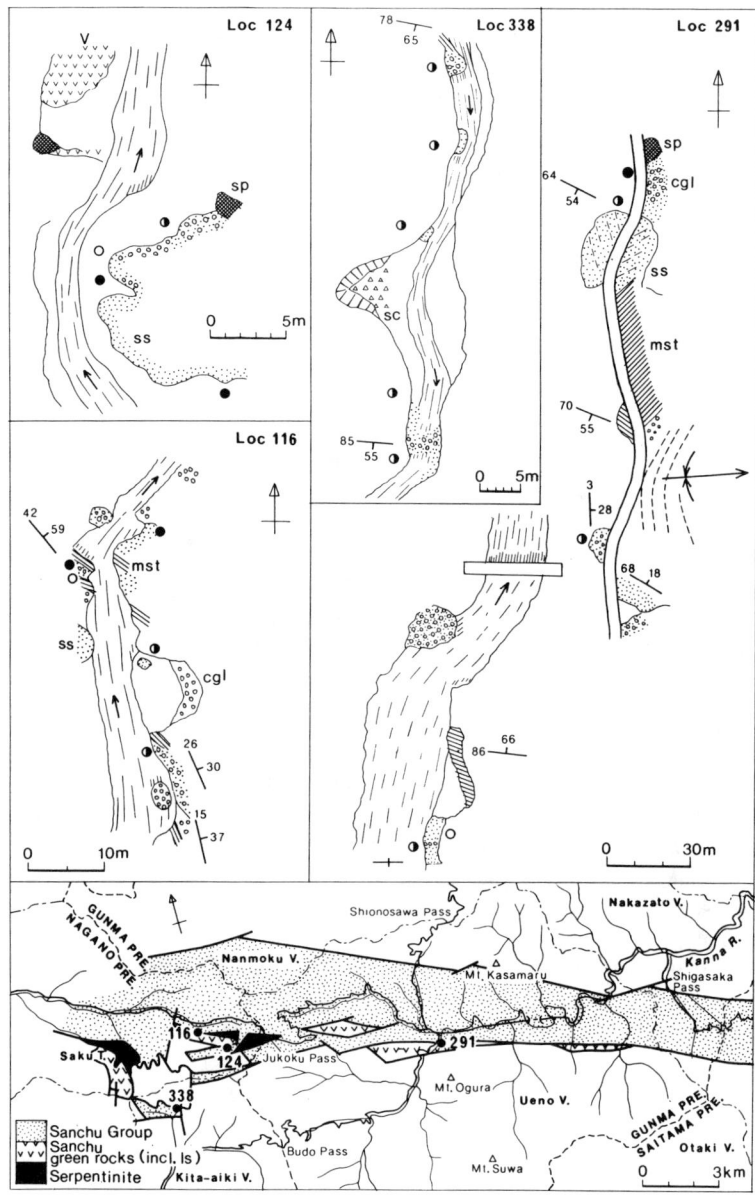

Figure 3. Occurrence of chromian spinel-bearing sandstones and their location map. Relative amount of chromian spinels are shown in following symbols; solid circle: abundant, half-solid circle: present, open circle: none.
mst:mudstone, ss:sandstone, cgl:conglomerate, sc:surface cover, sp:serpentinite, v:basalt.

All of "abundant" chromian spinel-bearing sandstones, inclusive of sandy matrix of conglomerate, collected from 6 localities belong to Unit 1. Among them, route maps for four localities (Locs. 116, 124, 291, 338) are shown in Figure 3. Serpentinite and/or basalt (Sanchu green rocks) occur very close to the sampling locality of sandstone at locs. 124 and 291, while sampling sites of locs. 116 and 338 are located within a few hundreds meters from now-exposed serpentinite (Figure 3). Common characteristics to sandstones containing "abundant" detrital chromian spinels are fine- to medium-grained, seemingly structureless and accompanied with conglomerate. Detrital chromian spinels are prolifically found in the sandy matrix of conglomerate, which includes more or less serpentinite cobbles [9]. They are rarely concentrated to make thin spinel-rich seams in conglomerate. Spinel particles can be microscopically classified into two types; coarse discrete clastic particles and fine euhedral crystals embedded in serpentine or chlorite particles.

CHEMISTRY OF CHROMIAN SPINELS

Chemical characteristics of detrital spinels

Chemical compositions of detrital spinels were determined with microprobe to estimate their provenances. The detrital spinels can be classified into two groups in terms of Cr# (Cr/(Cr+Al) atomic ratio)-TiO_2 relationship (Figure 4); one is low-Ti (<0.5 wt% of TiO_2) group, with relatively high Cr# (0.4 to 1.0, mostly 0.7-0.8), and the other is a high-Ti (ca. 1 wt% of TiO_2) group, with relatively low Cr# (0.5 to 0.6). High-Ti spinels tend to be slightly higher in Mg# (Mg/(Mg+Fe^{2+}) atomic ratio) than low-Ti spinels if compared at a given Cr#. Low-Ti detrital spinels are very similar in chemistry to spinels in serpentinites closely associated with the spinel-bearing sediments in the Sanchu graben [10].

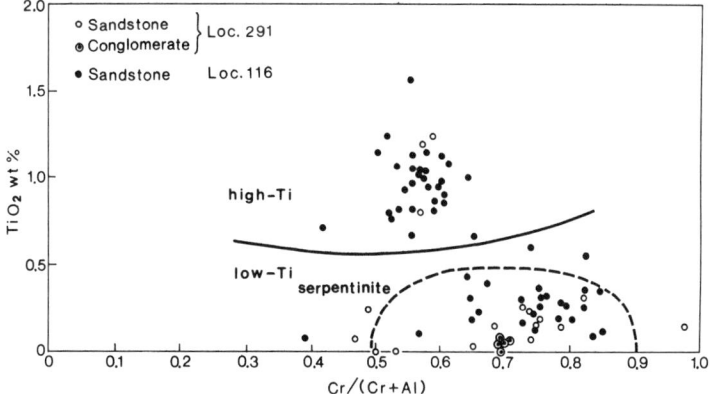

Figure 4. Relationships between Cr# and TiO_2 content of chromian spinels in sandstones and conglomerates of the Sanchu Group (Locs. 291, 116). See Figure 3 for sampling localities.
    serpentinite (surrounded by broken line): spinels in the Sanchu serpentinites.
    "Conglomerate": spinels from sandy matrix of the conglomerates.

Origins of detrital spinels

Low-Ti spinels had been most probably derived from serpentinites (or peridotites) essentially similar to now exposed ones. The original peridotites should be characterized by high-Cr# and low-Ti spinels, possibly originating from the arc mantle, especially from the fore-arc mantle (e.g., [26]).

High-Ti spinels, which are euhedral small grains embedded in chlorite or serpentine aggregates, were most probably derived from volcanic rocks. Small euhedral spinels are quite common in Mg-rich basalt, especially as inclusions in olivine phenocrysts (e.g., [3]). Olivine phenocrysts are easily altered into aggregates of chlorite or serpentine on alteration or weak metamorphism [27]. It is most probable, therefore, that the chlorite or serpentine aggregate including Ti-rich spinels was derived from olivine phenocrysts with euhedral spinel inclusions.

The chemistry of chromian spinel in volcanic rocks is an indicator of magma chemistry [2]. High Ti contents of the Sanchu high-Ti detrital spinels are clearly distinguished their magma(s) both from arc magmas, e.g. the Ryozen basalt, northeast Japan, and from N-type MORB (Figure 5). The magmas for the Sanchu high-Ti spinels are most probably of intraplate-type (Figure 5). Chromian spinels in the Sanchu green rocks (= weakly metamorphosed basalts) presently associated with the serpentinite

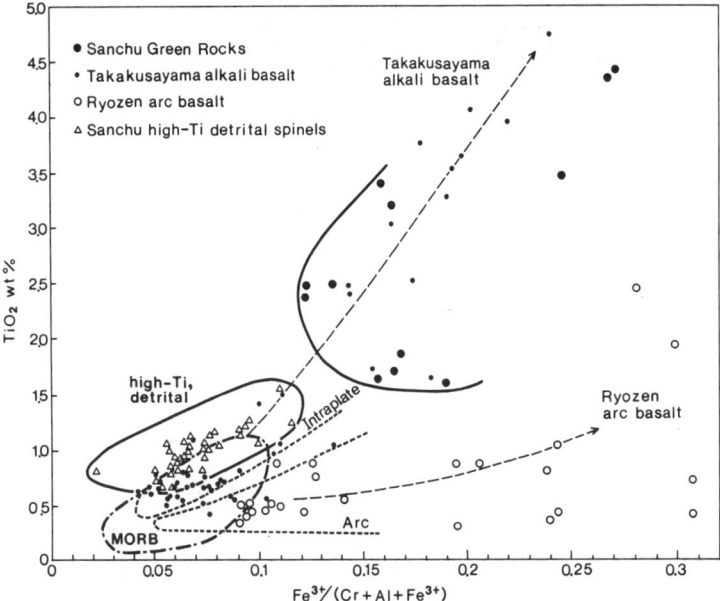

Figure 5. Relationships between Fe^{3+}# and TiO_2 content of chromian spinels in sandstones and conglomerates of the Sanchu Group (high-Ti group).
Discrimination lines are quoted from Arai [2]. Takakusayama alkali-basalt and Ryozen arc basalt exemplify intraplate and arc basalts, respectively. Arrows indicate the compositional variation of spinel during magmatic differentiation.

bodies (Figure 3) are very high both in Ti and Fe^{3+}, different from high-Ti detrital spinels [28] (Figure 5). The magmas for the Sanchu green rocks are also most probably of intraplate-type although highly evolved. The TiO_2-Fe^{3+} trends for the Sanchu green rocks spinels and high-Ti detrital spinels are roughly consistent with that for the Takakusayama alkali basalt, central Japan, a Miocene intraplate basalt (Figure 5). It is important that the magmas of the basalt associated with serpentinite had been of intraplate-type.

GENESIS AND EVOLUTION OF THE SANCHU FORE-ARC BASIN

General trend of the Sanchu fore-arc basin sandstones

The Chichibu Complex can be regarded as a part of the Mesozoic accretionary wedge which was developed along the eastern side of Asian continent (e.g., [29]). Figure 6 expresses a schematic profile of the accretionary wedge during Early Cretaceous. The Sanchu fore-arc basin was considered to be located landward on its accretionary wedge.

It is useful to compare fore-arc sandstones with melange ones in order to comprehend tectonics in the fore-arc region. The fore-arc basin sandstones (sandstones of Unit 1) and melange sandstones have respective characteristics with respect to mineral composition. The fore-arc basin sandstones are especially rich in chert and polycrystalline quartzose grains in comparison with melange sandstones (Figure 2). These grains are inferred to be derived from the uplifted melange ridge, which may be tectonically related more or less to the extrusion of basalt and the protrusion of serpentinite as suggested below. However, the fore-arc basin sandstones include scarce volcanic rock fragments. Takei [12] mentioned that dominant lithic fragments change from chert to granitic rocks in Cenomanian. Taking this shift into consideration, acidic magmatism of the volcanic arc in the arc-trench gap of the Mesozoic accretionary wedge may become more active as time proceeds [8].

Formation of the basalt-serpentinite complex

It is worthy of note that Unit 1 is more prolific in detrital chromian spinels than Units 2

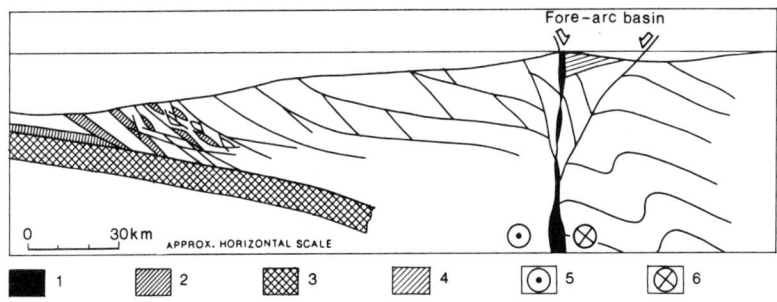

Figure 6. Schematic model of the Mesozoic accretionary wedge during Early Cretaceous.
1. basalt and serpentinite, 2. pelagic sediments, 3. ocean plate, 4. fore-arc basin sediments, 5. moving direction (coming), 6. moving direction (going).

and 3. Furthermore, the provenance of these chromian spinels can be regarded as undifferentiated basalt and serpentinite. These volcanic rocks and serpentinite may comprise a basalt-serpentinite complex, because detrital chromian spinels shed from both rocks form spinel particles seams in sediments. This interpretation is supported by now co-occurrence of the Sanchu green rocks and serpentinite (Figure 3). Recently, a preliminary K-Ar age (12 ± 6 Ma) for the Sanchu green rocks was reported by Ishida et al. [28]. The undifferentiated basalt can be regarded as a precursor of the "evolved" Sanchu green rocks. Thus, it is obvious that volcanic activity occurred immediately before and/or during the sedimentation of Unit 1.

Volcanic rocks containing analogous high-Ti spinels are absent in the neighboring areas. This indicates that the source basalts of the Sanchu high-Ti detrital spinels have entirely eroded. As indicated in Figure 5, there are some differences between spinels of the Sanchu green rocks and detrital chromian spinels of high-Ti group due to degree of magmatic differentiation. This spinel chemistry, however, suggests that both magma species are representative of intraplate-type. The chemistry of high-Ti spinel group and the Sanchu green rocks seems to exclude the possibility that these basalts originated from island-arc as magma species. Accordingly, this accretionary wedge can be inferred to shift into "intraplate-type" setting after earliest Cretaceous.

During Early Cretaceous, two oceanic plates, Izanagi and Farallon plates transcurrently acted along the eastern side of the Asian continent [30]. This plate boundary, therefore, may be assigned to transcurrent-type. A transcurrent fault in a fore-arc region can be possibly responsible for the extrusion of basalt and the protrusion of serpentinite. Serpentinite diapir is also considered to have facilitated greatly the protrusion, as demonstrated in the Mariana arc-trench gap (e.g., [31]).

The transcurrent faulting can produce a pull-apart basin on land (Figures 6, 7-1). Crowell [32] envisaged lava flow on a floor subsided by pull-apart. This lava flow may be genetically related to the extrusion of volcanic rocks as a precursor of the protrusion of serpentinite. The Sanchu sedimentary basin at the incipient stage was filled with shallow-water sediments of Unit 1, and was characteristically fringed by coarse-grained clastic rock facies on the southern border. This feature suggests that the basin has some characteristics of a half-graben which is bounded on the southern border by

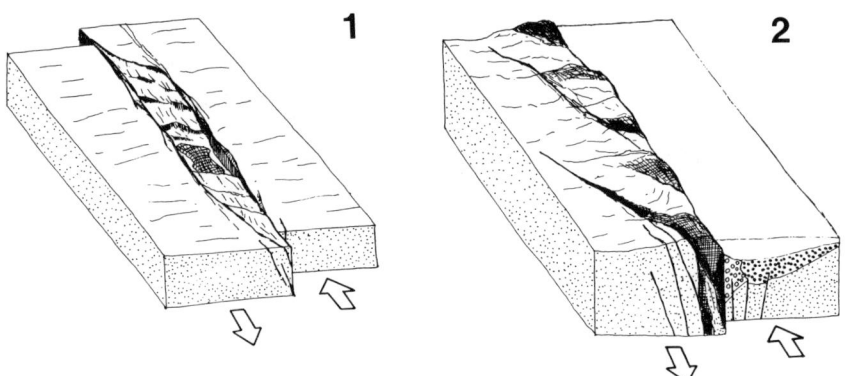

Figure 7. Formative model of the Sanchu sedimentary basin due to transcurrent faulting.
1. Formation of pull-apart basin and extrusion of basalt lava flow.
2. Protrusion of serpentinite and development of half-graben.

normal fault, though the transcurrent faulting continued (Figure 7-2).

CONCLUSION

Detrital chromian spinels were mainly found in fine-grained sandstone and matrix of conglomerate intercalated in the basalt part of the Sanchu Group. Two types of detrital chromian spinels such as low-Ti and high-Ti groups are recognized. Low-Ti spinels are considered to be derived from peridotitic rocks, possibly originating from an arc upper mantle. High-Ti spinels may be derived from intraplate-type magma. It is concluded that the transcurrent-type faulting occurred along the eastern side of the Asian continent during Early Cretaceous as testified by the chemistry of detrital chromian spinels in sandstones of the Sanchu Group. This transcurrent faulting followed the construction of the accretionary wedge, and is considered to originate from a change of subducting direction of oceanic plates; the shift from orthogonal to diagonal direction of subducting oceanic plate.

ACKNOWLEDGEMENTS

The authors would like to thank Professor H. Igo of University of Tsukuba and Professor T. Ishida of Yamanashi University for valuable discussion. Special thanks are extended to Mr. S. Ozaki and Miss T. Kawada of University of Tsukuba for their drafting and arranging the manuscript.

REFERENCES

1. T.N. Irvine. Chromian spinel as a petrogenetic indicator; Part I, Theory, *Canad. Jour. Earth Sci.*, **2**, 648-671 (1965).
2. S. Arai. Chemistry of chromian spinel in volcanic rocks as a potential guide to magma chemistry, *Mineral Magazine* **56**, 173-184 (1992).
3. S. Arai.. Petrological and geological significance of detrital chromian spinel and detrital serpentinite, *Mem. Geol. Soc. Japan*, no.38, 329-344 (1992).*
4. K. Hisada. Jurassic olistostrome in the southern Kanto Mountains, central Japan, *Sci. Rep., Inst. Geosci., Univ. Tsukuba (B)*, **4**, 99-119(1983).
5. K. Hisada, H. Ueno and H. Igo. Geology of the Upper Paleozoic and Mesozoic sedimentary complex of the Mt. Ryokami area in the Kanto Mountains, central Japan, *Sci. Rep., Inst. Geosci., Univ. Tsukuba (B)*, **13**, 127-151 (1992).
6. N. Iyota, K. Sashida, H. Igo and K. Hisada. The Ogochi Group of the Shimanto Terrane in the Kanto Mountains, central Japan, *Sci. Rep., Inst. Geosci., Univ. Tsukuba (B)*, 14 (in press).
7. K. Hisada and Y. Kishida. The Hamadaira Group in the western Kanto Mountains, central Japan - the developmental processes of the Jurassic-lower Cretaceous accretionary prism -, *J. Geol. Soc. Japan*, **92**, 569-590 (1986).*
8. K. Hisada. Sandstones from the Chichibu accretionary wedge, Kanto Mountains, central Japan, *Ann. Rep., Inst. Geosci., Univ. Tsukuba*, **17**, 63-68 (1991).
9. K. Hisada and S. Arai. Mode of occurrence of serpentinite and Cretaceous serpentinite- bearing conglomerate in the Kanto Mountains, central Japan, *J. Geol. Soc. Japan*, **92**, 391-394 (1986).*
10. S. Arai and K. Hisada. Detrital chromian spinels from the Ishido Formation of the Sanchu Cretaceous Formations, Kanto Mountains, central Japan, *J. Min. Petr. Econ. Geol.*, **86**, 540-553 (1991).*
11. K. Hisada and S. Arai. Detrital chromian spinels in Sanchu Cretaceous sandstone, central Japan: Indicator of serpentinite protrusion into fore-arc region, Palaeogeography, Palaeoclimatology, Palaeoecology (in press).
12. K. Takei. Development of the Cretaceous sedimentary basin of the Sanchu Graben, Kanto Mountains, Japan, *J. Geosci., Osaka City Univ. Art.1*, **28**, 1-44 (1985).

13. I. Obata, M. Matsukawa, H. Tsuda, M. Futakami and Y. Ogawa. Geological age of the Cretaceous Ishido Formation, Japan, *Bull. Natn. Sci. Mus.*, C 2(3), 121-140 (1976).
14. I. Obata, M. Matsukawa, K. Tanaka, Y. Kanai and T. Watanabe. Cretaceous cephalopods from the Sanchu area, Japan, *Bull. Natn. Sci. Mus.*, C 10(1), 9-37 (1984).
15. T. Kimura and M. Matsukawa. Mesozoic plants from the Kwanto Mountainland, Gunma Prefecture, in the Outer Zone of Japan, *Bull. Natn, Sci. Mus.*, C 5(3), 89-112 (1979).
16. M. Matsukawa. Stratigraphy and sedimentary environments of the Sanchu Cretaceous, Japan, *Mem. Ehime Univ., Sci.*, D 9, 1-50 (1983).
17. N. Yamagiwa, M. Tamura, K. Hisada and Y. Kamikawa. Early Cretaceous Hexacorallia collected from the Ishido Formation in the western Sanchu Graben, *Abstr. 38th Reg. Meet. Paleont. Soc. Japan* 8 (1989).**
18. F. Arai, K. Takei, H. Hosoya, S. Hayashi and K. Takahashi. Descriptions and some considerations concerning the ripple marks discovered in the Sanchu Graben, *Earth Sci. (Chikyu Kagaku)*, 40, 1-12 (1958).*
19. K. Takei. On the pebble of igneous rocks, metamorphic rocks and acidic tuffs in the Cretaceous conglomerate of the Sanchu Graben, Kwanto Mountains, *J. Geol. Soc. Japan*, 81, 247-254 (1975).*
20. Y. Hasegawa, T. Kase and S. Nakajima. Mega-vertebrate fossil from the Sanchu Graben, *Abstr. 91th Meet. Geol. Soc. Japan*, 219 (1984).**
21. Y. Saka and K. Koizumi. Paleocurrent of Turonian Sanyama Formation in the eastern part of the Sanchu Graben, Kwanto Mountains, central Japan, *J. Geol. Soc. Japan*, 83, 289-300 (1977).*
22. Y. Kamikawa, Y. Kishida, K. Kaiho and K. Hisada. Radiolarians obtained from the Cretaceous formations of the Sanchu graben in the Kanto mountains, central Japan, *J. Geol. Soc. Japan*, 94, 903-905 (1988).**
23. S. Sekiyama, T. Sato, F. Masuda and M. Matsukawa. Oblique-slip basin as an origin of the Sanchu Graben in Kanto Mountains, Ann. Rep., Inst.Geosci., Univ. Tsukuba 10, 113-116 (1984).
24. D.S. Cowan. Structural styles in Mesozoic and Cenozoic melanges in the western Cordillera of North America, *Geol. Soc. Amer. Bull.*, 96, 451-462 (1985).
25. W.R. Dickinson. Interpreting detrital modes of graywacke and arkose, *J. Sed. Petrology*, 40, 695-707 (1970).
26. S. Arai. Origin of ophiolitic peridotites, *J. Geogr. (Tokyo)*, 98, 231-240 (1989).*
27. T. Ishida, S. Arai and N. Takahashi. Metamorphosed picrite basalts in the northern part of the Setogawa belt, central Japan, *J. Geol. Soc. Japan*, 96, 181-191 (1990).*
28. T. Ishida, S. Arai, A. Ishiwatari, K. Hisada and M. Matsuzawa. Greenstones from the south marginal part of the Cretaceous Sanchu belt in the Kanto Mountains, central Japan, *J. Min. Petr. Econ. Geol.*, 87, 174-186 (1992).*
29. A. Taira and M. Tashiro. Late Paleozoic and Mesozoic accretion tectonics in Japan and eastern Asia. In : Historical biogeography and plate tectonic evolution of Japan and eastern Asia, A. Taira and M. Tashiro (eds). pp.1-43, Chap. 1, Terra Scientific Publishing Company, Tokyo (1987).
30. D.C. Engebretson, A. Cox and R.G. Gordon. Relative motions between oceanic and continental plates in the Pacific Basin, Geol. Soc. America, Spec. Paper 206, pp.1-59 (1985).
31. S.H. Bloomer. Distribution and origin of igneous rocks from the landward slopes of the Mariana trench:implications for its structure and evolution, J. Geophys. Res. 88, 7411-7428 (1983).
32. J. C. Crowell. Origin of late Cenozoic basins in southern California, In : *Tectonics and sedimentation.* W.R. Dickinson (ed.), Soc. Econ. Paleont. Mineral., Spec. Pub., .no.20, 190-204 (1974).

* in Japanese with English abstract.
** in Japanese.

(Manuscript received 7 December 1992; accepted 12 February 1993.)

Sedimentation and paleoenvironment in the uppermost Lower to Middle Miocene basin on the Japan Sea coast, Southwest Japan

TOSHIO MATSUMOTO

Kyushu Branch, Yachiyo Engineering Co., Ltd., Tenjin-Twin Bldg., Tenjin 1-6-8, Chuo-ku, Fukuoka 810, Japan

Abstract : In the backarc side of the Japanese Islands, many Miocene sedimentary basins link to surround the Japan Sea from the east. The uppermost Lower to Middle Miocene basin along the Japan Sea coast of the Southwest Japan is the target of this study, and is investigated to clarify the tectonics of basin genesis from a sedimentological aspect.

The sedimentary basin, about 20 km wide and 50 km long, is filled with thick clastic deposits associated with volcanogenic rocks. The clastic deposits consist of the lower conglomeratic sequence of alluvial fan and fan-delta origins, and the upper mudstone sequence deposited in an outer sublittoral to upper middle bathyal environments. The volcanogenic rocks developing in several horizons are mostly felsic and partly intermediate to mafic, showing a bimodal composition.

The sedimentary basin has the following characters ; (1) longitudinal elongation parallel to the general trend of the Southwest Japan arc, (2) thick accumulation of basin fills, more than 1,000 m, (3) high sedimentation rate attaining 500 m/m.y., (4) rapid deepening of the basin, which changes the sedimentary facies from brackish or inner neritic to upper middle bathyal, (5) bilateral supply of thick conglomerates of alluvial fan and fan-delta facies in the early stage and of numerous sandstone layers of submarine fan facies in the later stage, and (6) violent submarine volcanism of bimodal composition. These characters indicate that the sedimentary basin generated as a rift graben.

The sedimentological characters of the basin are similar to those of the contemporaneous rift basins in the backarc side of Northeast Japan, which were interpreted as a product of rifting related to the opening of the Japan Sea. The basin studied here also could have a genetic relationship to the opening.

Key words : *sedimentation, paleoenvironment, rifting, Japan Sea, backarc spreading*

INTRODUCTION

Many Miocene basins develop in the Japan Sea side of the Japanese Islands. Among them, this paper focuses on the uppermost Lower to Middle Miocene basin in the coastal belt of western Southwest Japan, i.e., the San'in district. The basin, about 20 km wide and 50 km long, elongates in the ENE-WSW direction, parallel to the general trend of the Southwest Japan arc. It is filled with clastic sediments and volcanogenic rocks more than 1,000 m thick. Their systematic, horizontal and vertical facies change is expected to provide an important

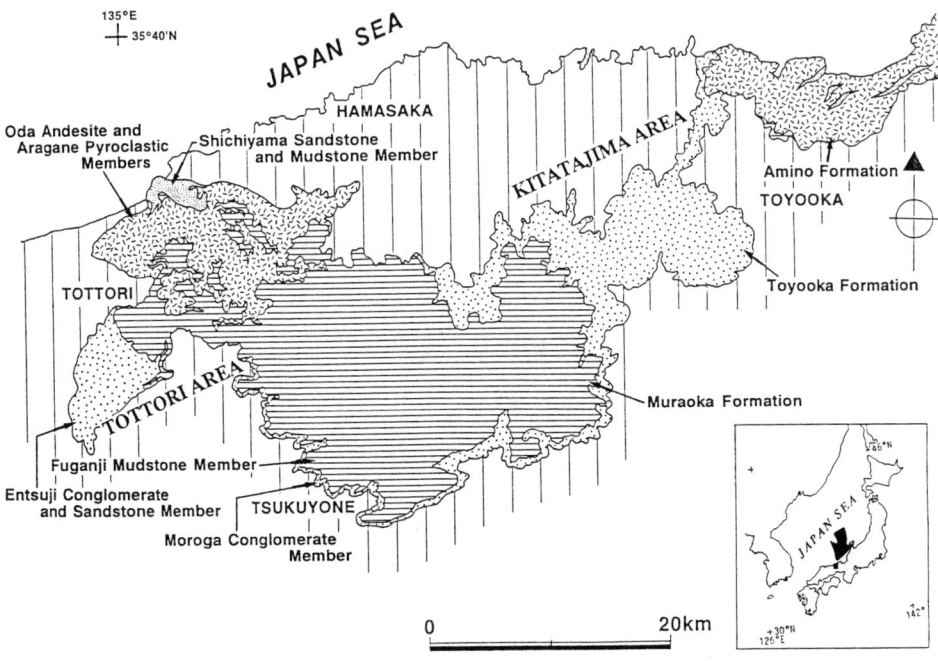

Figure 1. Distribution of the uppermost Lower to Middle Miocene series in the Tottori-Kitatajima area. Covering strata and intrusive rocks (Miocene to Recent) are omitted.

information on the tectonics which controlled the genesis of the basin. This paper, synthesizing sedimentological analyses, reconstructs the environmental change within the basin, and investigates the process of sediment supply. The information on paleoenvironment and provenance will reveal that the basin genesis was controlled by backarc rifting, as reported in the Japan Sea side of Northeast Japan [1, 2].

STRATIGRAPHY

The precise stratigraphy of the uppermost Lower to Middle Miocene sediments was clarified in Tottori and Kitatajima areas, the eastern San'in district (Figures 1, 2). The sediments in the Tottori area named the Iwami Formation, which is divided into the Entsuji Conglomerate and Sandstone, Moroga Conglomerate, Fuganji Mudstone, Oda Andesite, Aragane Pyroclastic and Shichiyama Sandstone and Mudstone Members [3]. The uppermost Lower to Middle Miocene sediments in the Kitatajima area consist of the lower Toyooka and the upper Muraoka Formations. The Toyooka Formation is subdivided into the Tsuji Conglomerate, Seto Volcanic, Kawae Volcanic, Ooka Conglomerate and Sandstone Alternation, Otani Sandstone and Conglomerate Members. The Muraoka Formation is also into the Ono-toge Sandstone, Shikada Tuff, Yubunegawa Shale and Myoken Tuff Breccia Members [4].

In both areas, gravelly sediments predominate in the lower sequence of the uppermost Lower to Middle Miocene sediments, and argillaceous ones in the upper sequence (Figure 2). The Entsuji and Tsuji Members, occupying the large part of the lower sequence, consist mainly of conglomerate, and are interpreted as deposits of alluvial fans or fan deltas [5]. The Moroga,

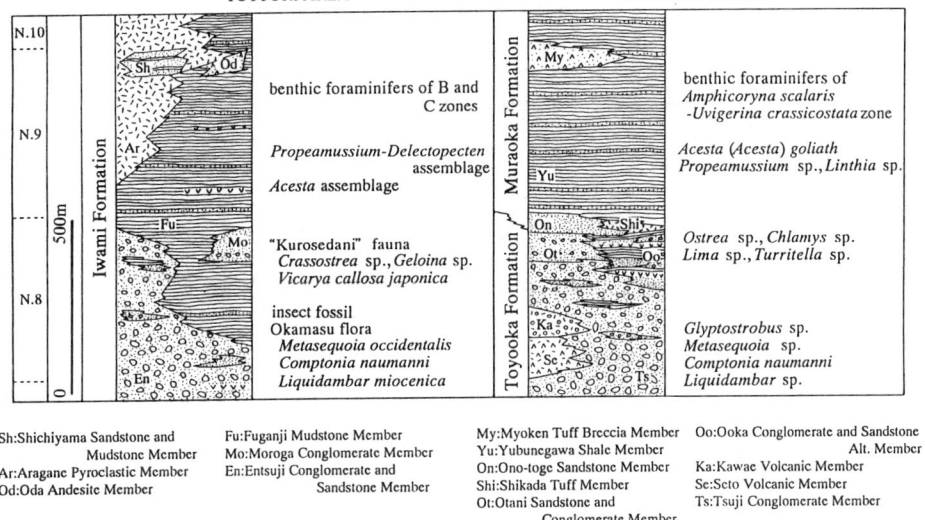

Figure 2. Lithostratigraphy and characteristic floral and faunal assemblages of the Iwami, Toyooka and Muraoka Formations in the Tottori and Kitatajima areas.

Ooka and Otani Members comprise an irregular alternation of conglomerates and sandstones, and yield brackish to inner neritic mollusca, such as *Vicarya, Geloina, Crassostrea, Anadara,* etc. [4, 6]. The Fuganji and Yubunegawa Members, occupying the large part of the middle to upper sequence, is made up of well-stratified mudstone with thin layers of sandstone. Well-preserved plant and insect fossils are yielded from the lowermost part, probably lacustrine [7], of the members [8-10], whereas outer neritic to upper middle bathyal mollusca and foraminifera occur abundantly in the middle to upper part of the members [6, 11-13]. The Oda, Aragane and Myoken Members are submarine volcanics consisting of pyroclastic- and lava-flow deposits, and are of bimodal composition as a whole [14].

PROVENANCE OF SEDIMENTS

Sedimentological analyses have revealed the provenance of coarse-grained sediments of the Entsuji, Moroga, Fuganji, Shichiyama Members in the Tottori area and of the Tsuji Members in the Kitatajima area [5, 15].

Gravels of the lower part of the Entsuji Member were derived mostly from the Lower Miocene volcanic rocks distributed on the southwest of the area. Those of the upper part were from the Mesozoic sedimentary rocks exposed to the south-southwest of the basin. Gravels of the Moroga Member were supplied from various pre-Neogene basement rocks distributed in the south areas of the basin [5]. Thus, the coarse-grained clastics of the lower part of the Iwami Formation were supplied mostly from the south. The sandstones intercalated in the upper part of the Fuganji Member are lithic wackes of which sand grains were derived probably from the basement rocks distributed in the southward of the Tottori area. In contrast, the sandstones of the Shichiyama Member, intercalated in the Aragane Pyroclastic Member of the upper Iwami Formation, are lithic wackes, and may have been derived mostly from the Cretaceous to Paleogene felsic volcanic and plutonic rocks exposed on the north of the Tottori area. In the Tottori area, thus, coarse clastics were supplied from the south, in large quantities, in the early stage, and bilaterally from the south and north in the late stage [5, 16].

The gravel compositions and size distributions of the conglomerates of the Tsuji Member in the Kitatajima area indicate that they were derived from two rows of basement highs on the northwest and southeast of the basin. The highs on the northwest were composed mainly of the Cretaceous to Paleogene felsic volcanic rocks and the Neogene andesite and basalt, and those on the southeast consisted mostly of the Cretaceous to Paleogene felsic volcanic rocks [15].

Although, the lateral supply from the south in the Tottori area, was much larger in volume and earlier in age than that from the north, the sediment supply into the sedimentary basin seems to have been essentially bilateral in the both areas.

DISCUSSIONS

The provenance analyses described in the previous section suggest that the sedimentary basin in question was bordered by the two rows of basement highs on both sides [16]. The southeastern high supplied a considerable amount of gravelly sediments of the Entsuji, Moroga and Tsuji Members and also the sandy intercalations of the upper Fuganji Member. The northwestern high provided a large amount of gravel of the Tsuji Member and the sandy sediments of the Shichiyama Member. The two basement highs elongated in northeast-southwest or eastnortheast-westsouthwest direction, parallel to the basin trend, and also to the general trend of Southwest Japan [16, 17].

The paleoenvironments and sediment supply in and around the basin during latest Early Miocene to Middle Miocene time are depicted in Figure 3, separated in three stages. In latest Early Miocene time, the two basement highs began to supply the conglomeratic sediments of the lower Entsuji and Tsuji Members to form alluvial fans at their foots (Figure 3-1). The lacustrine mudstone with plant and insect fossils of the lowest Fuganji and Tsuji Members accumulated in the marginal area of the fans. The subsidence of the basin and the transgression of earliest Middle Miocene age changed the major environment of the basin to brackish or inner sublittoral (Figure 3-2), as evidenced by molluscan fossils [4, 6] (Figure 2). Coarse-grained sediments from the basement highs formed fan deltas of which sediments are represented by the upper Entsuji and Moroga Members in the Tottori area, and the Ooka and Otani Members in the Kitatajima area. In early Middle Miocene time (Figure 3-3), the sedimentary basin rapidly deepened due to the advancement of the subsidence and the transgression. The rapid deepening is recorded through the drastic changes in fossil assemblage from brackish or inner sublittoral to outer sublittoral or upper middle bathyal [6, 11-13]. The sedimentary basin and the basin-ward margins of the basement highs were overlain by argillaceous sediments. The thickness of the accumulated mudstone attains more than 500 m. Arenaceous sediments were intermittently supplied from the retreated basement highs on the both sides of the basin, forming the sandstone layers of the Shichiyama and Fuganji Members. During this stage, the subaqueous volcanism of bimodal composition took place actively, and provided pyroclastic and lava flows in the basin [14].

Since the 1,000 m thick sediments accumulated in the basin during 1.5 million years ranging from N.8 to N.9., the averaged sedimentation rate is estimated to exceed 500 m/m.y.. Thus, the whole sedimentological characters of the basin are itemized as follows; (1) longitudinal elongation parallel to the general trend of the Southwest Japan arc, (2) thick accumulation of

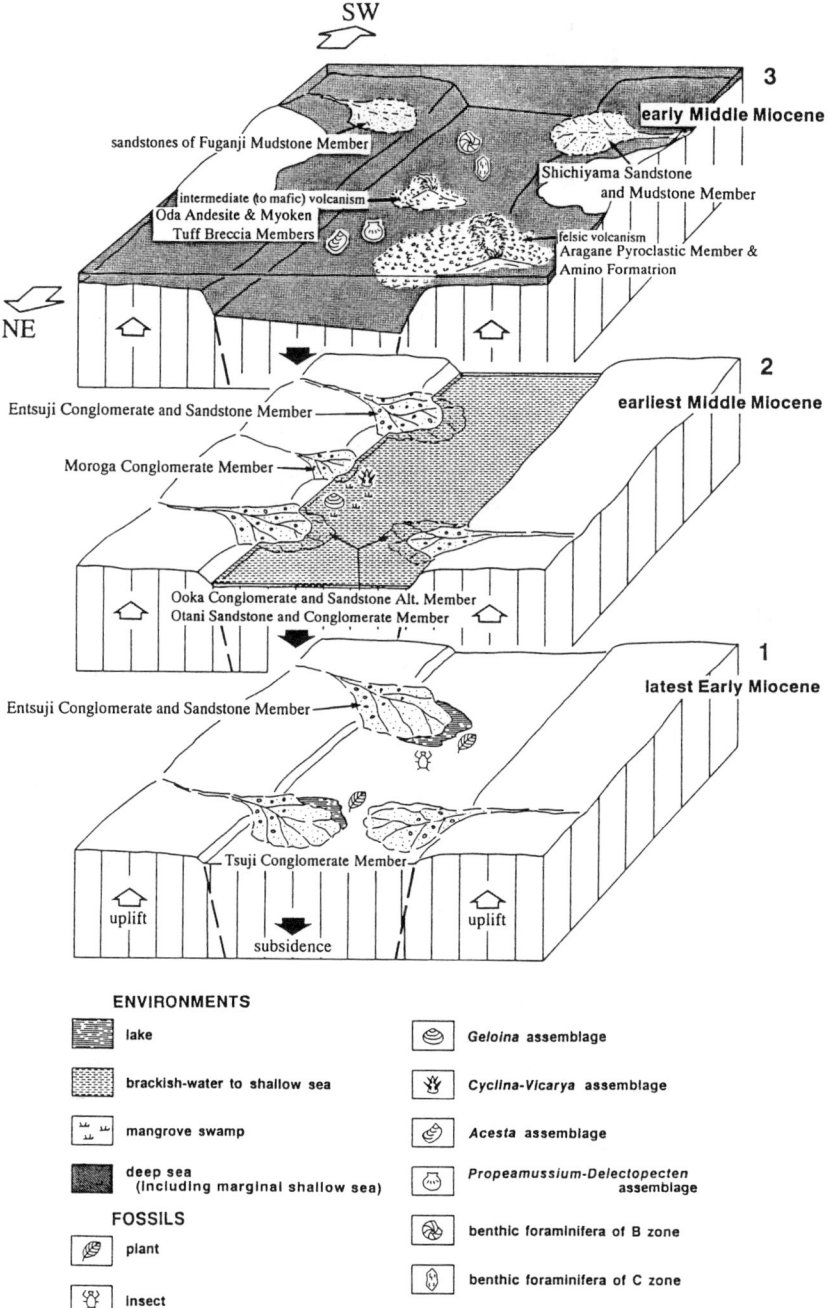

Figure 3. Block diagrams showing the sedimentary and tectonic history.
1. Depositional stage of the lower Entsuji and Tsuji Members (latest Early Miocene), 2. Depositional stage of the upper Entsuji, Moroga, Ooka and Otani Members (earliest Middle Miocene), 3. Depositional stage of sandstones of the Fuganji and Shichiyama Members (early Middle Miocene).

basin fills more than 1,000 m, (3) high sedimentation rate attaining 500 m/m.y., (4) rapid deepening of the basin, which changes the sedimentary facies from brackish or inner neritic to upper middle bathyal, (5) bilateral supply of thick conglomerates of alluvial fan and fan-delta facies in the early stage and of numerous sandstone layers of submarine fan facies in the later stage, and (6) violent submarine volcanism of bimodal composition. These characters are entirely identical with those of the contemporaneous rift basins in the backarc side of Northeast Japan [1, 2].

CONCLUSIONS

The sedimentological characters of the latest Early to Middle Miocene basin in the Japan Sea side of the Southwest Japan was analyzed in this paper. These characters indicate that the sedimentary basin generated as a rift graben.

The contemporaneous rift basins in the backarc side of Northeast Japan have the similar sedimentological characters. They were interpreted as a product of the initial rifting of the Japan Sea opening. The sedimentary basin analyzed in this article also could have a genetic relationship to the opening of the Japan Sea.

REFERENCES

1. N. Tsuchiya. Submarine basalt volcanism of Miocene Aosawa Formation in the Akita-Yamagata oil field basin, back-arc region of Northeast Japan, *Mem. Geol. Soc. Japan.* no. 32, 399-408 (1989).
2. A. Yamaji. Geology of Atsumi area and Early Miocene rifting in the Uetsu district, northeast Japan, *Mem. Geol. Soc. Japan.* no. 32, 305-320 (1989). (in Japanese with English abstract)
3. T. Matsumoto. Stratigraphical Study of the Miocene Series in the Eastern Part of Tottori Prefecture, Southwest Japan, *Jour. Sci. Hiroshima Univ. ser. C.* **9**, 199-235 (1989).
4. K. Wadatsumi and T. Matsumoto. The stratigraphy of the Neogene formations in northern Tazima -Study of the Neogene in the northwestern part of the Kinki district- part 1, *Jour. Geol. Soc. Japan.* **64**, 625-637 (1958). (in Japanese with English abstract)
5. T. Matsumoto. Latest Early to Middle Miocene paleogeography of the Tottori-Tsuyama area viewed from provenance study of clastic rocks, *Jour. Geol. Soc. Japan.* **97**, 817-833 (1991b). (in Japanese with English abstract)
6. I. Yamana. Molluscan assemblage from the Miocene Tottori Group in the eastern part of Tottori Prefecture, *Bull. Tottori Mus.* no. 14, 1-23 (1977). (in Japanese with English abstract)
7. S. Imamura, K. Hide, M. Nakano, M. Nishio and juniors of Hiroshima University. The Cenozoic strata and the basement rocks to the southeast of Tottori City, Tottori Prefecture, Southwest Japan, *Jour. Geol. Soc. Japan.* **68**, 414 (1962). (in Japanese)
8. I. Yamana. On the composition of Miocene Okamasu flora in Tottori Prefecture, Japan, *Bull. Tottori Mus.* no. 6, 1-5 (1968). (in Japanese)
9. Y. Hojo. Some Miocene plant fossils from Tottori and Shimane Prefecture, San-in District, *Mem. Fac. Sci. Kyushu Univ. ser. D, Geol.* **22**, 13-35 (1973).
10. K. Ueda. A Miocene fossil of long-armed scarabaeid beetle from Tottori, Japan, *Bull. Kitakyushu Mus. Nat. Hist.* **9**, 105-110 (1989).
11. Y. Tai. Miocene microbiostratigraphy of West Honshu, Japan, *Jour. Sci. Hiroshima Univ. ser. C.* **2**, 265-395 (1959).
12. T. Matsumoto. Stratigraphy of the Miocene series in the area southeast of Tottori City, Tottori Prefecture, Southwest Japan, *Jour. Geol. Soc. Japan.* **92**, 269-287 (1986). (in Japanese with English abstract)
13. R. Nomura and T. Matsumoto. Benthic foraminifera from the Tottori Group, *Studies of the San'in region -Natural environment-.* no. 6, 57-63. Center for studies of the San'in region, Shimane University (1990).

(in Japanese with English abstract)
14. T. Matsumoto. Stratigraphy of the Miocene series and the Middle Miocene volcanism in the area northeast of Tottori City, Southwest Japan, *Jour. Geol. Soc. Japan.* 9 7, 697-712 (1991a). (in Japanese with English abstract)
15. T. Matsumoto. Latest Early to Middle Miocene paleogeography of the eastern San'in district, Southwest Japan, viewed from a provenance study of conglomerate, *Mem. Geol. Soc. Japan.* no. 38, 205-216 (1992b). (in Japanese with English abstract)
16. T. Matsumoto. Latest Early to Middle Miocene paleogeography of the Tottori area viewed from volcanism and sedimentation, *Mem. Geol. Soc. Japan.* no. 37, 295-310 (1992a). (in Japanese with English abstract)
17. K. Ichikawa and N.Kitamura. Late Cenozoic sedimentary basins of Japan in relation to the basement structure. In. *Cenozoic Geology of Japan.* K. Huzita et al. (Eds). Prof. N. Ikebe Mem. Vol., 187-204 (1978). (in Japanese with English abstract)

(Manuscript received 3 February 1993; accepted 12 March 1993.)

Coexistence of provenance-reflected shallow-marine and deep-marine turbidite sandstones -sedimentation at the eastern margin of the Niigata Neogene backarc basin, northeast Japan-

SHUICHI TOKUHASHI

Fuel Resources Department, Geological Survey of Japan, 1-1-3 Higashi, Tsukuba, Ibaraki, 305 Japan

Abstract: The combination of the detailed stratigraphic study using many tuff marker beds and heavy mineral analysis of the turbidite sandstones disclosed that two kinds of turbidite sandstones, namely submarine-fan turbidite sandstone in the western part and shelf turbidite sandstone in the eastern part of the eastern margin of the Niigata backarc basin, were deposited during the deposition of the upper part of the Lower Pliocene Kawaguchi Formation. The heavy mineral composition of these sandstones reveals that the western submarine-fan turbidite sandstone was derived mainly from pre-Tertiary basement rocks and the eastern shelf turbidite sandstone was derived mostly from intrabasinal Neogene volcanic rocks.

Key words: heavy mineral analysis, shallow-marine turbidite, shelf turbidite, submarine-fan turbidite, Niigata backarc basin, Japan Sea

INTRODUCTION

Heavy mineral analysis is well known as a useful tool to estimate the characteristics of the provenance of the analyzed sediments. But it also plays a very important role to clarify the sedimentary process itself, especially when the sediments were supplied from different source rocks as shown in this study. In the active island arcs, the heavy mineral composition of sandstones is expected to be diverse as it includes minerals derived from not only the older basement rocks but also the younger volcanic rocks. This increases the importance of the heavy mineral analysis in the active margin areas compared to it in the other tectonically steady areas.
 In the Neogene backarc basins of the Japanese Islands along the Japan Sea the lowermost are mostly occupied with early to middle Miocene volcanic rocks related with the opening of the Japan Sea. In the Niigata backarc basin, one of the major oil-bearing basins around the Japanese Islands, these volcanic rocks are overlain by middle Miocene deep-marine mudstones and then late Miocene to early Pliocene turbidite-bearing sequences. During the late Pliocene to Pleistocene ages, Neogene backarc sediments suffered from a compressional field and were deformed to form many anticlines and synclines with axes trending NNE-SSW (Figure 1). This deformation formed the many anticlinal turbidite traps of oil and gas in the Niigata basin. According to the previous studies, many turbidite bodies supplied from various source rocks were formed widely in the central area of the basin during the late Miocene to early Pliocene ages [1-4].

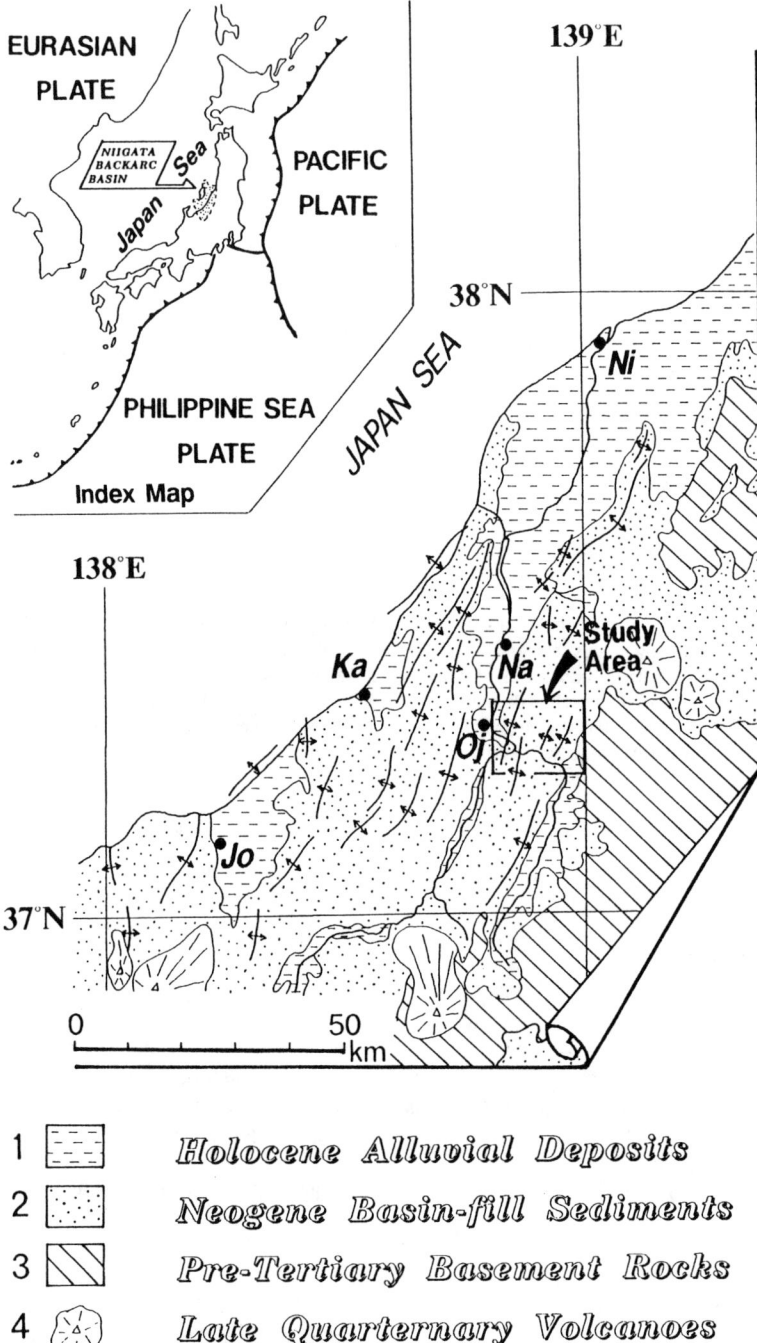

Figure 1. Geologic outline and the location of the study area.
Jo: Joetus, Ka: Kashiwazaki, Na: Nagaoka, Oj; Ojiya

STRATIGRAPHY IN THE STUDY AREA

The recent detailed stratigraphic work using many volcanic ash layers as useful marker beds at the eastern margin of the Niigata basin disclosed the dynamic lateral change from deep-marine sediments to shallow-marine sediments [5]. In this area, andesitic to dacitic volcanic rocks dominate in the early to middle Miocene age, but in the late Miocene to Pliocene age, normal sediments, deep-marine to shallow-marine sediments in ascending order, dominate as shown in Figure 2. The basement rocks are distributed widely east of the basin and comprise of Mesozoic rocks mostly composed of chert and shale, and partly of sandstone and green rocks, and Cretaceous granitic rocks intruding those Mesozoic rocks.

In the western part of the study area, the lower Pliocene Kawaguchi Formation (850 m in maximum thickness), the target of this study, is composed of flysch-type alternation of sandstone (turbidite) and mudstone, and conformably overlies the late Miocene Araya Formation (more than 400 m in thickness), mostly formed of deep-marine mudstone, and conformably overlain by the Ushigakubi Formation (350m in maximum thickness) characterized by the upper bathyal mudstones. The Shiroiwa Formation (350 m in maximum thickness), mostly composed of shallow-marine sandy siltstones, conformably overlies the Ushigakubi Formation and overlain by the Wanazu Formation (200 m in maximum thickness) made up of shallow-marine sandstones. The shallow-marine sandstones of the Wanazu Formation is widely overlain by the Plio-Pleistocene Uonuma Formation (600 to 430 m in thickness), characterized by terrigenous conglomerate and near-shore (transitional from marine to non-marine) sand and mud. Based on the tuff marker beds traced through the study area, western deeper-water sediments are gradually replaced by the shallower-water sediments to the east. In the study area, these formations are repeatedly distributed under the control of north-south trending folding structures.

KAWAGUCHI FORMATION

The Kawaguti Formation, in the western area of the study area, is divided into the lower and upper parts. The Lower Kawaguchi Formation (Kl) is characterized by the sandstone (turbidite)-dominated sandstone-mudstone alternation. The sandstone beds vary in thickness from ten meters to a few centimeters, and they have a tendency to occur more or less collectively in the vertical succession forming depositional tongues. Trace fossils are rarely observed in the vertical succession. The Lower Kawaguchi Formation is nearly exclusively distributed in the western part of the study area.

The Upper Kawaguchi Formation, on the other hand, is characterized by mudstone-dominated mudstone-sandstone alternation, and divided into a western part and eastern part. The western Upper Kawaguchi Formation (Ku1) is composed of mudstone-sandstone alternation, while the eastern Upper Kawaguchi Formation (Ku2) is composed of sandy mudstone-sandstone alternation. The boundary between the western and eastern Upper Kawaguchi Formations is gradual. The turbidite sandstone beds in the western area occur more or less collectively as like as those in the Lower Kagaguchi Formation, but those in the eastern area occur more randomly in vertical succession. In the western area the vertical succession yields rare trace fossils but in the eastern area it yields abundant trace fossils. The turbidite sandstone beds in the western area attain about five meters in maximum thickness and sandstone beds more than one meter in thickness are common, but in the eastern area they can be more than two meters but are usually less than several tens of centimeters in thickness. Therefore, in the Upper Kawaguchi Formation the turbidite sandstone-mudstone alternation with few trace fossils in the western area (Ku1) changes gradually to the turbidite sandstone-sandy mud-

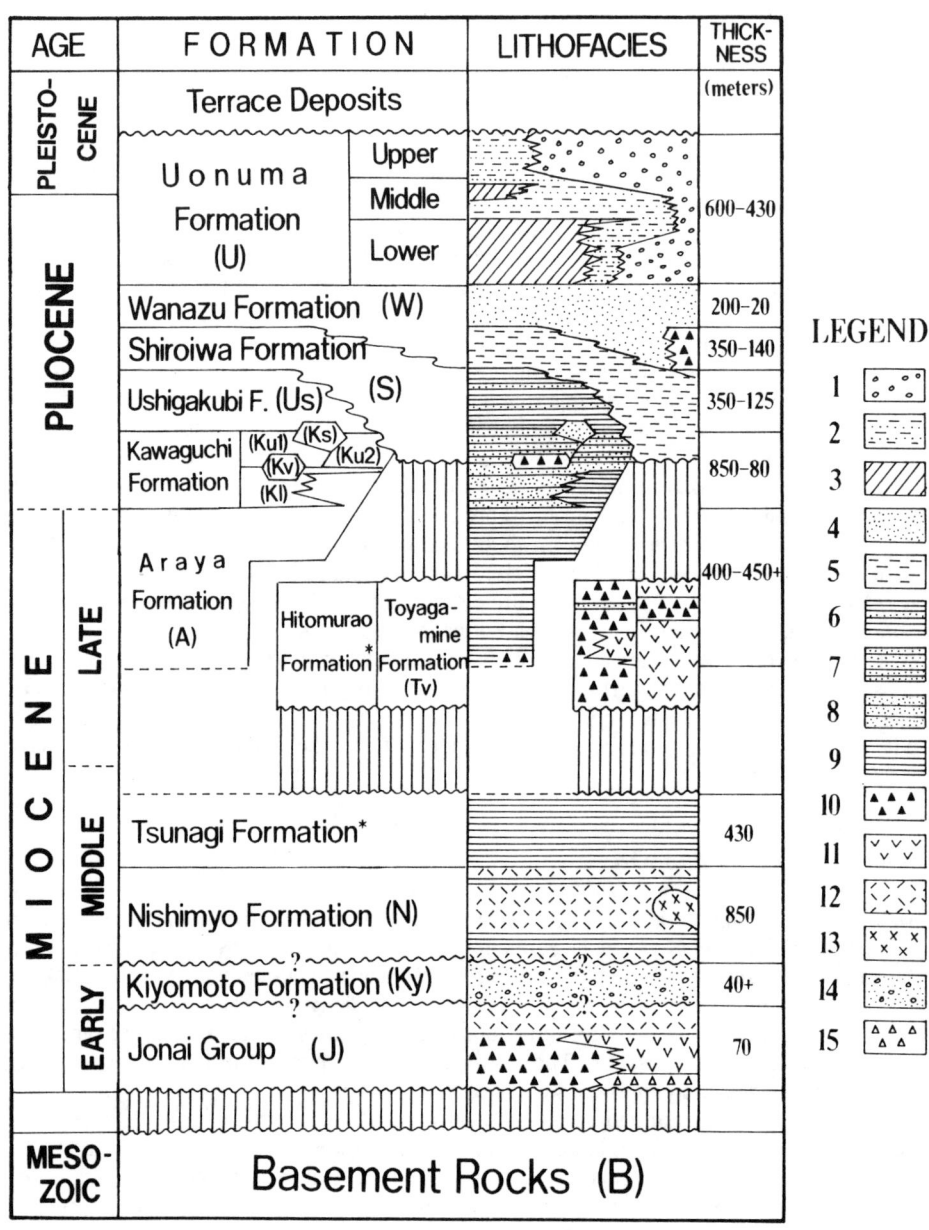

Figure 2. Stratigraphy of the eastern part of the Niigata backarc basin. After [5].
Legend for lithofacies; 1. gravel, 2. sand and silt, 3. marine silt and sand, 4. sand, 5. sandy silt, 6. mudstone, 7. mudstone-rich alternation of sandstone and mudstone, 8. sandstone-rich alternation of sandstone and mudstone, 9. mudstone, 10. andesitic and dacitic volcanic breccia, 11. andesitic and dacitic lava, 12. dacitic pyroclastic rocks, 13. rhyolitic lava, 14. sandstone and conglomerate, 15. andesitic pyroclastic rocks.

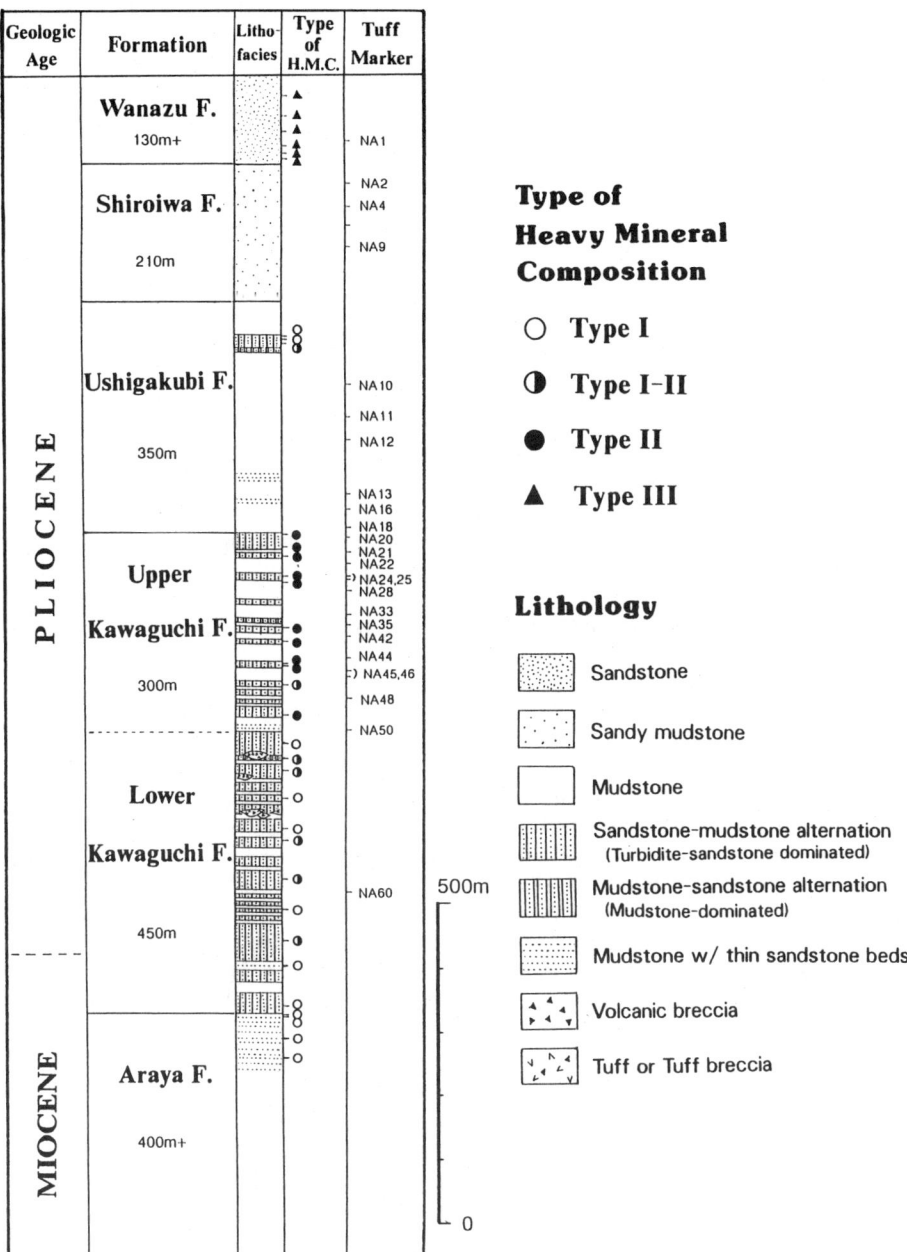

Figure 3. Stratigraphic distribution of heavy mineral assemblages of sandstones shown in the geologic column along the Nobegawa River in the western part of the study area.

stone alternation with abundant trace fossils in the eastern area (Ku2).

The paleocurrent of the turbidite sandstone beds of the Kawaguchi Formation both in the western and eastern parts of the study area indicates a westward direction.

RESULTS OF HEAVY MINERAL ANALYSIS

Previously the heavy mineral analysis of sandstones collected from the Araya Formation to Wanazu Formation in ascending order along the Nobegawa River, type locale of these formations in the western part of the study area, was tried [6]. Except for the shallow marine sandstones in the Wanazu Formation, sandstones of other formations are all turbidites. As a result, the heavy mineral composition of these sandstones is classified at least into four types, i.e., type I, type I-II, type II, and type III.

In type I, opaque minerals dominate and such basement-origin minerals as garnet, zircon, epidote, zoisite and hornblende occur as other minor minerals. In type II, hornblende dominates and is more abundant than opaque minerals. The same basement-origin minerals as those in type I also occur as minor minerals. Type I-II is the intermediate of type I and typeII, but opaque minerals are still more abundant than hornblende. In type III, hypersthene, augite and hornblende dominate. Type III is basically different from other types, as hypersthene and augite occur nearly exclusively in type III.

The basic characteristics of these types are simplified as follows:
Type I :Opaque.>>Gar.,Zir.,Epi.,Zoi.,Hor.
Type I-II :Opaque.>Hor.>Gar.,Zir.,Epi.,Zoi.
Type II :Hor.>Opaque.>Gar.,Zir.,Epi.,Zoi.
Type III :Hyp., Aug.,Hor.>>Epi.,Opaque>Gar.,Zir.,Zoi

Type I is recognized in thin sandstones of the uppermost part of the Araya Formation and sandstones of the Lower Kawaguchi Formation. Type I-II is also recognized mainly in the Lower Kawaguchi Formation. Turbidite sandstones of the Upper Kawaguchi Formation are characterized mostly by type II and partly part I-II. Turbidite sandstones partly intercalated in the upper part of the Ushigakubi Formation are composed of type I and type I-II. Shallow-marine sandstones in the Wanazu Formation, on the other hand, are all composed of type III. Therefore, type-III heavy mineral composition is recognized only in the shallow marine sandstones of the Wanazu Formation, and not recognized in the turbidite sandstones of the underlying formations along the Nobegawa River in the western part of the study area (Figure 3).

The results of heavy mineral analysis of sandstones collected from many localities both in the western and eastern parts of the study area are summarized in Figure 4. As shown in this figure, the heavy mineral composition of the sandstones of the Lower Kawaguchi Formation, exclusively distributed in the western part of the study area, are composed of type I and type I-II similar to those along the Nobegawa River. Turbidite sandstones of the Upper Kawaguchi Formation in the western part of the study area are mostly composed of type II like those along the Nobegawa River, but those in the eastern area are characterized by type III.

DISCUSSION

Heavy mineral analysis of the turbidite sandstones of the Upper Kawaguchi Formation firstly discloses that their heavy mineral types are basically different from each other between the western area and the eastern area. This means that turbidite sandstones of the Upper Kawaguchi Formation in the western area and in the eastern area are both supplied from the east

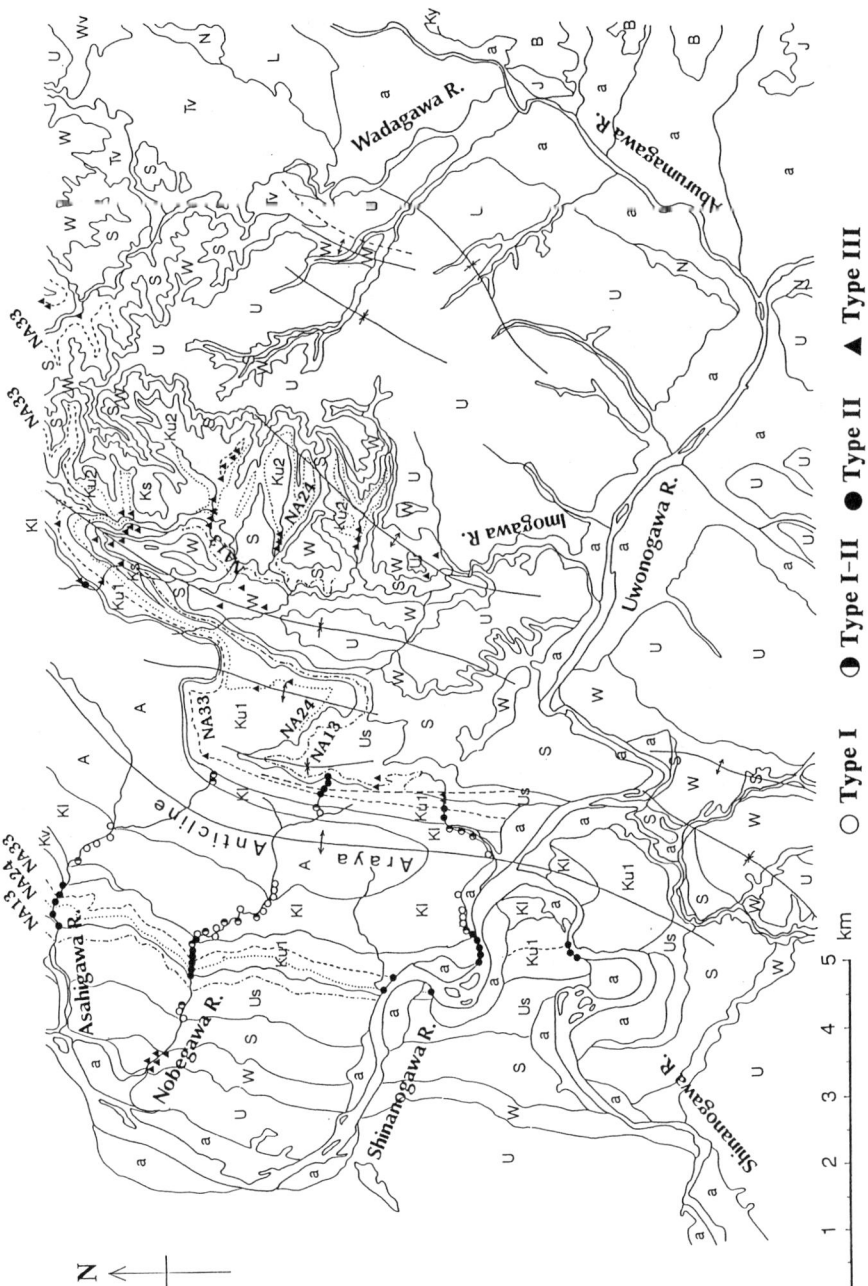

Figure 4. Distribution of heavy mineral composition of sandstones in the study area. Each formation is represented by the same alphabetic symboles shown in Figure 2. Dashed line with NA number means a tuff maker bed listed in the colum of Figure 3.

○ Type I   ◐ Type I-II   ● Type II   ▲ Type III

based on the paleocurrent data, but from a different source and through different routes. Therefore, during the deposition of the Upper Kawaguchi Formation, two kinds of turbidite sandstone beds separated each other were formed in different environments at the eastern margin of the Niigata backarc basin, and resulted in the formation of the two kinds of sandstone–mudstone alternations (Ku1, Ku2) as previously mentioned.

The western turbidite sandstone beds are inferred to be deposited on the submarine fan environments, as these turbidite sandstone beds are intercalated in the finer-grained mudstone with very few trace fossils in its vertical succession as suggested by the previous study[7]. The eastern turbidite sandstone beds, on the other hand, are estimated to be deposited on the shelf area as they are intercalated in the coarser sandy mudstone with abundant trace fossils on the vertical succession. Therefore, during the deposition of the Upper Kawaguchi Formation, two kinds of turbidite sandstone beds, submarine–fan turbidite sandstone beds in the western area and shelf turbidite sandstone beds in the eastern area, were formed concurrently in the study area [8].

As augite and hypersthene in the type–III turbidite sandstones often have mechanically-broken edges or hacksaw edges, and volcanic–origin oxy–hornblende are commonly included in the type–III turbidite sandstones, the shelf turbidite sandstones in the eastern area are inferred to be supplied mostly from Tertiary volcanic rocks, that is, from intrabasinal rocks. The fact that near–shore shallow–marine sandstones of the Wanazu Formation are also composed of type–III sandstones strongly suggests that relatively smaller rivers, which mainly drained through the Tertially volcanic–rock areas, supplied type–III sands westward to the coast. Then these coastal or near–shore sands were transported farther westward to the shelf area intermittently through many small channels forming many small depositional tongues when some events such as huge storms occurred.

The submarine–fan turbidite sandstones in the western area, on the other hand, are inferred to be from the pre–Tertiary basement rocks, that is, Mesozoic sedimentary rocks and Cretaceous granitic rocks, because very few volcanic–origin minerals such as hypersthene, augite and oxy–hornblende are included in these sandstones. The composition of gravels sporadically included in these sandstones, especially those in the lower part of channelized sandstones, also supports this interpretation. Therefore, it is supposed that the relatively larger river, which mainly drained through the basement areas, supplied sand to the mouth of the river and formed there a delta–like sand body, i.e., delta–fan or delta. The submarine–fan turbidite sandstones in the western area must have been supplied from such a delta–like sand body through the submarine canyon or valley and deposited on the basin floor forming depositional tongues as shown in Figure 5.

CONCLUDING REMARKS

The combination of detailed stratigraphic work using many tuff marker beds and the heavy-mineral analysis of turbidite sandstones reveals the new fact that two kinds of turbidite sandstone beds, submarine–fan turbidite sandstone beds and shelf turbidite sandstone beds, were concurrently deposited at the eastern margin of the Niigata Neogene backarc basin. As this fact could not be disclosed without the heavy mineral analysis or mineralogical study, heavy mineral analysis has proven to be very important not only for the estimation of the provenance but also the investigation of the sedimentary process of the sandstones itself. Especially in active margins including many volcanic rocks, heavy mineral composition is more diverse and therefore heavy mineral analysis is confirmed to be more efficient and important tool.

ACKNOWLEDGEMENTS

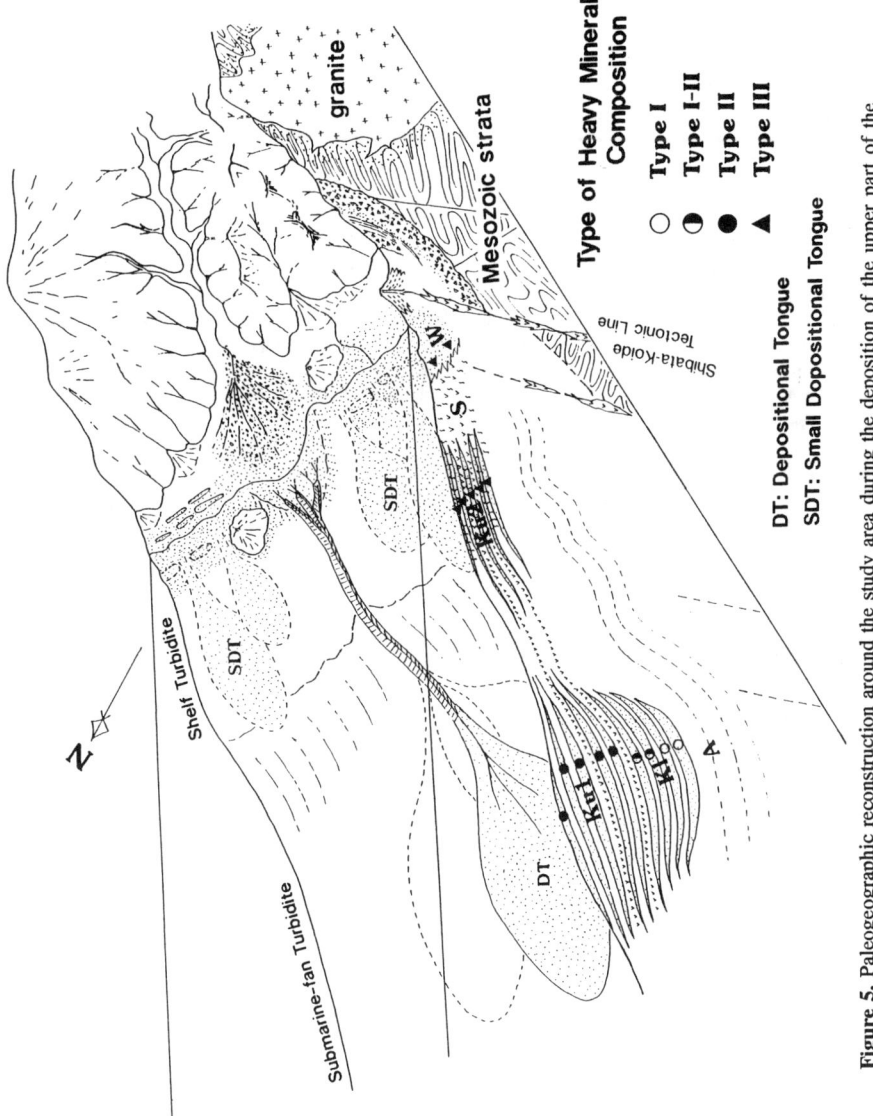

Figure 5. Paleogeographic reconstruction around the study area during the deposition of the upper part of the Lower Pliocene Kawaguchi Formation. Some alphabetic symbols in the cross section correspond to the formations shown in figure 2.

The author expresses his deep gratitude to Dr. Fujio Kumon, and Dr. Kang Min Yu for their active organization and promotion of this proceeding. He is grateful to Dr. Philip A. Jarvis, a guest researcher of the Geological Survey of Japan, for his refinement of the manuscript.

REFERENCES

1. K. Kageyama and Y. Suzuki, The paleogeographic reconstruction of northern Fossa Magna region, *Rept. Geol. Surv. Japan*, **250**, 285-306 (1974).*
2. M. Tateishi, H. Irino, T. Minezaki, and M. Endo, Submarine fan sediments in the Niigata active marginal basin, central Japan, *Jour. Res. Gr., Clastic Sed. Japan*, no. 3, 41-56 (1984).
3. U. Suzuki, Geology of Neogene basins in the eastern part of Sea of Japan, *Mem. Geol. Soc. Japan*, **32**, 143-183 (1989).*
4. I. Kobayashi and M. Tateishi, Neogene stratigraphy and paleogeography in the Niigata region, central Japan, *Mem. Geol. Soc. Japan*, **37**, 53-70.*
5. Y. Yanagisawa, I. Kobayashi, K. Takeuchi, M. Tateishi, K. Chihara and H. Kato, *Geology of the Ojiya district*, Geol. Surv. Japan (1986).*
6. S. Tokuhashi, Sedimentological and mineralogical study on the late Miocene to Pliocene sandstones occurring in the southern area of the Higashiyama region, Niigata Prefecture, northeast Japan − heavy mineral composition of sandstones exposed along the Nobegawa River −, *Jour. Geol. Soc. Japan*, **96**, 745-758 (1990).*
7. S. Tokuhashi, A preliminary study on turbidite sandstone beds in the southern part of the Higashiyama oil belt, Niigata Prefecture, central Japan, *Bull. Geol. Surv. Japan*, **36**, 611-635 (1985).*
8. S. Tokuhashi, Sedimentological and mineralogical study on the late Miocene to Pliocene sandstones occurring in the southern area of the Higashiyama region, Niigata Prefecture, northeast Japan −sedimentological relationship between submarine-fan turbidite sandstone and shelf turbidite sandstone −, *Jour. Geol. Soc. Japan*, **98**, 355-372 (1992).*

* in Japanese with English abstract.

(Manuscript received 28 December 1992; accepted 20 January 1993.)

Petrography of turbidite sandstones in Niigata basin, northernmost part of Fossa Magna, central Japan

MASAAKI TATEISHI[1], ADEL A.A. El HABAB[2], and MITSUO SHIMAZU[3]

[1] Dept. Geol. Mineral., Fac. Sci., Niigata Univ., Ikarashi-2, Niigata 950-21, Japan
[2] Geol. Dept., Fac. Petrol. Min. Eng., Suez Canal Univ., Suez, Egypt
[3] Niigata Prefectural Women's College, Ebigase, Niigata 950, Japan

Abstract: Neogene turbidite sandstones constitute the important reservoir rocks for hydrocarbon resources in the Niigata sedimentary basin located in the north of Fossa Magna. The turbidite sequence in the basin was deposited during two periods; Middle Miocene and latest Miocene to Pliocene times. This paper attempts to clarify the petrographical features and the source rocks of these turbidite sandstones deposited in the southwestern part of the basin.
  Although the modal composition of the sandstones generally indicates the quartzose recycled orogen to transitional arc provenances, the late Miocene to Pliocene turbidite sandstones are derived from the provenance richer in the factor of volcanic arc than the Middle Miocene turbidite sandstones. On the basis of the heavy mineral composition, the sandstones are classified into two types of assemblages. One is characterized by the abundance of zircon and garnet grains, while the other is by the abundance of hornblende grains. The former one suggests the provenance of plutonic and metamorphic rocks. On the other hand, the latter does the main supply from the volcanic and metamorphic rocks. The Middle Miocene turbidites include only the former assemblage, while the late Miocene to Pliocene ones contain both types of heavy mineral assemblages. From the chemical composition, garnet grains included in these turbidite sandstones are considered to have been derived from the sedimentary and metamorphic rocks distributed to the west of the Itoigawa-Shizuoka Tectonic Line, and from the Cenozoic volcanic rocks distributed in the environs. Conclusively the provenance, which is located to the west through the Itoigawa-Shizuoka Tectonic Line and is composed of older sedimentary, metamorphic, acid plutonic and volcanic rocks, continued to supply the detritus into the Niigata basin during Middle Miocene to Pliocene time. The Cenozoic volcanic rocks distributed to the south and southeast of the basin began to supply the abundant detritus at late Miocene time.

Key words: Niigata basin, provenance, petrography, sandstone, modal composition, heavy mineral composition, chemical composition, garnet.

INTRODUCTION

This work concerns with the petrographical characteristics of turbidite sandstones of Middle Miocene to Pliocene age in the Niigata sedimentary basin, northern Fossa Magna. Petrographical characteristics of the sandstones give the information in relation to the tectonic movement of the provenance which supplied coarse detrital grains.
  Heavy mineral composition of clastic sedimentary rocks plays an important role in the presumption of source rocks which supplied detrital grains [1,2]. Clastic rocks deposited in mobile belts, however, have experienced complex tectonic movements, and contain various kinds of heavy minerals. Therefore, we find difficulty in presuming the source rocks of this kind of sedimentary rocks solely from the heavy mineral composition. Garnet grains have high potential in the provenance studies among the various kind of heavy minerals, because they are not only stable and common in sandstone, but also they show a wide

variety of chemical compositions, mainly with six principal end-members (pyrope, almandine, spessartine, grossular, andradite, and uvarovite). Their chemical composition is controlled by the chemical and physical conditions of original magma and rocks. The chemical composition of detrital garnet grains is expected to bring the important information on the petrographic character of source rocks [3,4,5].

The provenance of the turbidite sandstones deposited in the Niigata sedimentary basin during Middle Miocene to Pliocene time, is discussed in this report, on the basis of the modal composition, heavy mineral assemblage, and the chemistry of garnet grains. Depositional system of the turbidite sequences in the Niigata sedimentary basin has been discussed so far from stratigaphical and sedimentological viewpoints by many investigators [6,7], because the turbidite sandstones are important hydrocarbon resources in Japan. However, the petrographical study of these sandstones has just started [8]. A part of the present study was already reported [9].

GEOLOGICAL SETTING AND SAMPLES

The Niigata sedimentary basin, located in the northernmost part of the Fossa Magna (Figure 1), is buried by a thick Neogene to Quaternary sequence up to 6,000 m in total thickness.

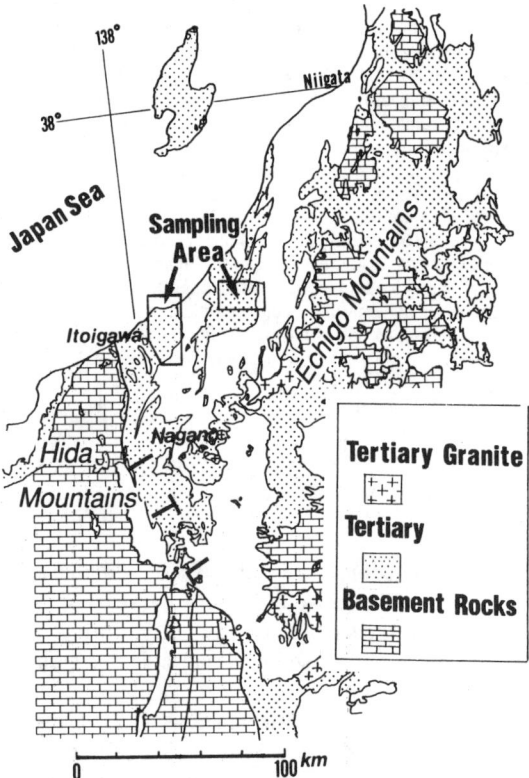

Figure 1. Index map showing the sampling areas for petrographical examination of the Middle Miocene to Pliocene turbidite sandstones in the southwestern part of the Niigata sedimentary basin. I T L means Itoigawa-Shizuoka Tectonic Line.

Table 1. Stratigraphy of the Neogene sequences distributed in the southwestern part of the Niigata sedimentary basin.

Age		West		East	
Pliocene		Nadachi F.	Massive mudstone	Higashigawa F.	Sandy siltstone
		Kawadume Formation	Sandy to muddy flysch with Conglomerate & mudstone	Tamugigawa F. / Shobu F.	Sandy to muddy flysch with thick-bedded sandstone / Mudstone
Miocene Late		Nodani Formation	Muddy flysch & Bedded Shale	Sugawa Formation	Massive mudstone
				Taruda Formation	Muddy flysch
				Matsunoyama F.	Dacitic volc. rocks
Miocene Middle		Nambayama Formation	Sandy to Muddy Flysch with massive mudstone & thick-bedded sandstone		
		Tsugawa F.	Rhyolitic vol. rocks		

It is an intra-arc basin with the preceded Paleogene to Early Miocene volcanic ridge called the Sado-Pohan belt in the back-arc side [10]. The Niigata sedimentary basin is bordered on the west by the Hida Mountains through Itoigawa-Shizuoka Tectonic Line, and on the east by the Echigo Mountains. Although this basin appeared associated with violent terrestrial volcanism during Oligocene to Early Miocene time, the real subsidence of the basin and the marine transgression into the basin took place associated with submarine bimodal volcanism in latest Early to early Middle Miocene time. The upper Lower to lower Middle Miocene in this basin is mainly composed of fluvial to deltaic sedimentary rocks and submarine bimodal volcanic rocks. The basin continued to subside and were generally covered by bathyal black shale and muddy flysch during Middle to Late Miocene time. Thick turbidites and their associated sediments called the Nambayama Formation, however, were deposited in the soutwestern part of the basin during Middle Miocene time (Table 1). The tectonic movement of the basin changed and the basin began to shallow and to differentiate at latest Miocene time. Turbidites and their associated sediments were deposited in several various places of the basin during latest Miocene to Pliocene time. They in the southwestern part of the basin are called the Kawadume and Tamugigawa Formations (Table 1). The upper bathyal to lower neritic mudstone and sandy siltstone were deposited during Pliocene time. Through the depositional history of the basin, the turbidites and their associated sediments were deposited at two stages, that is, at the stage of the rapid subsidence and deepening and at the stage of the shallowing and differentiation of this basin.

The Middle Miocene Nambayama Formation is lithologically divided into A, to H Members, in ascending order [6]. Their paleocurrents show the south-southwestern longitudinal and east-southeastern lateral supplies of the detritus [6]. The latest Miocene to Pliocene Kawadume and Tamugigawa Formations also show the south-southwestern longitudinal and east-southeastern lateral paleocurrents [7,11]. That is, the provenance of their turbidte sandstones of the two stages cann't distinguished sedimentologically. The numbers of sampled sandstones for the petrographical examination are 29, 13, and 14, respectively from the Nambayama, Kawadume, and Tamugigawa Formations.

PETROGRAPHICAL DESCRIPTION OF TURBIDITE SANDSTONES

The sandstones collected are generally fine-grained, and several samples from the Tamugigawa Formation are medium-grained.

In order to get modal composition of these sandstones, 500 points were counted on a stained thin section by the Gazzi-Dickinson method [12,13] under polarized microscope. The same samples were crushed sufficiently to break up the lumps into individual grains, and were sieved. The heavy minerals were concentrated by using the bromoform liquid (Specific Gravity 2.87) from the sieved fine-grained sand to coarse-grained silt (1/4 to 1/32 mm) fraction. 200 heavy mineral grains were identified and counted under polarized microscope. Then, garnet grains picked up from the heavy mineral concentrate under stereoscope were analyzed chemically by electron probe micro-analyzer.

The chemical components analyzed were SiO_2, Al_2O_3, total oxide of iron, MgO, CaO, MnO, Na_2O_3, and K_2O.

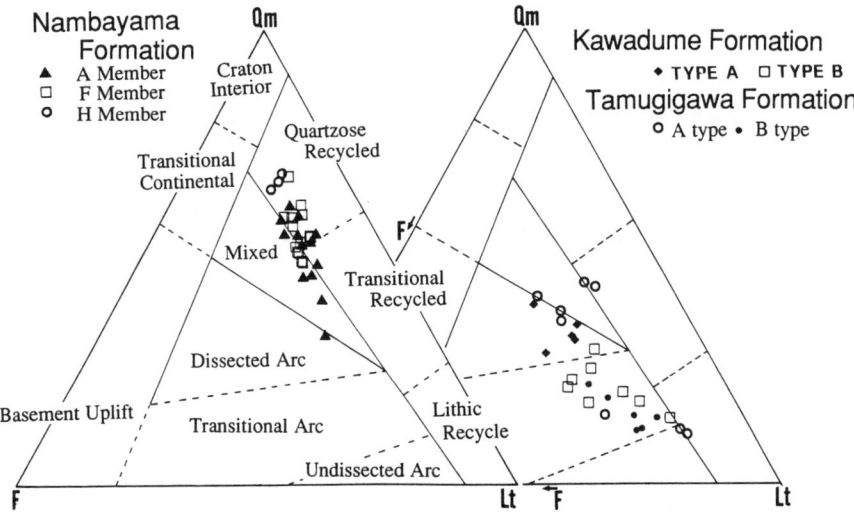

Figure 2. Qm-F-Lt diagrams of turbidite sandstones from the Middle Miocene Nambayama and Late Miocene Kawadume and Tamugigawa Formations. The Middle Miocene turbidite sandstones are separated into samples from three members, that is, A, F, and H Members. The Late Miocene to Pliocene turbidite sandstones are separated into two types based on the heavy mineral composition (Figure 3).

Modal composition

The modal composition of the turbidite sandstones from the Nambayama, Kawadume, and Tamugigawa Formations are shown in the Qm-F-Lt diagram (Figure 2). Matriceous materials smaller than 30 μm, and cementing materials of silica and carbonate are totally more than 15 % in all thin sections. Therefore, most of them belong to lithic wacke-type of sandstone, and a part of them to feldspathic wacke-type [14]. In Figure 3, the Tamugigawa and Kawadume sandstones are separated into two types based on the heavy mineral composition described later. Two types of sandstones from the Kawadume Formation are able to distinguish also in the modal composition, that is, Type A is richer in monocrystalline quartz grains than Type B. While two types of sandstones from the Tamugigawa Formation are overlapped on the plotted area. This overlap is caused by the content of a lot of volcanic rock fragment in three samples among the Type A of sandstones, plotted in the area rich in the rock fragment.

Potash feldspar grains are included more than plagioclase grains in the Nambayama sandstones, while the former is less than the latter in the other sandstones.

Chert fragments are included more than volcanic rock fragments in the Nambayama sandstones, while the former is less than the latter in the other sandsones.

Heavy mineral compostion

Heavy mineral grains are included about 1 to 3 % in total framework grains of the sandstones. Opaque minerals are more than non-opaque minerals. About twenty kinds of non-opaque heavy minerals are found in the sandstones, but ten kinds of minerals among of them are included in appreciable amount. Biotite and chlorite grains are generally abundant in the heavy mineral concentrate. Their grains were excluded from the diagram, which shows relative proportion of the non-opaque heavy minerals, because their density is similar

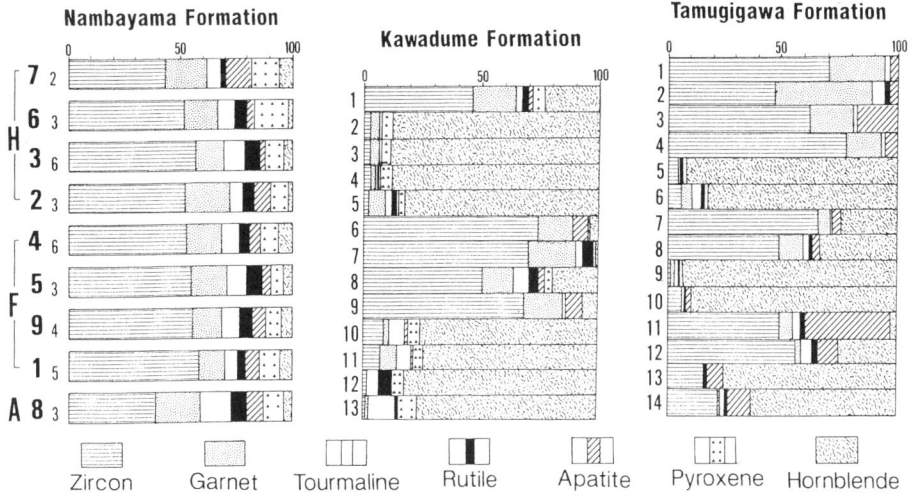

Figure 3. Heavy mineral composition of the turbidite sandstones of the Nambayama, Kawadume, and Tamugigawa Formations.

to the specific gravity of bromoform liquid, and they have not been separated thoroughly from the light minerals.

Zircon grains included abundantly are mostly colorless, but rarely yellow to pale pink colored. They show mostly euhedral prism shape with pyramid termination, partly elongate to elliptic shape with well worn termination. Some grains contain small rod-shaped inclusions of other minerals and small bubble-like inclusions of liquid or gas, by which they look a dirty appearance.

Garnet grains included commonly are pale pink or apricot yellow in color. Most of them are angular with sharp corner, but there are some colorless, well rounded grains. Surface etching, very regular and sharp "hillocks" [15], is a common feature of these garnet grains.

Hornblende grains are predominant in some samples from the Kawadume and Tamugigawa Formations. The grains are usually in the shape of elongated plate, and exhibit green to brown colored.

Tourmaline, showing conspicuous feature of prismatic striation and strong pleochroism, are yellow to brown colored. Inclusions are common in the brown-colored grains. Most

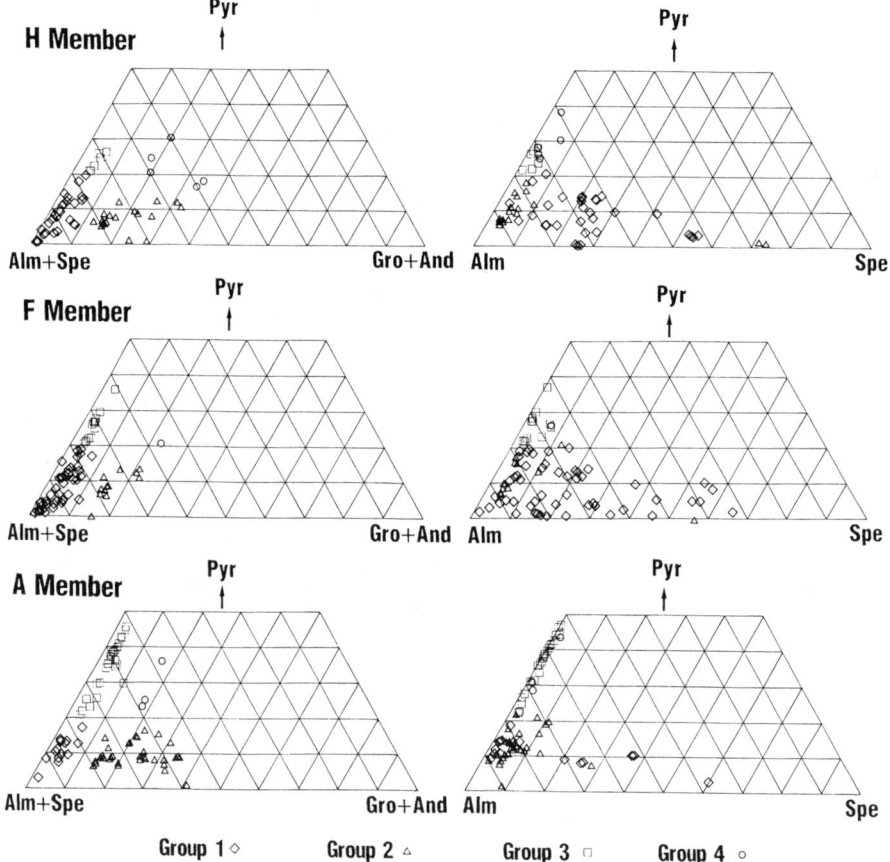

Figure 4-1. (Alm+Spe)-Grand-Pyr and Alm-Spe-Pyr diagrams of garnet grains in the turbidite sandstones of the Nambayama Formation. On the basis of the chemical feature, their garnet grains are subdivided into five groups, that is, Groups 1 to 4.

common inclusions are black, opaque materials which are in the form of irregular masses, elongated rod, or dust.

Rutile grains show three distinctive color varieties, yellow, wine red, and chocolate brown. Short, stumpy, and subhedral grains are predominant, but some ones are prismatic and rather rounded.

Colorless apatite grains occur mostly in rounded and anhedral shape, rarely in euhedral prismatic shape. They are generally fresh and unaltered, but a few grains exhibit a cock comb structure. Inclusions are generally numerous and sometimes arranged in row.

The non-opaque heavy mineral composition (Figure 3) shows the characteristics as follows: 1) There is not the difference of heavy mineral composition in the Nambayama sandstones, which are mainly composed of zircon and garnet grains with tourmaline and rutile. 2) Two types of heavy mineral assemblages are clearly recognized in the Tamugigawa and Kawadume Formations. The one is characterized by abundant zircon and garnet grains, that is, similar to the heavy mineral composition of the Nambayama Formation. The other type includes abundantly hornblende grains, and merely zircon and garnet grains. The former is

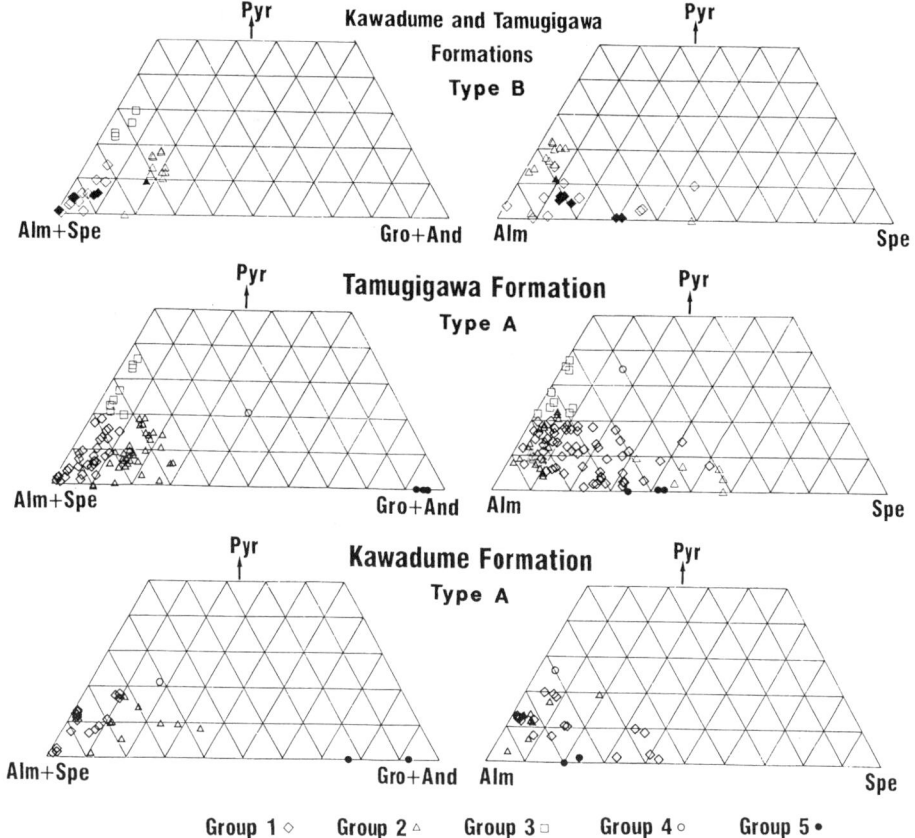

Figure 4-2. (Alm+Spe)-Grand-Pyr and Alm-Spe-Pyr diagrams of garnet grains in the turbidite sandstones of the Kawadume and Tamugigawa Formations. The garnet grains are subdivided into Groups 1 to 5, based on the chemical feature. In the sandstones of Type B, solid symbols show the garnet grains from the Tamugigawa Formation, and open symbols are from the Kawazume Formation.

called Type A of sandstone, and the latter is Type B.

Chemical composition of garnet grains

The chemical feature of garnet grains are shown on the (Alm+ Spe)-Gran-Pyr and Alm-Spe-Pyr diagrams (Figure 4). In general, the garnet grains in these sandstones show a variety of chemical composition.

The garnet grains are tentatively classified into the five groups on the basis of their chemical compositions (Figure 4). Group 1 is almandine with low content of pyrope and grandite component. Group 2 is almandine with low content of pyrope and modelate content of grandite. Group 3 is almandine with moderate contnent of pyrope and low content of grandite. Group 4, recognized only in the Nambayama sandstone, is characterized by moderate contents of pyrope and grandite component. Group 5, recognized in the Tamugigawa and Kawadume Formations, is grandite. In terms of content of spessartine, Groups 2, 3, 4, and 5 are characterized by the content less than 10 %, while Group 1 of garnet grains are mostly more than 10 %, and partly less than 10 %. Garnet grains with more than 10 % of spessatine component are called Group 1a, while those with less than 10 % of spessartine component are Group 1b.

SOURCE ROCKS AND PROVENANCE OF TURBIDITE SANDSTONES

The modal composition of the sandstones (Figure 2) suggests that the source rocks were distinctly different between the sandstones of the Middle Miocene Nambayama Formation and ones of latest Miocene to Pliocene Tamugigawa and Kawadume Formations. On the basis of the discriminative diagram of sandstone [11, 16], the sandstones of the Nambayama Formation are assigned to the quartzose recycled orogen to mixed provenances. On the other hand, the sandstones of the Tamugigawa and Kawadume Formations show the dissected to transitional arc provenance. These presumptions are generally coincide with the relative proportion of potash feldspar and plagioclase and the proportion of chert fragment to total rock fragments.

Abundant content of fresh zircon and garnet grains with a small quantity of tourmaline and rutile grains (Figure 3) suggests that acid plutonic and metamorphic rocks consisted a part of the provenance of the Nambayama Formation. Adding the presumption from the modal composition, the provenance of the Nambayama Formation consisted of various kinds of older rocks, that is, older sedimentary, acid plutonic, and metamorphic rocks of orogenic terrane. On the other hand, the heavy mineral composition suggests that the sandstones of the latest Miocene to Pliocene formations were supplied from two types of provenances different in source rocks. One contains a large amount of zircon and garnet grains, and is similar to the provenance of the Nambayama sandstone, though it is accompanied by volcanic rocks, and the supply from the recycled sedimentary rocks was decreased, as suggested from the modal composition. The other provenance contains a huge mass of volcanic rocks, as shown by the content of a lot of hornblende and volcanic rock fragment.

The chemical composition of garnet grains is expected to give more detail information for the source rocks. The distribution area of basement rocks around the southwestern part of the Niigata sedimentary basin, are divided into several tectonic terranes (Figure 5). Furthermore, the Paleogene to Early Miocene volcanic rocks are distributed in the environs. The chemical composition of garnet grains in them are shown in Figure 6 after many reports.

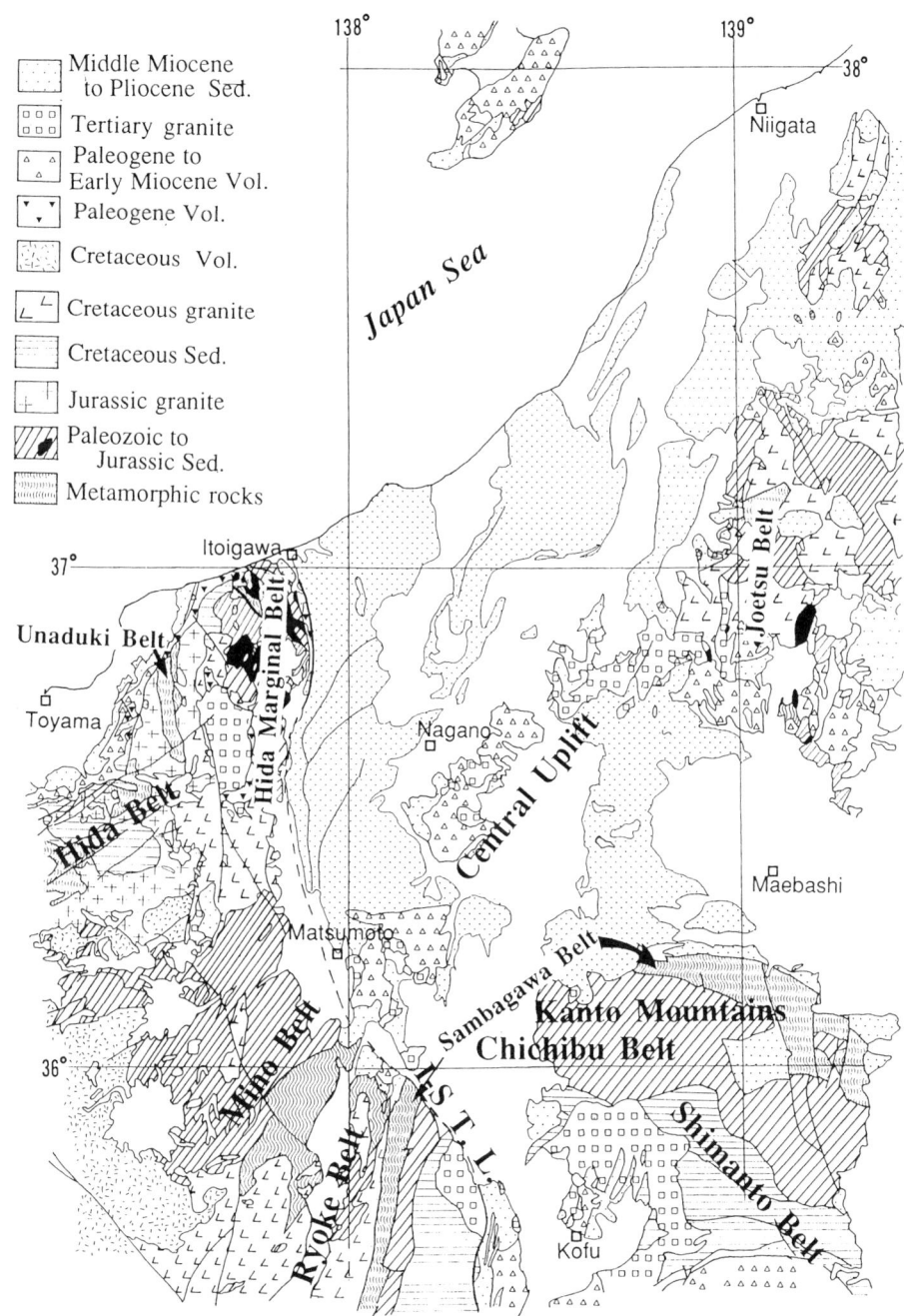

Figure 5. The distribution of the Paleozoic to Mesozoic terranes and Cenozoic plutonic and volcanic rocks around the southwestern part of the Niigata sedimentary basin. I-S T. L. means Itoigawa-Shizuoka Tectonic Line.

The Hida Mountains to the west of the Itoigawa-Shizuoka Tectonic Line consists of the Hida Terrane, the Hida Marginal Belt, the Jurassic sedimentary rocks, and the Cretaceous sedimentary and granitic rocks. The Hida Terrane comprises the Hida Gneiss Zone, the Unazuki Zone [17], and the Trias to Jurassic Granitic rocks called the Funatsu Granite. Garnets in the pelitic schist of the inner Hida Gneiss Zone are almandine with low content of pyrope and spessartine component, while those of the outer Hida Gneiss Zone are almandine with up to 28 % of pyrope and less than 10 % of spessartine component [18, 19, 20, 21]. Almandine in the Unazuki pelitic schist is characterized by more than 10 % of spessartine component [22]. Calcareous gneiss in the Hida Gneiss Zone includes grandite with high Ca component [23]. Garnets in the pelitic and psamitic schists occurring as blocks in melange of the Hida Marginal Zone are almandine with less than 10 % of spessartine and more than 10 % of grandite components [24]. Sandstones in the Mino Terrane include garnets characterized by high pyrope, low spessartine and grandite components [3]. Garnets in Cenozoic acid volcanic rocks are almandine with low spessartine, pyrope, and grandite components [25].

Compared with these chemical features of garnet grains included in the basement rocks and volcanic rocks, each groups of garnet grains of the turbidite sandstones analyzed in this study are inferred to have been supplied from the terranes as follows. The Group 1a of garnet grains, which are almandine characterized by spessartine component more than 10 %, are from the Unazuki Belt, while the garnet grains in Group 1b having spessartine component less than 10 % is from the Hida Gneiss and the Cenozoic acid volcanic rocks. Group 2 is from the Hida Marginal Belt and the Cenozoic acid volcanic rocks, and Group 3 is from the sandstones of the Mino Belt. The supply of detritus from the Mino Belt decreased as the depositional age became younger. Grandite of Group 5 is from the calcareous gneiss of the Hida Gneiss Zone. On the other hand, the garnet grains showing the composition similar to that of Group 4 of garnet grains has not been reported so far from the basement rocks and the volcanic rocks in the surrounding area. Their composition is near that of the garnet included in granulite to eclogite facies of metamorphic rocks [26, 27], and generally implies

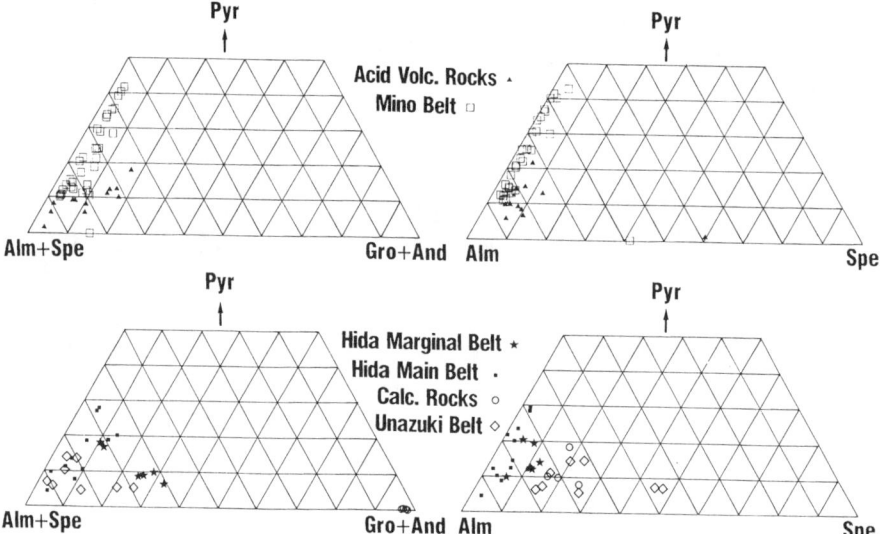

Figure 6. (Alm+Spe)-Grand-Pyr and Alm-Spe-Pyr diagrams of garnet grains in the basement rocks and Cenozoic volcanic rocks. Their data are from the many reports referred in the text.

the high pressure type of metamorphism [28]. High-pressure schists are distributed in the Hida Marginal Belt [29]. Group 4 of garnet grains is probably supplied from this provenance, though the metamorphic rocks including these garnet grains have been eroded out.

The paleocurrents of the turbidite sandstones suggest that their detritus are transported by the south-southwest and east-southeast currents [6, 7, 11]. The southeastern area in which the basement rocks presently crop out, however, is inferred to have submerged at least in Middle Miocene time [30]. Therefore, the main provenance of the Nambayama Formation is reasonably attributed to the western mountains mainly composed of sedimentary and metamorphic rocks with acid plutonic and volcanic rocks. On the other hand, the provenance of the Kawadume and Tamugigawa Formations are complex, because the southeastern mountains began to uplift at late Miocene time [30]. The southeastern mountains, which are composed of sedimentary and metamorphic rocks with the Cretaceous granite, are widely covered by the Early Miocene volcanic rocks, and are intruded by the Tertiary granite (Figure 6). Their sedimentary and metamorphic rocks has been correlated to them of the Mino and Hida Marginal Belt [31]. The calcareous metamorphic rocks including grandite, however, are not found in their metamorphic rocks. The modal composition and the amount of Group 3 of garnet grains suggest that the supply from the older sedimentary rocks such as the Mino Belt was almostly stoped at the stage of the Kawadume and Tamugigawa Formations. Although the detritus of volcanic rocks were abundantly supplied from the southeastern mountains, a part of the sandstones, especialy of the Type A, are considered to have been still supplied from the Hida Terrane and the Hida Marginal Belt.

CONCLUDING REMARKS

The petrograhical analysis of the Neogene turbidte sandstone deposited in the southwestern part of the Niigata sedimentary basin resulted in the following conclusion for the provenance of them.

1) The Middle Miocene turbidite sandstone are mainly supplied from the orogenic terrane with acidic plutonic and volcanic rocks, distributed to the west of the Itoigawa-Shizuoka Tectonic Line.

2) The Late Miocene to Pliocene turbidite sandstones are derived from the Cenozoic volcanic rocks and the basement rocks same as the Middle Miocene ones, but except the Mino Belt. Although two types of sandstones were formed at this stage, the difference of provenance remains unsolved.

REFERENCES

1. H. Blatt, G.V. Middleton and R.C. Murray. *Origin of Sedimentary Rocks*. Prentice Hall, New Jersey (1980).
2. M.E. Tucker. *Sedimentary Petrology*. Blackwell Sci. Pub., London (1991).
3. M. Adachi and M. Kojima. Geology of the Mt. Hikagedaira area, east of Takayama, Gifu Prefecture, central Japan. *J. Earth Sci., Nagoya Univ.* No. 31, 37-67 (1983).
4. A.C. Morton. Heavy minerals in provenance studies. In:*Provenance of Arenites*. G. Zuffa (Ed.), pp. 249-277. Reidel Pub. Co., Dordrecht (1985).
5. A.C. Morton. A new approach to provenance studies: electron microprobe analysis of detrital garnets from Middle Jurrasic sandstones of the northern North Sea. *Sedimentol.* 3 2, 553-566 (1985).
6. O. Takano. Depositional processes of trough-fill turbidites of the upper Miocene to Pliocene Tamugigawa Formation, northern Fosa Magna, central Japan.*J. Geol. Soc. Japan*. 9 6, 1-17 (1990). (in Japanese with English abstract).

7. M. Endo and M. Tateishi. Reconstruction of the Miocene Nambayama submarine fan, northern Fossa Magna, central Japan. *J. Geol. Soc. Japan.* 96, 193-209 (1990). (in Japanese with English abstract).
8. S. Tokuhashi. Sedimentological and mineralogical study on the late Miocene to Pliocene sandstones occurring in the southern area of the Higashiyama Region, Niigata Prefecture. *J. Geol. Soc. Japan.* 98, 355-372 (1992). (in Japanese with English abstract).
9. M. Tateishi, A.A.A. El Habab and M. Shimazu. Source rocks of Miocene- Pliocene turbidite sandstones in the Kubiki area, northern Fossa Magna, Japan. *Mem. Geol. Soc. Japan.* No. 38, 181-190 (1992). (in Japanese with English abstract).
10. M. Shimazu, S. Yoon and M. Tateishi. Tectonics and volcanism in the Sado-Pohang Belt from 20 to 14 Ma and opening of the Yamato Basin of the Japan Sea. *Tectonophys.* 181, 321-330 (1990).
11. M. Endo and M. Tateishi. Upper Neogene in the Nishi-Kubiki area, Northernmost part of the Fossa Magna - with reference to the sedimentary environmnet of the Tsunako Conglomerate -. *Contri. Dept. Geol. Mineral., Niigata Univ.* No. 5, 33-48 (1985). (in Japanese with English abstract).
12. W.R. Dickinson. Interpreting detrital modes of graywacke and arkose. *J. Sed. Petrol.* 40, 695-707 (1970).
13. G.G. Zuffa. Optical analysis of arenites: influence of methodlogy on compositional results. In: *Provenance of arenites.* G.G. Zuffa (Ed.). pp. 165-189. Reidel Pub. Co., Dordrecht (1985).
14. H. Okada. Cllasification and nomenclature of sandstones. *J. Geol. Soc. Japan.* 74, 371-384 (1968). (in Japanese with English abstract).
15. G.S. Simpson. Evidence of overgrowths on, and solution of, detrital garnets. *J. Sed. Petrol.* 46, 689-693 (1976).
16. W.R. Dickinson, L.S. Beard, G.R. Brakenridge, J.L. Erjavec, R.C. Ferguson, K.F. Inman, R.A. Knepp, F.A. Kindberg and P.T. Ryberg. Provenance of North American Phanerozoic sandstones in relation to tectonic setting. *Geol. Soc. Amer. Bull.* 94, 222-235 (1983).
17. M. Suzuki, S. Nakazawa and T. Osakabe. Tectonic development of the Hida Belt - with special reference to its metamorphic history and late Caboniferous to Triassic orogenies-. *Mem. Geol. Soc.. Japan.* No. 33, 1-10 (1989). (in Japanese with English abstract).
18. M. Suzuki. An occurrence of "eclogitic rock" in the Hida metamorphic belt. *J. Japan. Assoc. Mineral. Petrol. Econ. Geol.* 68, 371-382 (1973). (in Japanese with English abstract).
19. M. Suzuki. Granulite-facies rocks from the Odori-river area of the western province of the Hida metamorphic belt. *The basement of Japanese Islands.* Memorial papers of Prof. H. Kano. 71-84 (1979). (in Japanese).
20. M. Asami. Pelitic metamorphic rocks from the Arashima-dake area, Toga area and Wada-gawa area of the Hida metamorphic belt. *The basement of Japanese Islands.* Memorial papers of Prof. H. Kano. 41-49 (1979). (in Japanese).
21. K. Suwa. Biotite schist and leptite from the Kitamata-dani and Kasadani of the upper Katakai-gawa area, Toyama Prefecture. *The basement of Japanese Islands.* Memorial papers of Prof. H. Kano. 15-27 (1979). (in Japanese).
22. Y. Hiroi. Progressive metamorphism of the Unazuki pelitic schists in the Hida Terrane, central Japan. *Contrib. Mineral. Petrol.* 82, 334-350 (1983).
23. A. Okui. Polymetamorphism in the Hida metamorphic rocks of upper Katakai river area, Toyama Prefecture, central Japan, with special reference to the effect of intrusion of the Funatsu granitic rocks. *J. Japan. Assoc. Min. Petr. Econ. Geol.* 80, 382-397 (1985). (in Japanese with English abstract).
24. M. Nakamizu, M. Okada. T. Yamazaki. and M. Komatsu. Metamorphic rocks in the Omi-Renge serpentine melange, Hida Marginal Tectonic Belt, central Japan. *Mem. Geol. Soc. Japan.*, No. 33, 21-35 (1989). (in Japanese with English abstract).
25. M. Shimazu. Trace elements of garnets in volcanic rocks. Abstract of papers presented at the Fall Joint Meeting of Japan. *Assoc. Mineral. Petrol. Econ. Geol., Volcanol. Soc., and Mining Geol. Japan.* A-9. (1988). (in Japanese).
26. R.G. Coleman, D.E. Lee, L.B. Beatty and W.W. Brannock. Eclogites and eclogites: the differences and similarities. *Geol. Soc. Amer. Bull.* 75, 483-508 (1965).
27. Y. Karakida, A. Miyachi, H. Yamamoto and T. Oshima. The Kurosegawa Tectonic Zone in Kyushu, with special reference to garnet amphibolites of Tsubokinohana in the Yatsusiro district. *The basement of Japanese Islands.* Memorial papers of Prof. H. Kano. 369-396 (1979) (in Japanese with English abstract).
28. A. Miyashiro. *Metamorphism and metamorphic belts.* George Allen and Unwin, London (1973).
29. S. Banno and T. Nakajima. Metamorphic belts in Japan. *Episodes*, 14, 280-285. 30. I. Kobayashi and M. Tateishi. Neogene stratigraphy and paleogeography in the Niigata region, central Japan. *Mem. Geol. Soc. Japan*, No.37, 53-70 (1992). (in Japanese with English abstract).
31. M. Komatsu, M. Ujihara and K. Chihara. Pre-Tertiary basement structure in the Inner Zone of Hoshu and the north Fossa Magna region. *Contrib. Dept. Geol. Mineral., Niigata Univ.,* No. 5, 133 -148 (1985). (in Japanese with English abstract).

(Manuscript received 6 March 1993; accepted 13 April 1993.)

EVAPORITE AND DESERT ENVIRONMENT

Editors
Y. Watanabe and A. Motamed

CONTENTS

Preface	199
Geology and climate evolution of Central Iran *A. Motamed*	201
Cluster and factor analysis of geochemical data of Najmah Formation, Upper Jurassic, Western Desert, Iraq *A.K. Jamil and M.A. Al-Hilaly*	221
Shortite formation in Turkey and its geochemical properties *F. Suner*	237
Mineral assemblages and formation of the Kestelek and Sultançayiri borate deposits *C. Helvaci*	245
Hydrogeochemical controls on the formation of primary dolomite in some ephemeral lakes in the Coorong region of South Australia *R. Ahmad and P.B. Hostetler*	265
Biomineralization of the mirabilite deposits by the exemplification of the Barkol Lake *W. Dongyan, L. Zhenmin, D. Xiaoling, and X. Shaokang*	283
Halogenic basins: facial and paleotectonic models *G.A. Belenitskaya*	289
An occurrence of primary inyoite at Lagunita Playa, Northern Argentina *C. Helvaci and R.N. Alonso*	299

Preface

The Desert Environment represents a waste area of our planet, about 1/3 of total surface of the globe; equals to 25 million square kilometers. The desertification is increasing, specially in third world and developing countries. Many organizations, domestic or international, contribute to battle the desertification.

These organizations have the comprehensive biological concept by "diminution of biological potential of land" or "reduction of biomass" and "destruction of ecological system". This concept covers the sail rasalimty, erosion and deforestation in which the "anthropological effect" is the main responsible leading ultimately to desert condition; specially where the human pressure represents a major factor in desertification by more intensive use of agriculture resources.

But the geological concept of desertification is somewhat different; the geologists behave that the desertification is essentially due to geodynamic process which produces some conditions for "evaporites" such as halite, gypsum, phosphate, or deliquescent salts as "sedimentary ore deposit".

Each actual desert environment has his history in the past, present and future by evolving system of desertification factors like important ancient geological desert and evaporitic environments in Precambrian, Triassic, and Eocene systems or south central Asian Neogene system. The geologic conditions are provided by 1) tectonic movement with increasing sea level which spreads the salt water in decreased depth area being subject to evaporation and erosion, 2) volcanism and lithologic conditions in which alkaline rocks or sediments provide saline materials, 3) topographic condition such as depressed zone, back shelf or flood plain and playa, 4) ground water near the surface with high salinity and "efflorescence" salt by capillarity, and 5) climatic situation in Norse latitude which is changing periodically, or even year to year.

These factors make it possible to consider the different types of "desert environment" as very dry, dry (arid) and semi-arid zones, in which the anthropologic effect has only a secondary importance The region itself, off course, has a potential for "desertification" equivalent of "rheexestasie" period in geologic history, in contrast with "biostasy" periods.

The aim of this treaty is to summarize briefly the relationship between these factors and the dynamics of each region, in each period, for its physical and diagenetic process and the classification of these regions, submitted to these geological and mineralogical factors of desertification.

The desertification is present in two important ways. Firstly given as a mechanical one by surface to surface denudation, sand transport and accumulation, yarding and deflation surface form (landscape). In the second one, the environmental geology in desert area can present an inventory of natural resources and the technology for gainful utilization of "desert resource" by transfer of the scientific and technological knowledge from laboratory to the field for the benefit of the community and founding a "desert industry".

We hope that some improvement should be possible by combined geological and biological concepts to 1) save the decreasing life in some desert areas by treatment in soil, water, plant, animals, etc. and monitoring the desert factor. 2) understanding the dynamics of the landscape and determine its tendency to change as a result of different factors of geology. 3) to drive the history evolution of each region and varying climatic and topographic conditions, specially in Quaternary for understanding its "susceptibility" for desertification. 4) elaborate the plan action to combat the desertification.

This publication consists of papers presented at the Evaporite and Desert Environment Symposium at the 29th International Geological Congress held in Kyoto on 26 August 1992. We are very gratefull to the Organizing Committee for having our session possible and fruitful.

A. Motamed
Professor at the Department of Geology
The Tehran University
Tehran 11
IRAN

Y. Watanabe
Fuel Resources Departmant
Geological Survey of Japan
Tsukuba 305
JAPAN

Geology and climate evolution of Central Iran

A. MOTAMED
Professor of Tehran University, Department of Geology. Faculty of Sciences. Tehran 11, IRAN

Abstract-- The increasing desertification of Central Iran is a joint effect of geology and climatic change in Late Quaternary. Peripheric highs, mountains, central volcanic and salty plugs isolated like lakes depressions. Intense erosion accumulated detrital and evaporitic materials as "Glacis " in piedmont, and in form of annual cyclic sedimentation in playa, with some diagenetic minerals the climatic change from late Holocene increased the aridity.
    The evolution of desertification is present by changing of ancient brackish lake to salty playa.

Key words--Desertification, Climatic change, Diagenesis, Subsidense, Quaternary, Biostasy, Rheexistasy.

INTRODUCTION

The central Iranian plateau is bounded by three young mountain ranges: Elborz in the north, Zagros in the southwest, and Eastern mountain ranges in the east. The mountains with summits about 3000m high isolate an endoreic and internal basin which is dissected by some volcanic, sedimentary and salty ranges, providing some isolate depression zones in North, East and Southern parts, evolving to arid playas; such as Houze Sultan, Daryaye-Namak, Dasht-e-Kevir, Lut, Saghand, Gavkhuni Damghan, *etc.* (Fig. 1).
    This study concerns with the origin of these central deserts and the present sedimentary process to establishing the effect of the climatic and geological evolution. The data are provided from aerial photos, satellite images, field works and laboratory analyses.

GEOGRAPHIC SITUATION

The Central Iran lies between 26 to 40°N and 50 to 58°E. It is surrounded by Alborz, Zagros West, and Eastern ranges mountains. The volcanic and sedimentary ranges and salt domes divide the region into some individual depression, mostly elongated in NW-SE or W-E directions.

GEOMORPHOLOGY

The morphological units in each one of the depressed basins are as follow: high elevation mountain(s), foothills, alluvial fans, bajada and finally accumulated flat zones, playa or kevir (Fig. 2).
    The high relief, generally Eocene sedimentary or volcanic rocks or post-Eocene salty domes, overlook the depressions by thrust or gravity faults in the form of horst or graben systems which are subject to intensive mechanical erosion. The foothills, usually post-Eocene or Neogene salty sediments or Pleistocene conglomerates, are generally folded or faulted and

are subject also to severe denudation (Fig. 3). The piedmonts and alluvial fans generally grouped together as "glacis form"; are drained by new network system, which at least 3 or 4 terraces that were recorded on the alluvial plain or bajada, due to the oscillation of base-level through Quaternary period (Fig. 4).

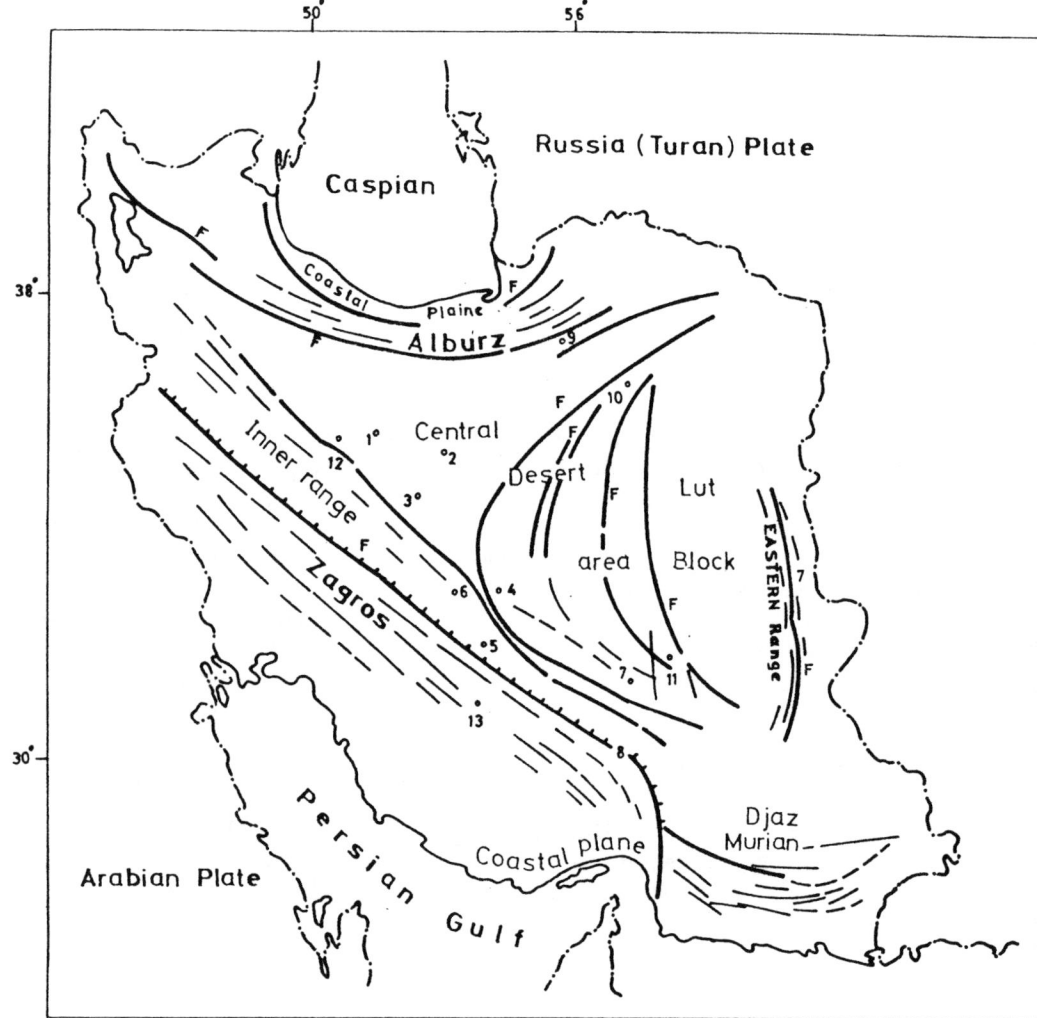

Fig. 1. Geographical situation and structure map of central Iran.
Principal depression units of Iranian Plateau are 1: Houze-Soltan, 2: Daryaye-Namak, 3: Ardestan desert, 4: Ku-siahku of Yazd, 5: Gaw khoni, 6: Abar Ghow, 7: Shar-e-Badak, 8: Sirjan desert, 9: Molla Ali, 10: Dasht-e-Kavir, 11: Shahaad-Lut, 12: Turgool Arak, and 13: Bakhtegan.

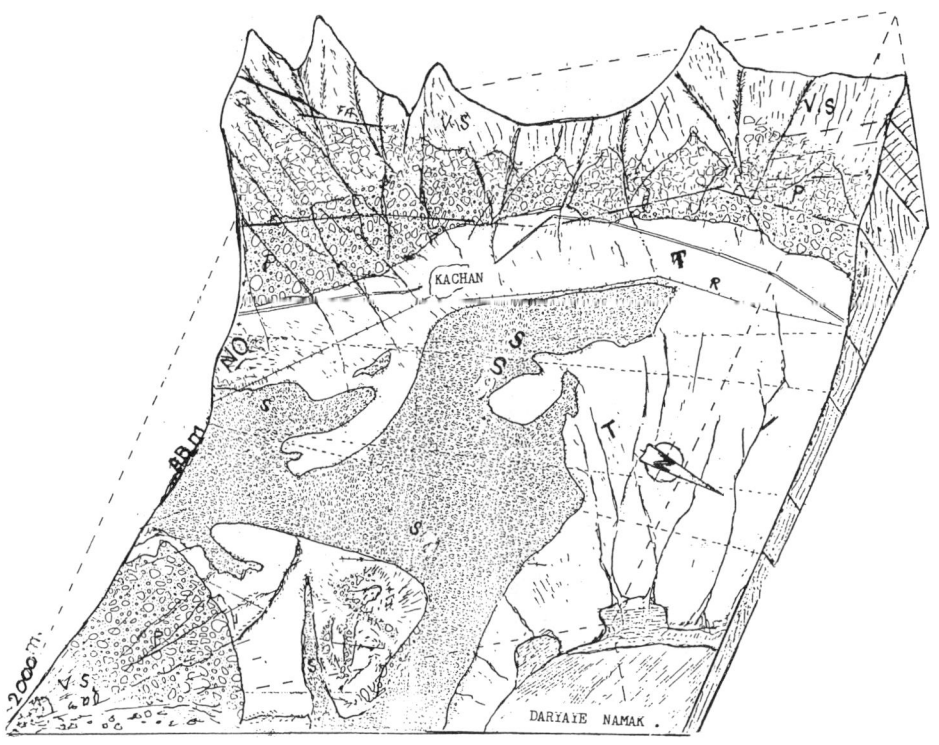

Fig. 2. Block diagram of Kachan Mountain and Daryay Namak playa.
S: Sand dune (f: fixed), R: Railway, T: Alluvial plain, ABm: Alluvial plain (covered by recent sand dune), P: Pediment sediment, NO: Neogene Sediment (Navab), V.S: Volcano-sedimentary, high Eocene land.

Fig. 3. Volcanic and sedimentary rocks (Eocene - in back).

Fig. 4. Piedmont increased by stream in "terrace".

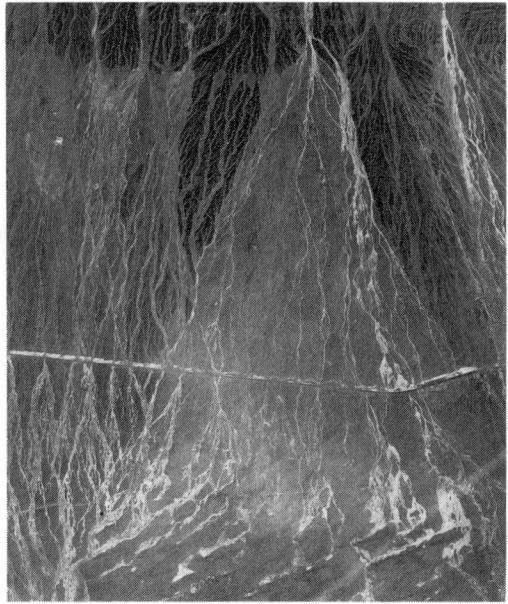

Fig. 5. A general dereal view of Alluvial plain in central Iranian Desert, with saline "efflorescence" [17].

All depression zones of central Iran are generally depositional areas. The surrounding mountains provide the detrital materials and soluble salts. The aerial photos, in eastern part, present a total erosion of a Neogene marly sandy anticlinal structure with 5 km length and 2 Km height [8]. The alluvial fan and piedomont areas with coarse materials present fining up layers with cut and fill sedimentary structure and winnowing process. The straight channel systems in mountains become braided or meandering with ephemeral flow, not too far from the fault line scarps and incise their own alluvial debris. The alluvial plains with terrestrial system show a very high density narrow drainage pattern without deep thalweg in which the saline "efflorescence" present a whitish dendritic divide in aerial photos (Fig. 5). The end member of this depressed area is generally flat zones, covered by clay or salt accumulations in dry seasons with poffy zone in marginal parts (Fig. 6).

The water table is deep in foothills and alluvial fans but can reach the surface in playas zone. The ground water is carbonaceous in piedmonts, but gypsiferous and saline in alluvial plains and playas.

One of the most important recent geomorphological features of Iranian Deserts is windy denudation, sand accumulation, and salt depositions. Windy structures in alluvial

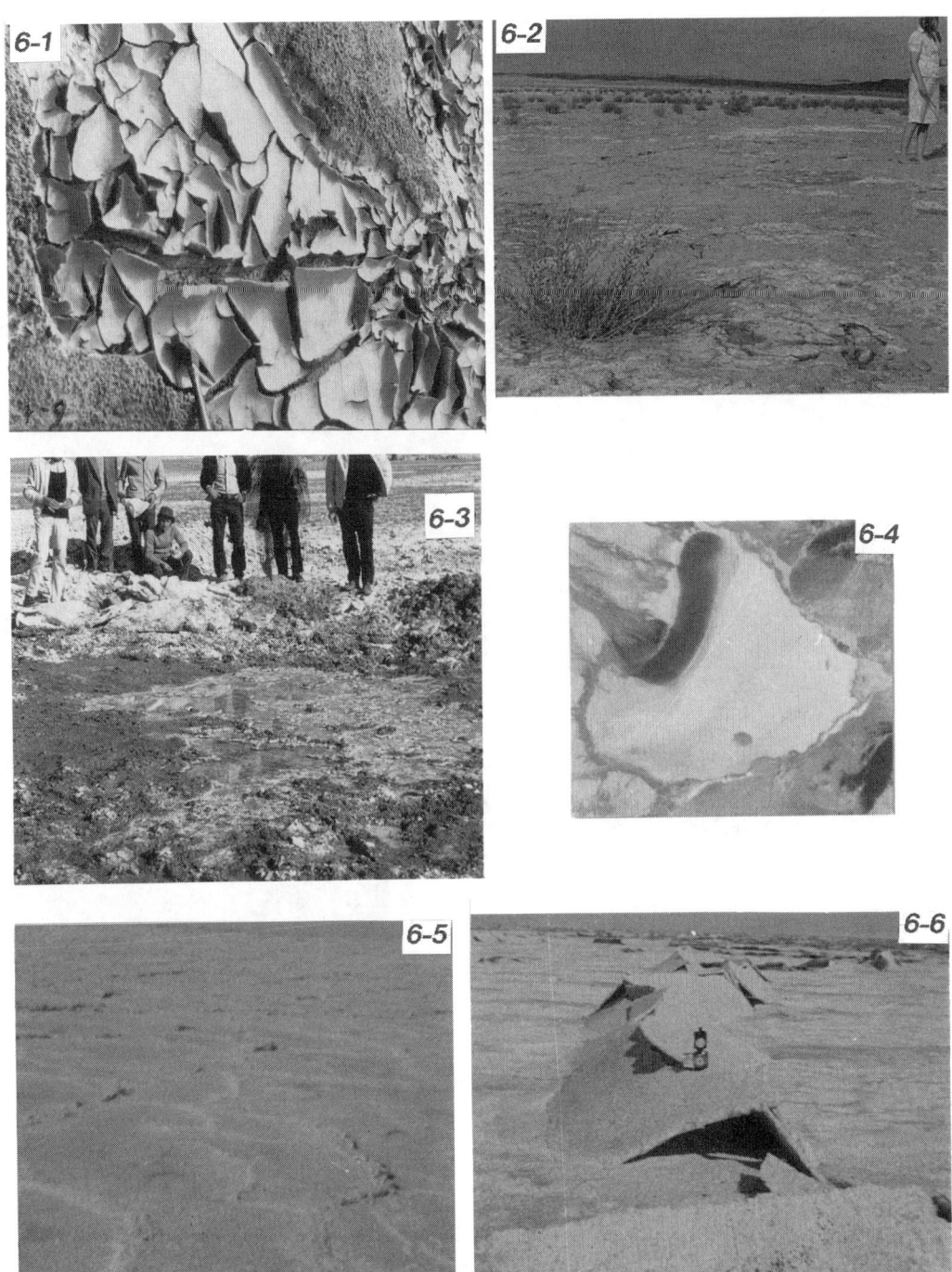

Fig. 6. 1: Alluvial salty-clayey plain. 2: Marginal puffy zone, 3: Central salty crust of playa with saline water, 4: Satellite imagery of Daryaye-Namk playa with saline water in N and W due to tilting in western part, 5: Polygon in halite crust, and 6: Uplifting of salt crust.

Fig. 7. Sand dunes. 1: Ancient sand "dune": Late Pleistocene, 2: Recent "multched" sand dune, 3: Individual batkhans, 4: New sand dunes covering the recent sediment (B), and 5: Wind deflation as Yardang - (Kaluto).

plains are: desert, sand moving and sand accumulation. Sand dunes are settled in low and middle terraces or ancient dried playas surface or in ancient alluvial valleys. The sand duned as 250m high in Lut to 50m high in Kachan, and the individual moved batkhans dunes are increasing recently (Fig. 7). The salt deposition in depression generally formed primary by dissolved materials, running from surrounding detrital and evaporative marly layers, or secondarily by capillarity effect of ground water as gypsum or saline efflorescent or salt crust at or near the surface (Figs. 5 and 6).

CLIMATE

The climate of Central Iran is classified as arid and semiarid; average rainfall is less than 150 mm per year which is very irregular during a year (Fig. 8). The temperature is high, with a mean value of about 25°; reaching the maximum 45° in August, and the minimum is -10° in January.

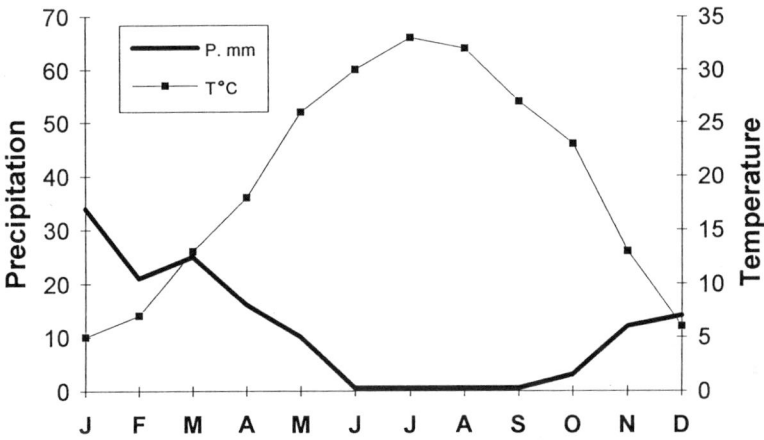

Fig. 8. Schematic climatogram of Iranian Central Desert.

The relative humidity is low and the intention of evaporating and sun radiating is high in summer. The weather is usually windy, with lack or light density of vegetation and very low biomass production is important in all over the region.

The E/p ratio of De Marton (evaporation/precipitation) over the area is less than 5 and the aridity index is less than 6. The temperature decreases with relief; in high periphery mountains there are the remains of glacial valleys.

Thermoclastic errosion in both volcanic and sedimentary rocks. The detrital materials are carried with steady rainfall, the fine particles are moved by wind. The wind blows in many directions but the dominant direction generally is through the graben valleys, SE-NW, like the important sand accumulations from Kachan to Yazd depression or, NW-SE in WLW and SN-NE in East Lut. The mean speed of wind is about 10 to 50 km/h, or 22-25 "Not"; some one can reach even 80 to 100 km/h.

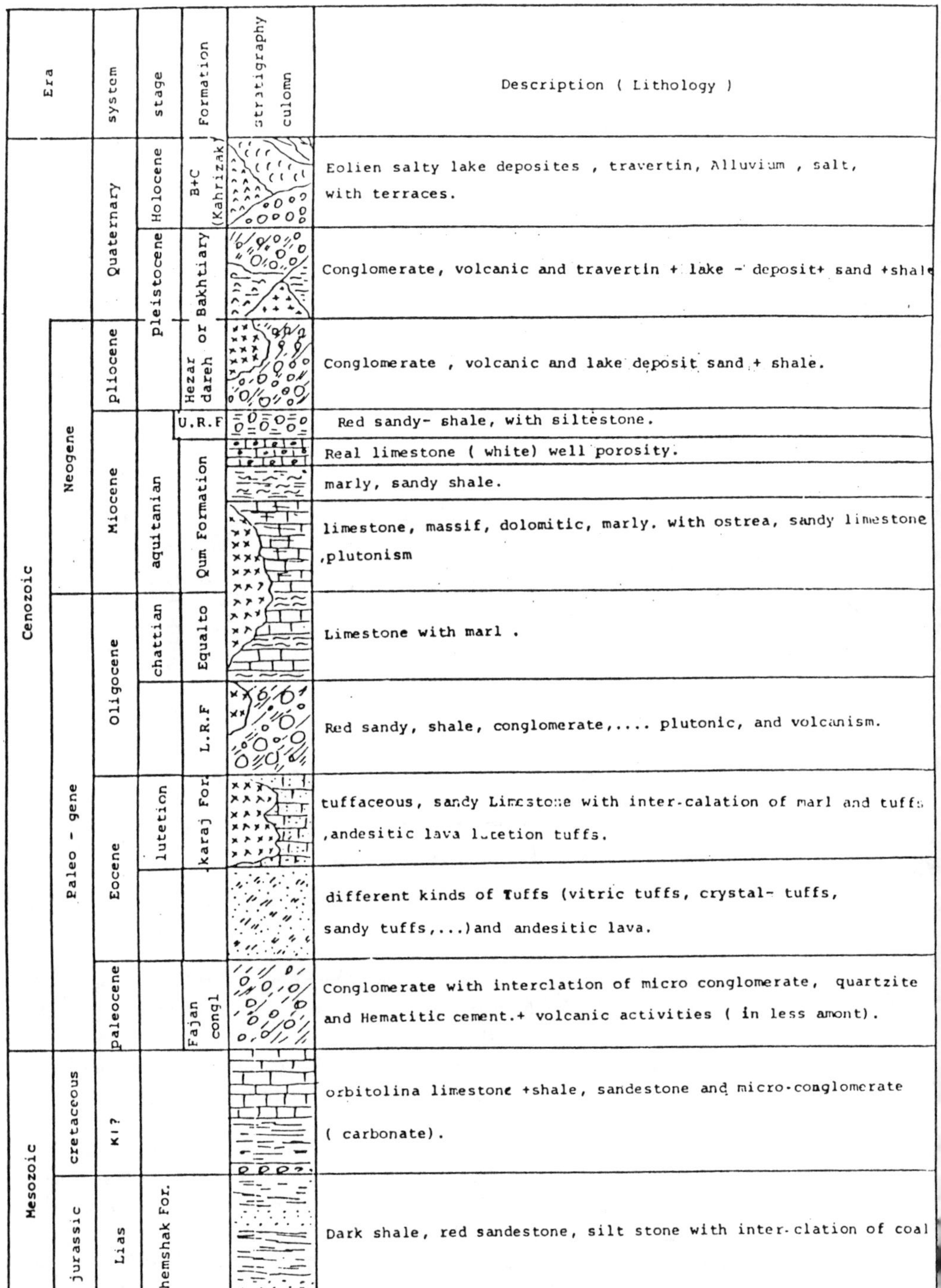

Era	system	stage	Formation	stratigraphy column	Description (Lithology)
Cenozoic	Quaternary	Holocene	B+C (Kahrizak)		Eolien salty lake deposites, travertin, Alluvium, salt, with terraces.
Cenozoic	Quaternary	Pleistocene	Bakhtiary		Conglomerate, volcanic and travertin + lake - deposit + sand + shale
Cenozoic	Neogene	pliocene	Hezar dareh or Bakhtiary		Conglomerate, volcanic and lake deposit sand + shale.
Cenozoic	Neogene	Miocene	U.R.F		Red sandy- shale, with siltestone.
Cenozoic	Neogene	Miocene	Qum Formation		Real limestone (white) well porosity.
Cenozoic	Neogene	Miocene aquitanian	Qum Formation		marly, sandy shale.
Cenozoic	Neogene	Miocene aquitanian	Qum Formation		limestone, massif, dolomitic, marly. with ostrea, sandy limestone ,plutonism
Cenozoic	Paleo-gene	Oligocene chattian	Equalto		Limestone with marl.
Cenozoic	Paleo-gene	Oligocene	L.R.F		Red sandy, shale, conglomerate,.... plutonic, and volcanism.
Cenozoic	Paleo-gene	Eocene lutetion	karaj For.		tuffaceous, sandy Limestone with inter-calation of marl and tuffs ,andesitic lava lutetion tuffs.
Cenozoic	Paleo-gene	Eocene	karaj For.		different kinds of Tuffs (vitric tuffs, crystal- tuffs, sandy tuffs,...) and andesitic lava.
Cenozoic	Paleo-gene	paleocene	Fajan congl		Conglomerate with interclation of micro conglomerate, quartzite and Hematitic cement.+ volcanic activities (in less amont).
Mesozoic	cretaceous	K I ?			orbitolina limestone +shale, sandstone and micro-conglomerate (carbonate).
Mesozoic	jurassic	Lias	shemshak For.		Dark shale, red sandestone, silt stone with inter-clation of coal

Fig. 9. Stratigraphy of central Iran.

The new urban developments in marginal parts of desert, create a problem of water management; the traditional water exploitation "Kanta" is subject to severe decrease of discharge.

The grazing affect of some desert domestic animals such as camel and goat is also important, but it seems that the increase of the aridity and salinity clue to geological and climatological processes is the origin for breaking the biostasy equilibrium and caused the rheexistasy phenomenon in this region. However, some desert and salty plants persist in Central Iran, such as: Halaxylon, Atriplex, Zygophyllum, Acacia, Calligonum, Salsola Zyphus, Tamarix, Nitraria, Artemesia, Aristida, Agrophyllum, Stripilgrostis, Coudinia, Eucaliptus, Salix, Aristida, Agreophyllum.

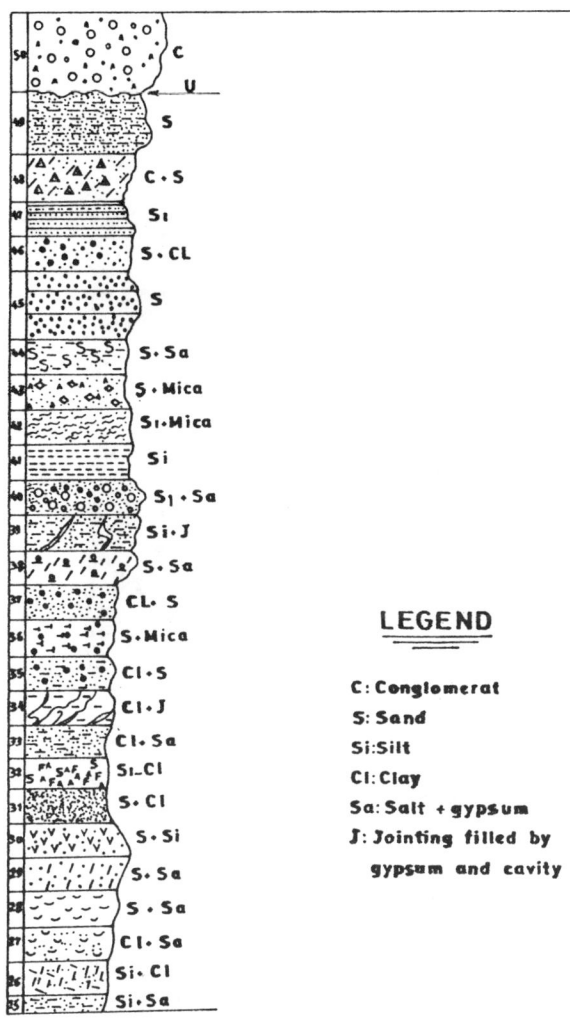

Fig. 10. Geological section of the Neogene sediments in central Iran..

GEOLOGY

Stratigraphy of Central Iran

The region present a vast spectrum of rocks from early Paleozoic to Quaternary. The Paleozoic rocks generally detrital, with occasional ore deposits, can be seen in central or southern part (Yadz), but it is more metamorphic in eastern part.

The Mesozoic rocks are more sandy to limy or silty with coal deposit (Jurassic) and are abundant in south and eastern part.

The development of early Eocene rocks, mostly volcanic agglomerate with tuff or some Lutetien Nummulitic limestones is very important in northern part (Fig. 10).

The post Eocene sediments generally deposited in shallow eater, as limestone, sandstone, (Chattiall) or in lagoonal environments (Upper Miocene - early Pliocene) mainly consist of marly (Figs. 11 and 12), sandy or evaporative rocks with new volcanism which are present over the area: in the north the volcanic episodes decrease on Quaternary; detrital and evaporative facies are abundant in piedmonts and playa.

Fig. 11. 1: A faulting in the Neogene sediments, **2:** Spectacular erosion in the Neogene (see the stratigraphic column).

Structure of Central Iran

The Iranian plateau is, lying between Arabian and Russian plates; 3 major marginal orogenic trends limiting the Central Iran, which is dissected into multiple depression zones by secondary mountains ranges, mostly volcanic or metamorphic and sedimentary (Fig. 3). The high land presents a large success fold system like Jurassic, but more faulted and thrusted, with horst and graben form and many important fault lines, even in depressed zone (Doroneh, Dehshir, etc.). The graben systems form parallel generally depressions in which the elongated playas develop.

Fig. 12. A salt diapir in Central Iran (Gom Area).

The diapirism is very important in most area specially ill northern areas where some 50 post-Eocene salt plugs are grouped in broken or conic domes. The post-orogenic erosion is severe and form a thick coarse sediments deposition in piedmonts; and more fine evaporative deposition in depressions.

Geological evolution; origin of desertification

The geological evolution of central Iran, from shallow water of post Eocene to recent environment to be presented as follows (Fig. 11).

In Eocene, the Central Iran was occupied by sea, in which submarine volcanism was active, depositing tiff and limestone.

In Oligocene, by mid-Alpine orogeny, lagoonal environment developed and the detrital and evaporative depositions progressed, so-called Lower Red Formation (L.R.F.).

In early Miocene some limited inner parts became deeper, depositing limestone, marl and gypsum as Qum formation (Q.F.).

In late Pliocene a new lagoonal facies established and new red marl and sandstone facies is deposited called Upper Red Formation (U.R.F.) with volcanism (Fig. 3)

In Pliocene and early Pleistocene, orogenic movements and diapirism, folding, thrusting and faulting created a new depresses areas. The region was subject to intensive erosion, low rainfall, acute solar energy and salt accumulation in inner zones with increasing capillarity, the secondary salinity, provided more desert area with very reduced biomass production; there is the evidence of climatic change, the aridity increase (Fig. 6).

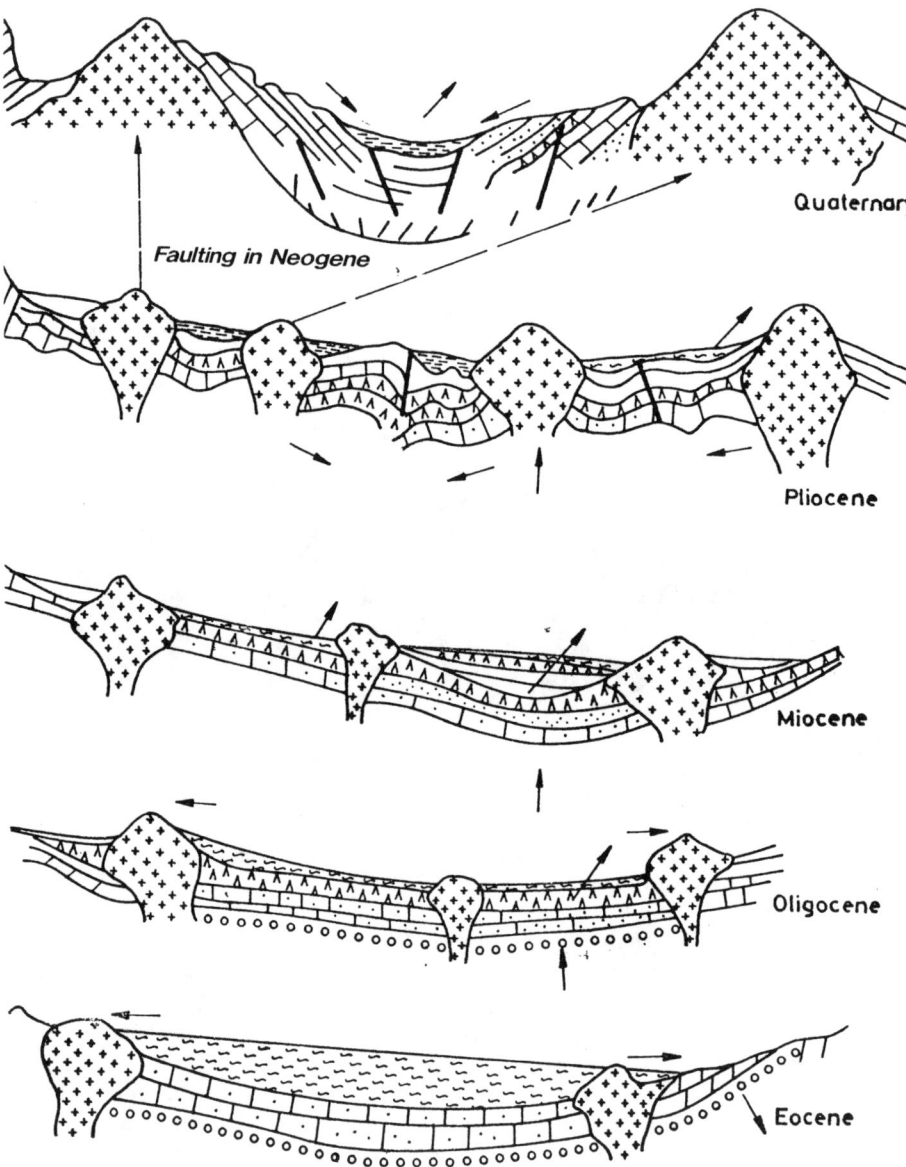

Fig. 13. Evolution of Central Iran Desert from Eocene to Quaternary.

Sedimentation and diagenesis

The fine and soluble materials are carried to alluvial plains, in which the ground waters is near the surface. The wind erosion can remove these fine grains when dried. Sedimentation in the playa systems shows clear changes both horizontally and vertically (Fig. 6).

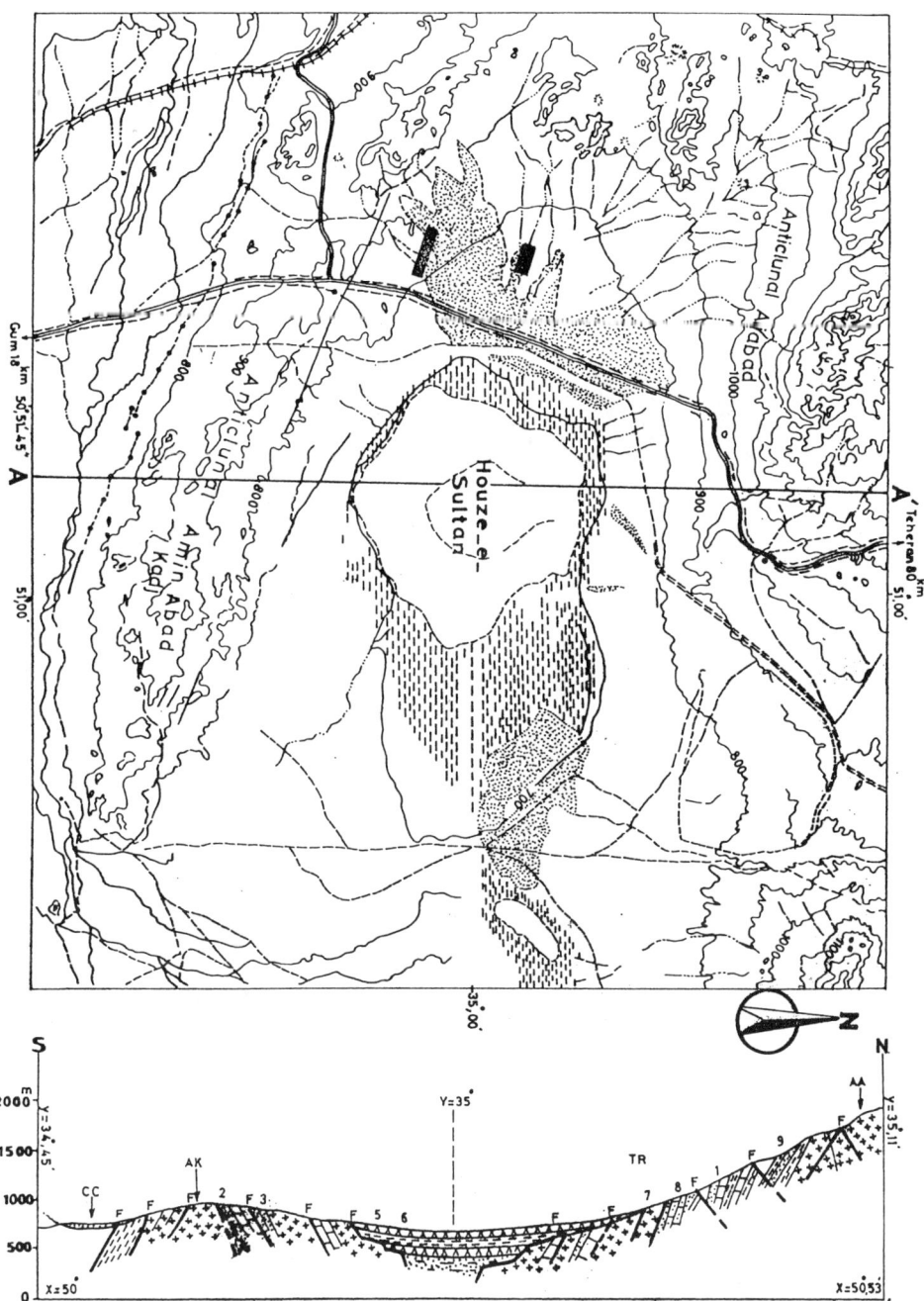

Fig. 14. Geology and topographic section of Houze Soltan Playa and its boundary mountains.
CC: Charachai River, AK: Amin Abad-Kadj Anticline, AA: Ali Abad Anticline, L.F.: Lower Red Formation, G: Ghom, F: Fault, 1&10: Andesite (Eocene), 2: Scutella Limestone (Ghom Fm. a), 3: Marl (G. Fm. b), 4: Reef Limestone (G. Fm. f), 5-7: Playa, 7: Proluvial Bahada, 8: Eocene (Sandy tuff), and 9: Eocene (Limestone with Numulites).

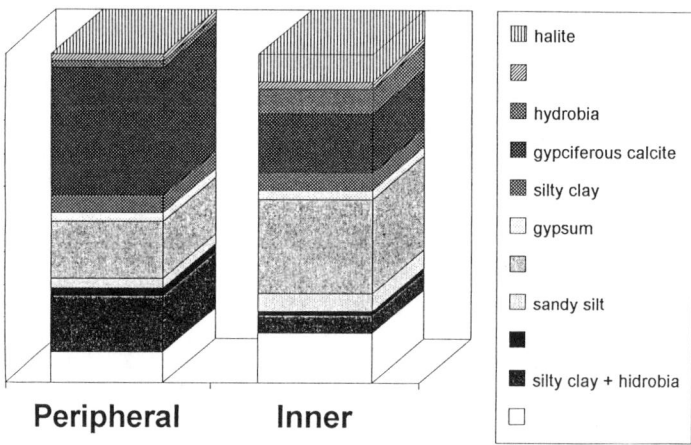

Fig. 15. Comparison and correlation of the deposits between peripheral and inner zone of Houze soultan.

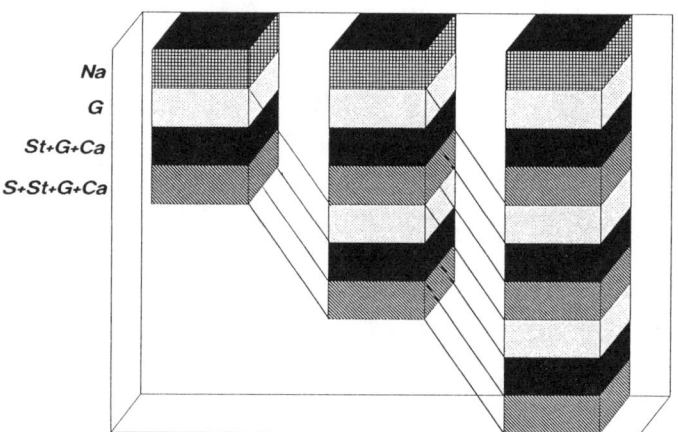

Fig. 16. Diagram showing annual sequence of sedimentation in Houze sultan playa with three annual successive cycles. Na: halite, g: gypsum, Ca: calcium carbonate, St: silt and clay, and S: sand.

a) Horizontally: the peripheral zone of playa is calcareous marl. It is immediately followed by gypsiferous clay toward the center, in which the gypsum efflorescence in fibrous form; bending in the wind direction. Halite crust is located in the central part and a narrow wet zone separates this halite zone from the margin. The thickness of the halite crust reaches to 10 to 40 cm in different playas and in different distances of marginal which is extracted mechanically where the thickness is enough to be exploited. By breaking this crust, one can visualize the recrystallization of NaCl over the saturated water table. Some playas are covered by sodium sulfate (South of Tehran).

b) Vertically: Some wells drilled in playas (Hauze - sultan, Daryay-e-Namak, Sirdjan, etc.) present a cyclic sedimentation with sand, marl and halite sequence. In wet seasons, the fine sand grains, the suspended materials and soluble salts are transported to playas by temporary dried rivers. The sand is deposited first (February-March), then the suspended materials generally with carbonate is decanted as marl (March-April). At the end of this wet period, the brackish water accumulated in the playa becomes clear and the evaporation makes it more saturated, in which Hydrobia (mollusca) can survive and in stagnant less saturated water, green algae can develop (Fig. 15).

At the beginning of dry season, the intensive evaporation at surface make more concentrated water and the gypsum is deposited at the first step, filling also the pores of marl, previously deposited. In warmer temperatures, the supersaturated water redeposit halite whose crust becomes compact and thicker by capillarity effect. In drier months (July to September), polygons are formed, then the edges grow, break and bend toward the interior. The gypsum present very quick and early diagenesis following to evaporation, the pore space

Fig. 17. Different forms of authigenic gypsum in Houze Soultan playa. 1: Saccheroidal, 2: Lens form, 3: Monoclinical form, 4: Twinning (Macle) lanceole, and 5: Halite.

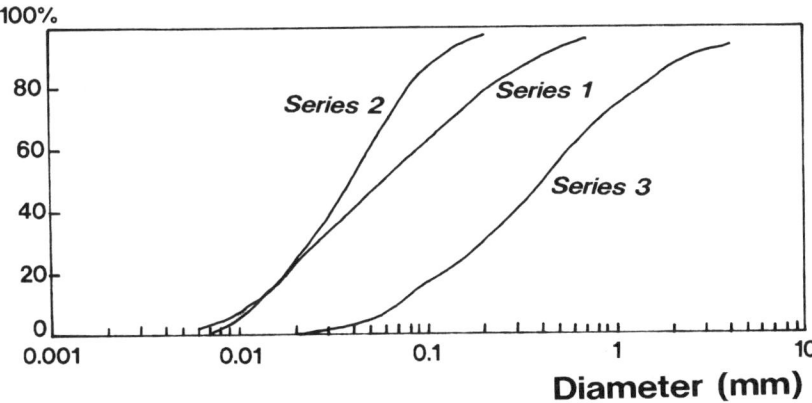

Fig. 18. Cumulative curve of detrital sediment.

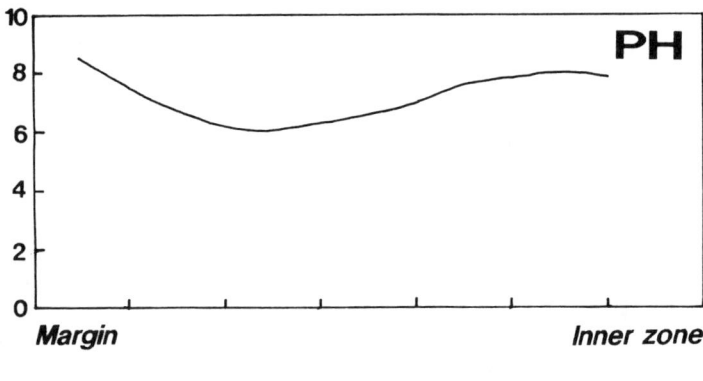
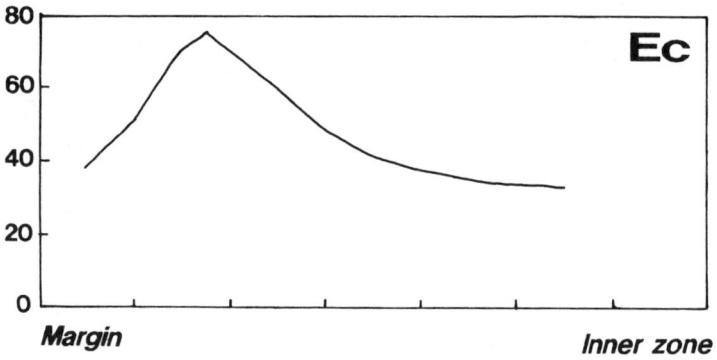
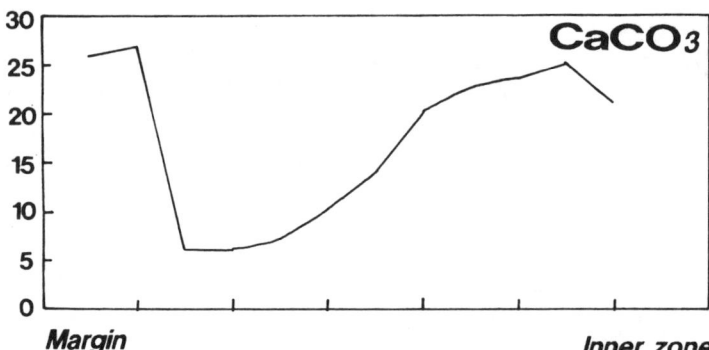

Fig. 19. Diagrams showing the change in pH, CaCO3, and Ec in playa.

and the low value of NaCl. In the marly layer the macro-monoclinical crystals of gypsum are abundant. The compaction of marl reduce the size, which is generally twiny (Macle) or Lallceolate.

In more saturated water, the saccharoidal form of gypsum is produced. It seems that the gypsum is not present in "supersaturated water". Above 240 g/l concentration, halite is deposited in the last step and become thick and crusty by evaporation on the surface. It should be mentioned that in northern playa, potassium is not present and Ca+Mg/Na ratio is less than 10 and the SO_4^{2+}/Cl is about 1/4, except in sulfuric playas. This seasonal or periodic sequence is truncated in the new wet seasons. New runoff less saturated redissolve the previous salt crust, and the sedimentary sequence is renewed on the previous saccharoidal gypsum or marl containing new sandy layer, marl with Hydrobia, gypsum and new salt crust, that will be thicker than previous period.

However some salt crusts can persist and remain untouched, when 1) this crust becomes so thick that the wet season is not able to dissolve it totally; 2) the climate changes into more arid condition, in which the water supply is not enough to remove the salt crust; or 3) the influx of detrital materials is very important as their deposition with high amount can save the salt crust.

It is worth to mention that this recent evaporative truncated cycles are comparable with Oligocene-Miocene facies in Central Iran containing sandy-gypsiferous marl and gypsum layers. Halite is irregularly presented in some thick layers after 50 or 100 truncated cycles; e.g., halite layer in 110 m depth of Daryaye Namak (Report of N.I.O.C.) and in the depth of 2 to 8 m in Hauze-Sultan (Motamed, geophysical exploration of House-Soultan).

MINERALOGY OF INSOLUBLE MATERIALS IN IRANIAN PLAYAS

Detrital materials in these depressions consist of quartz, feldspars, mica, and few heavy minerals (tourmaline, pyroxene, etc.) and also clay minerals. The amount of the clay minerals reaches 20 to 30% of silty, marly layers. They are essentially constituted from illite, smectite and rarely kaolinite. They show that the degradation of igneous minerals dose not produce these undeveloped precursor minerals; then they are not neoformed. They have been transported as "heridity " minerals to playa systems in which the "neoformation " seems very inconsistent. In other words, the presence of alkaline ions and high PH, generally higher than 7.5 to 8, makes the formation of kaolinite difficult, but smectite can be produced by saturation of illite from Mg and Ca (Fig. 20).

The sandy materials are generally quartz with mica, feldspar or lithic volcanic and heavy mineral grains. The clay minerals are essentially illite (alluminous) and smectite; these minerals are common in alluvial plains. A trace of kaolinite and chlorite is also observed in X-RAY diffraction These minerals are "Heredity " and not neoformed phase.

COMPARISON IN PAST AND PRESENT ENVIRONMENTS

The comparison of present situation and paleoenvironment in late Pleistocene and early Holocene provide some evidence for salinity and aridity, increasing toward the time. Some authors have mentioned that more humid condition precedes this actual arid situation [2, 8, 12-14]. These evidences are present in biological, sedimentological and archeological data; such as spore and pollen, the remains of wet organisms in the desert and the presence of

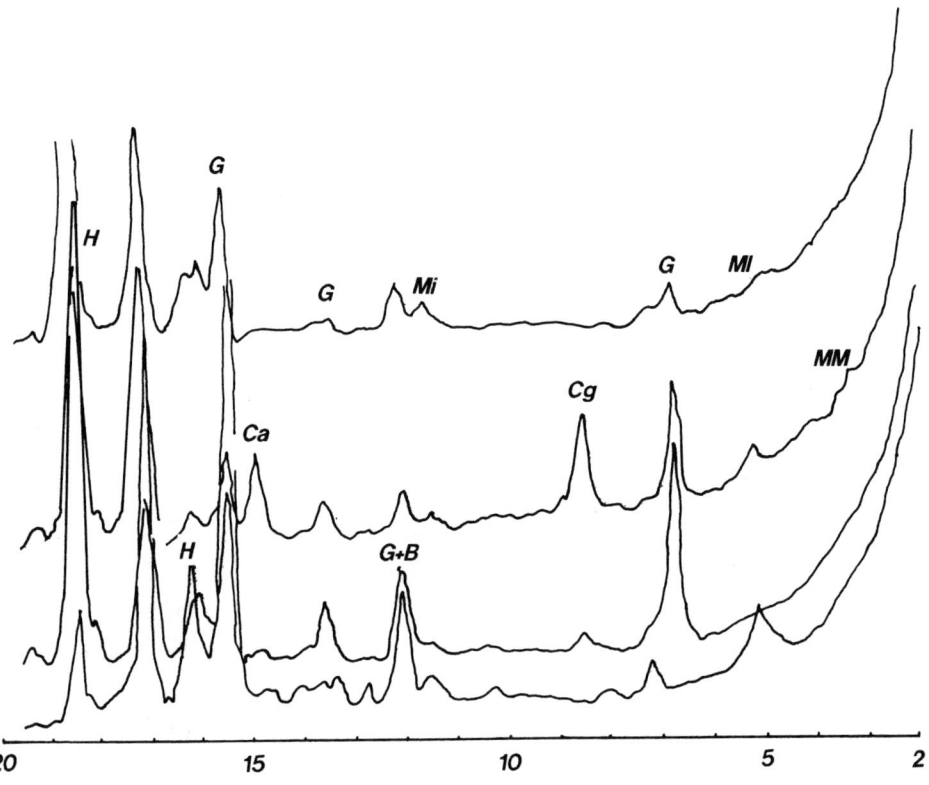

Fig. 20. X-ray diffractogram of clay minerals in central playa.

The silty marly sediments, with less salinity in lake like areas may be correlated with "Atlantic" period which had more rainfall and humidity in internal pluvial basins. The recent sand dunes formed on alluvial plains and valleys or on ancient dried lake surfaces or on terraces of the rivers are also the evidences of more humid climate of the past time.

According to Erhard's theory [4], there is recently an evolution to Rhexistasy (in contradiction to Biostasy) in central Iranian plateau, by increasing salinity and aridity and biomass decreasing. This critical situation made difficult the arid zone management and make the new conception of large "Potential for desertification of Central Iran".
The sandy materials are generally quartz with mica, feldspar or lithic volcanic and heavy mineral grains. The clay minerals are essentially illite (alluminous) and smectite; these minerals are common in alluvial plains. A trace of kaolinite and chlorite is also observed in X-RAY diffraction These minerals are "Heredity " and not neoformed phase.

CONCLUSION

The central Iranian plateau is dissected into some elongated small basins by mountain forming, volcanism, diapirism and large network of faults and thrusts, engendering a horst and graben system. All this complex evolve to more wilderness condition.

The aridity and salinity are increasing through Holocene. The evaporative and detrital materials carried from surrounding mountains are deposited and accumulated in depressing zones. A relative slow subsidence affects these depression zones in which the annual fine detrital-evaporative cyclic sedimentation reaches about 30 cm per year.

The coarse detrital materials are deposited in piedmonts near fault line scarps forming the glacis and the medium pebbles and gravel are carried to flood plains in which the multiple fine drainage networks appear with evaporite efflorescence. In wet and puffy zone the gypsum crust is formed within the saline soil by pedogenetic process and under the capillarity effect; in this case the sandy silty materials are cemented by gypsum.

The diagenesis of gypsum is quick. The saccaroidal form (2 mm) appear when the salinity increases in brine by evaporation; the long isolated or twined monoclinic crystals or lanceolate forms (1-3 cm) develop in silty marl within pores filled with saturated waters; the fibrous from is developed by capillarity effect and the lens from (2-4 mm) grows in more compact marl.

The sedimentation cycle comprises: sandy layer from December to February gypsum and halite or sodium sulfate from March to June, halite or sodium sulfate from the superficial crust developing by evaporation and capillary phenomena. The halite crust of previous cycles are dissolved in next wet months when the playas are flooded by ephemeral rivers, then the new cycle is redeposited above the previous truncated cycle. The salty crust on surface from polygons in which the edges are bent towards the interior of the polygon. This sedimentation cycle is similar to Oligocene and Miocene deposition (L.R.F. and U.R.F.) in Central Iran.

The central Iran presents some evidences of more wet climate in the past with less saline depositions and larger brackish lakes. In recent situation, the intensive erosion, evaporation and the increase of salinity present the very quick evolution to desertification with decreasing the biomass; this may lead to rheexistasy period according to Erhard's theory.

At the end it should be mentioned that these should be a deep consideration towards the capability of central Iran for desertification in all management plans.

ACKNOWLEDGMENT

I present my thanks to University of Tehran for financial support, to Dr. Y. Watanabe, for his collaboration in "Evaporite and Desert Environment" session of 29th I.G.C., Kyoto 1992, and to Dr. Esfandiari, M. Saffarzadeh and Ms. Gorbani for their kind help.

REFERENCES

1. Berberian, M., Seismotectonic map of IRAN.1/2500000, Geol. Surv. Iran, (1976).
2. Bobek, H., Features and formation of the great kewir and Massileh. *University of Tehran, Arid Zone research Center Publ.* **2,** 63p (1961).
3. Furer, M.A. and P.A. Soder, The Oligogocene formation in the Gom Region (Central Iran). *Proceding of the Fourth World Petroleum Collgress.* (1955).
4. Gansser, A., New Aspect of the Geologicum Central Iran, Petroleum Congress S.I.A.S, A1956, Proceeding 4, (1956).
5. Ghorbanni, M. and J. Lamblimon, Premier apercu de la zonation de la vegetation Halogypsophile du lac Ghom (provinnce de Tehran, Iran), Lejeunia Ser., 29, (1978).
6. Geological Survey of Iran, Geological map of Iran 1/2,500,000, (1983).
7. Charkoff, F., Report on Nakhlak Geology, Geological Survey of Iran, (1983).
8. Krinsley, Geomorphology study and paleo-climate of playa Iran, Geography Survey of Iran, 2, (1971).

9. Emami, M.H., Geologie de la region de Gom - Aran (Iran) (Volcanisme), These Doctrorat., Grenobles, France, (1978).
10. Mohafez-Reyr, Une premiere contribution des accords N.I.O.C. - E.R.AF a la connaissance geologique de l'Iran, Revue de l'Institut Francais du petrole, (1976).
11. Rieben, H., Geological observation on alluvial deposits of Tehran region; Arid zone research, Tehran University, No. 4, (1966).
12. Motamed A. pour Motamed F., Contribution a l'etude geologique de House Sultan Quar, Bull. Fac. Sci., Tehran Univ., 9, No. 3-4, (1978).
13. Motamed, A., Biwa, M. et al., Geophysic exploration of Houze Sultan, Bull. Fac. Sci., Tehran Univ., 2, No. 15, (1976).
14. Motamed, A., Contribution a l'etude depression de Lut, Bull. Geograph., Univ. Tehran, No. 11, (1971).
15. National Iranian Oil Company, Geological map of Iran, 1/10,000,000, (1977).
16. Nogale-Sadat, M.A., Les zones decrochements et les regions structurals en Iran, These. Doctral, Grenobles, France, (1978).
17. LANDSAT Image, Gom Area, (1978).

Cluster and factor analysis of geochemical data of Najmah Formation, Upper Jurassic, Western Desert, Iraq

A.K. JAMIL and M.A. Al-HILALY
Department of Geology, College of Science, University of Baghdad, Iraq.

Key word- *Najmah Formation, R- and Q-mode cluster and factor analyses, geochemical facies.*

INTRODUCTION

Najmah Limestone Formation of Malmian age represents the calcareous neritic and lagoonal facies of the Upper Jurassic period in the region especially in Iraq. Its type section was first described by Dunington [10] (in Van Bellen [30]) in well No.29 on the Najmah structure at the foothill zone and giving a thickness of about 330 m reaching almost its maximum. In the western desert, outcrops of the formation were first recognized and reported by Karim and Ctyroky [20] as few isolated patches located at Wadi Horan about 130 Km NE of Rutba Town (Fig 1). Since then several investigations were carried out on the formation dealing with its age, nomenclature, geographic extension and stratigraphic position [4,5,15,16,18,19].

Bassi *et al.* [6] were the first to describe a subsurface section of the formation west of Euphrates River in key hole KH 12/7, located about 50 Km south of Anah Town (Fig. 1). They regarded the succession from 550 to 902 m depth, according to its micro-paleontological features as Najmah Formation.

This paper is a mineralogical and geochemical study of two surface sections N1 and N2 at wadi Horan and one subsurface section in KH 12/7 of Najmah Formation in the Western Desert (Fig. 1). It aims to characterize the formation in terms of geochemical parameters using R and Q mode cluster and factor analyses.

The results, no doubt, will be utilized as a possible tool in stratigraphic correlation in the region, vertically and laterally, as well as in microfacies differentiation. It is important to note that the nomenclature and stratigraphic position as suggested by Jassim *et al.* [16] and Al-Sinjari [5] were adopted in this study where Najmah formation is considered as the fifth major sedimentlogical cycle (Malmian age) among the six well known cycles of the Jurassic System in the western desert.

SAMPLING AND METHODS

26 rock samples at about 2m interval were collected from two composite surficial sections, N1 and N2 at Wadi Horan, and 71 core samples were selected from KH 12/7 at 5-10 m interval (Fig. 1). All samples were examined by X-ray diffraction using Cu Kα radiation to study rock

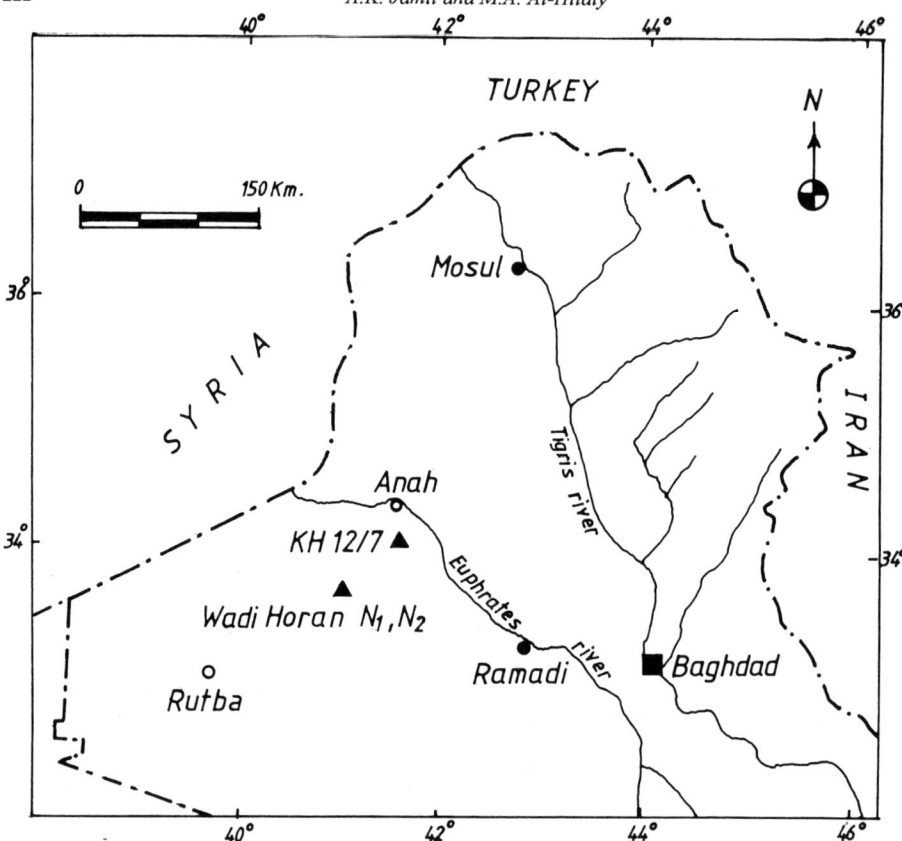

Fig.1. Location map of the studied sections.

mineralogy and were chemically analyzed for major components and few trace elements. Thin sections of different intervals were made for petrographic examinations and a number of samples were selected for the identification of mineral content of the non-carbonate fraction or insoluble residue. In the chemical analyses, the I.R. was determined using dilute hydrochloric acid [14,27], CaO and MgO by the EDTA method, Na_2O and K_2O by flame photometery, Fe_2O_3, MnO and SrO by atomic absorption, Al_2O_3 by Autoanalyzer and the trace elements Cr, Ni and V by emission spectrography. The precision of the chemical analyses were determined following the methods of Stanton [14], Maxwell [27] and Rose et al. [26], and were within the acceptable limits with high analytical accuracy.

MINERALOGY

At Wadi Horan, the formation is composed entirely of calcite with little effect of dolomitization indicating a relatively deep environment of deposition. In KH 12/7, dolomite is very common, sometimes dominant particularly in the upper part of the section indicating a relatively shallowing upwards depositional environment. Few samples, restricted to the bottom of both sections contained detectable amounts of quartz. In both sections the rocks are characterized by their low content in I.R. except samples of detrital nature. The I.R. consisted mainly of quartz, kaolinite, orthoclase, gypsum and minor ammounts of heavy minerals such as

Table 1. Summary of geochemical results for Wadi Horan sections (N1 and N2).

Element	No. of samples*	Maximum	Minimum	Average	Standard Dev.
IR (%)	24	5.81	0.07	0.89	1.38
CaO (%)	24	55.72	51.8	54.93	0.88
MgO (%)	24	0.30	0.09	0.20	0.16
Al_2O_3 (%)	24	1.72	0.03	0.20	0.36
Fe_2O_3 (%)	24	0.060	0.016	0.102	0.14
Na_2O (%)	24	0.049	0.013	0.023	0.031
LOI (%)	24	43.66	40.9	43.06	0.65
MnO (ppm)	24	135	37	61	24
SrO (ppm)	24	127	40	71	21
Ca/Mg (M)	24	703	480	459	72
Fe/Mn (M)	24	29.8	1.41	6.50	6.71
Sr/Ca (10^3A)	24	0.128	0.039	0.07	0.02
Na/Ca (10^3A)	24	1.28	0.26	0.62	0.86
Fe/Ca (10^3A)	24	5.65	0.15	0.97	0.37

* two samples of detrital nature were excluded.
Notes: K_2O, Cr, V, Ni, Pb, Cu, Ga and B were below detection limit. M : Molar ratio, A: Atomic ratio.

Table 2. Summary of geochemical results for KH 12/7 section.

Element	No. of samples	Maximum	Minimum	Average	Standard dev.
IR (%)	71	26.49	0.18	3.42	6.69
CaO (%)	71	55.32	29.14	48.09	7.49
MgO (%)	71	21.60	0.21	4.4	6.43
Al_2O_3 (%)	71	2.09	0.047	0.201	0.38
Fe_2O_3 (%)	71	2.30	0.024	0.294	0.48
Na_2O (%)	71	0.36	0.023	0.154	0.06
K_2O (%)	71	0.60	0.01	0.05	0.09
LOI (%)	71	47.22	32.01	42.13	3.88
MnO (ppm)	71	525	17	64	78
SrO (ppm)	71	640	41	222	132
Ni (ppm)	71	34	>5	8	8
Cr (ppm)	71	52	>5	43	11
V (ppm)	71	180	>5	21	14
Ca/Mg (M)	71	301	1.63	57	56
Fe/Mn (M)	71	137.9	4.72	18.9	26.1
Sr/Ca (10^3A)	71	0.87	0.072	0.256	0.17
Na/Ca (10^3A)	71	9.69	0.66	5.23	1.94
Fe/Ca (10^3A)	71	39	0.25	3.7	6.8

Note: Cu, Pb, Ga and B were below detection limit. M: Molar ratio, A: Atomic ratio.

iron oxides, tourmaline, zircon, rutile, staurolite and kyanite which were restricted in the detrital samples of Wadi Horan.

GEOCHEMICAL DATA

The range, arithmetic mean and standard deviation of all analyzed major and minor components for Wadi Horan and KH 12/7 sections are given in Tables 1 and 2, respectively. Geochemical interpretations and discussions are based completely on cluster and factor analysis as they are most widely used multi-variant statistical methods in recent geochemical studies. Davis [8] describes cluster analysis as a statistical method used to place the variables into groups (R-mode) or the samples (Q-mode) in a way where they have strong interrelations, while factor analysis, anther statistical method, used to find interrelations in a matrix of correlation between variables (R-mode) and interrelations between samples (Q-mode). In this study both analyses are applied to the geochemical data using programms given by Davis [8] with the aid of H.P 3000 computer at the computer centre of the S.E. of Geoloaical Survey and Mining.

CLUSTER AND FACTOR ANALYSIS

In R-mode cluster analysis the value of $r = \pm 0.25$ were taken as a minimum value to measure the strength of correlation between the variables while all values less than 0.4, positive or negative, were considered ineffective on the deduced factors in the R-mode factor analysis [1]. On the other hand, in Q-mode cluster analysis the value of d (relative eqlidious distance) = 6 was taken as the minimum value to measure the interrelationships between samples, while all deduced factors with eigen values less than 1 were considered ineffective [7,13].

RESULTS AND DISCUSSION

R-mode:
The cluster analysis revealed three major groups of components for each section with very similar distribution of most components but different intensities as shown in Figs. 2 and 3.

For KH 12/7, group1 includes the I.R. and its contents of clay minerals (Al_2O_3, K_2O, and Na_2O), trace elements and iron oxides (Fe_2O_3, Fe/Mn) in addition to depth, SrO, Sr/Ca, Fe/Ca and Na/Ca. Group 2 includes L.O.I. and MgO and group 3 includes CaO and Ca/Mg indicating an increase in the dolomite content of the rocks of this section, i.e. increasing effect of dolomitization. For Wadi Horan, group 1 is very similar to that of KH 12/7, group 2 includes L.O.I., CaO, and Ca/Mg indicating the high limestone content of these rocks while group 3 includes MnO referring to the little effect of dolomitization in this section. The relative difference in the intensity of each component within a group is generally due to the relative difference in the environment of deposition or type of later diagenetic processes that have affected the rocks in the two sections.

R-mode factor analysis for the two sections revealed four main interpretable factors covering 77.38% and 89.12% of the total variables for KH 12/7 and Wadi Horan sections respectively as shown in Tables 3 and 4.

In KH 12/7 section:
Factor 1: represents the leaching of clay minerals and iron oxides within the I.R. during acid digestion of the samples. It is considered a pure experimental factor.
Factor 2: represents the degree of closure of the dominant diagenetic system which caused a relative increase in the Sr and Mn content [21,24] and that dolomitization intensity relatively decreases with depth.
Factor 3: indicates the high effect of dolomitization on these rocks.
Factor 4: represents a limited depositional environment or an increase in the salinity of the solutions that have affected the rocks [22]. In Wadi Horan. it seems that factor 1 is similar to that of KH12/7 except the negative influence of CaO and LOI, representing an increase in the non-carbonate content in this section downwards. Factor 2 represents a relative opening in the dagenetic system which caused a decrease in the Sr content [21].

Cluster and factor analysis of Najmah Formation 225

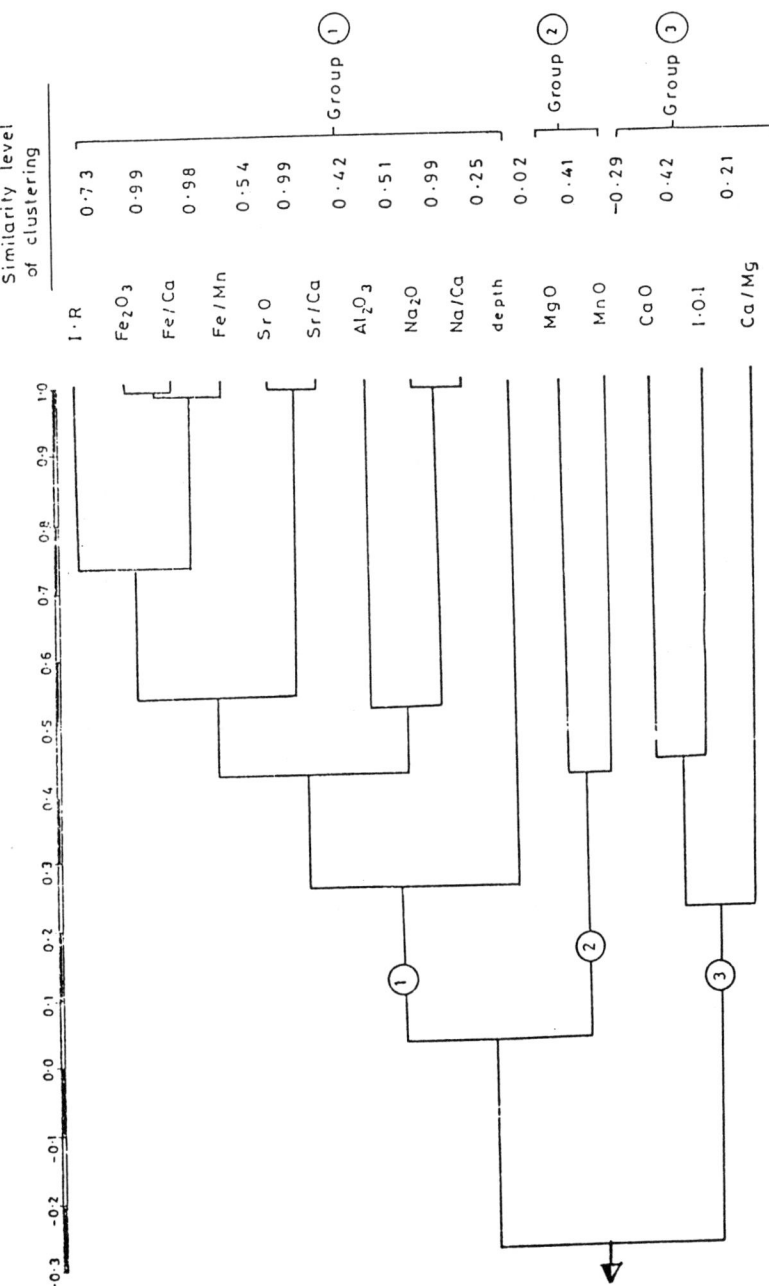

Fig. 2. R-mode cluster analysis of the carbonate rocks, Najmar Formation, KH 12/7 section.

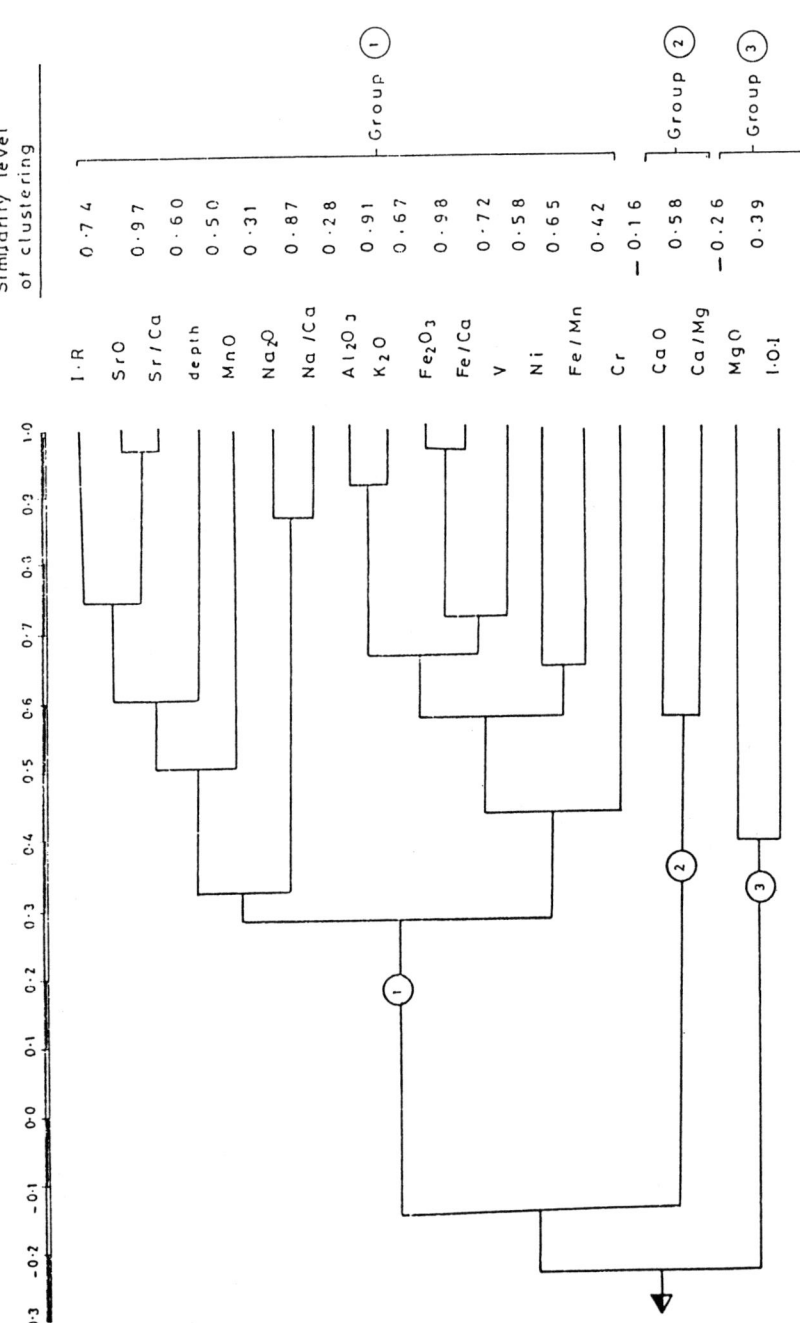

Fig. 3. R-mode cluster analysis of the carbonate rocks, Najmar Formation, Wadi Horan.

Table 3. R-mode factor analysis for the variables of the carbonate rocks of KH 12/7 section, Najmar Formation.

Element\Factor	F-1	F-2	F-3	F-4	Communality
IR	.472	.800			.843
CaO			-.895		.909
MgO			.819		.933
Al_2O_3	.813				.969
Fe_2O_3	.872				.919
Na_2O				.984	.981
K_2O	.862				.852
LOI	-.551	-.740			.864
MnO		.541			.828
SrO		.925			.909
Ni	.739				.672
Cr	.617				.469
V	.802				.774
Ca/Mg			-.794		.701
Sr/Ca		.953			.928
Na/Ca				.858	.963
Fe/Ca	.826				.819
Fe/Mn	.791				.859
Depth		.751			.652
Effect %	39.93	16.78	12.86	7.79	77.38% total
Eigen Value	7.588	3.189	2.443	1.481	percent of trace

Factor 3: may represent a more oxidizing depositional envlronment and more effect of open circulation which causes the precipitation of iron as iron oxides on the account of Mn which usually remains in solution in oxidizing conditions [12].

Factor 4: represents the low effect of dolomitization on the rocks of this section.

Q-mode:

Due to the limited capacity of the applied programme, 50 representative samples were selected from the two sections: 24 from Wadi Horan and 26 from KH 12/7 including all petrographically examined samples.

Q-mode cluster analysis revealed two maior groups: A included all the samples of KH 12/7 section while B included those of Wadi Horan as shown in Fig. 4. Such bimodal distribution is largely due to the different effect of dolomitization on the two sections in addition to the relative difference in their depositional environments. It is observed from the dendrogram of Fig. 4 that the samples are actually distributed into six secondary groups which in turn are made up of eight subgroups or geochemieal facies, 5 for KH 12/7 and 3 for Wadi Horan which differ in their chemical and mineralogical composition as shown in Table 5. A brief description of each facies is given below [9].

Geoehemical facies of KH 12/7 section:

Facies A-1: It is composed of dolostone rocks and the effect of intensive dolomitization caused a relative decrease in their trace element content [23].

Facies A-2a: A bioclastic packstone microfacies (Plates 1-1 and 1-2) containing oolites; partial effect of dolomtization is evicent.

Facies A-2b: A Pelletal packstone-grainstone microfacies showing very little effect of dolomitization (Plate 1-3).

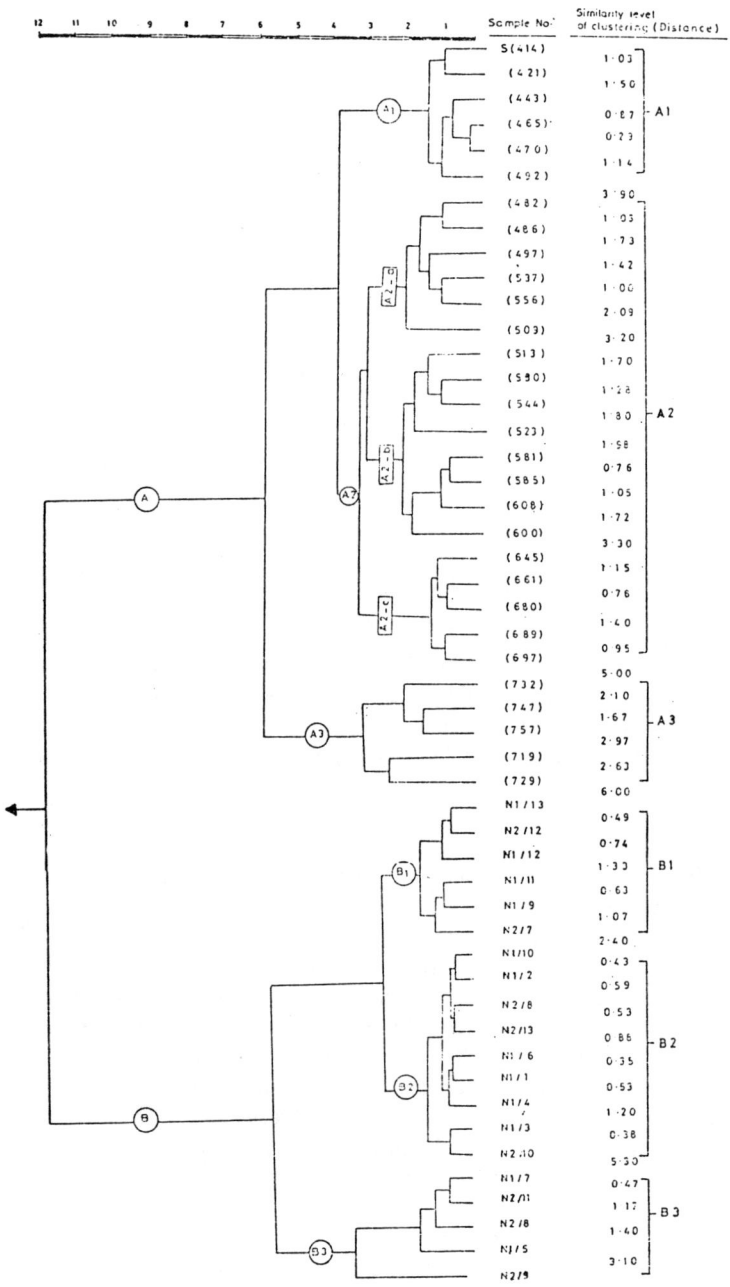

Fig. 4. Q-mode cluster analysis of the carbonate rocks, Najmah Formation.

Table 4. R-mode factor analysis for the variables of the carbonate rocks of Wadi Horan section, Najmah Formation.

Element\Factor	F-1	F-2	F-3	F-4	Communality
IR	.955				.977
CaO	-.922				.918
MgO				.940	.940
Al2O3	.940				.940
Fe2O3			.959		.995
Na2O	.721				.825
LOI	-.896				.841
MnO	.602		.476		.778
SrO		.977			.992
Ca/Mg		-.466		-.765	.945
Sr/Ca		.976			.994
Na/Ca	.677				.722
Fe/Ca			.953		.994
Fe/Mn			.935		.957
Depth	.400				.996
Effect %	45.67	18.86	15.22	9.36	89.12 % total
Eigen value	5.948	2.453	1.979	1.216	percent of trace

Facies A-2c: A dolomitic pelletal packstone facies (Plate 1-4); effect of dolomitization is highly evident.

Faceis A-3: Highly dolomitized and recrystallized microfacies.

It is characterized by its high I.R. content which caused a high content of minor components (Al2O3, Fe2O3, K2O) and trace elements [29]. Quartz grains are very common in this facies (Plates 1-5 and 1-6), in addition to clay minerals and iron oxides, as they form the main constituents of I.R.

GEOCHEMICAL FACIES OF WADI HORAN SECTION

They are characterized by their similar chemical and mineralogical composition, in particular their low content in trace elements and the relative decrease in Ca/Mg from facies B-1 to B-3, associated with an increase in Sr/Ca ratio.

Facies B-1: It is a bioclastic lime wackstone microfacies (Plates 2-1 and 2-2).

Facies B-2: An oolitic bioclastic wackstone-packstone mlcrofacles, partly recrystallized (Plate 2-3).

Facies B-3: A peloidal bioclastic packstone microfacies (Plates 2-4 and 2-5).

On the application of Q-mode factor analysis to the same samples, three main factors, F1, F2, and F3 with eigen values greater than 1 were revealed, covering 98.68% of the total effect as shown in Table 6, which also exhibit the effective components of each factor and their interpretations.

On examining the influence of the deduced factors in the two groups of samples, or the eight geochemical facies, where the influence of the three factors on each sample is projected as shown in Fig. 5, it is very evident that each group has been influenced very differently by each factor while such difference diminishes or become less evident within the facies of the same group. Karim (personal communication) have shown that the two sections have relative resemblance in their petrographic characters and microfossil content.

Nevertheless, according to this study, the cluster and factor analyses of their geochemical data revealed results of remarkable differences which can be taken as an indication that the rocks of Wadi Horan section (Najmah Formation) does not necessarily

Table 5. Average geochemical and mineralogical data of geochemical facies of KH 12/7 and Wadi Horan sections, Najmah Formation.

Facies	A-1	A-2a	A-2b	A-2c	A-3	B-1	B-2	B-3
No. of samples	6	6	8	5	5	6	9	5
IR (%)	0.33	1.05	0.89	0.44	14.46	0.21	0.39	1.01
CaO (%)	32.02	53.61	54.42	51.91	41.30	55.34	55.40	54.79
MgO (%)	19.72	1.94	0.68	3.01	3.74	0.101	0.129	0.47
Al_2O_3 (%)	0.068	0.098	0.075	0.066	0.617	0.061	0.094	0.13
Fe_2O_3 (%)	0.122	0.180	0.063	0.071	0.590	0.051	0.045	0.062
Na_2O (%)	0.13	0.16	0.20	0.14	0.21	0.017	0.024	0.031
K_2O (%)	0.015	0.030	0.020	0.020	0.150	bdl	bdl	bdl
LOI (%)	46.12	43.22	43.43	43.89	36.34	43.37	43.17	43.29
MnO (ppm)	42	35	31	38	234	51	49	59
SrO (ppm)	70	168	226	214	455	51	68	97
Ni (ppm)	5	9	7	4	8	bdl	bdl	bdl
Cr (ppm)	7	11	8	11	13	bdl	bdl	bdl
V (ppm)	6	12	10	4	64	bdl	bdl	bdl
Ca/Mg (M)	1.97	52.3	102	23.3	23	566	515	222
Fe/Mn (M)	16.58	30.8	9.6	8.24	19.9	4.5	4.3	5.3
Sr/Ca (A)	0.115	0.175	0.225	0.224	0.604	0.05	0.067	0.097
Na/Ca (A)	5.75	4.42	5.15	3.76	7.3	0.46	0.62	0.83
Fe/Ca (A)	2.61	1.70	0.60	0.70	10.00	0.48	0.42	0.60
Calcite (%)	0.25	81.3	95.4	19.6	27.4	-	-	-
Dolomite (%)	99.75	16.6	4.2	80	59.8	100	100	99

Note: bld = below detection limit, M = Moloar ratio, A = Atomic ratio.

belong to Najmah Formation as it is supposed. Al-Hilali [5] have shown that the rocks of this section show great resemblance in the mineralogical, geochemical, petrographic and facial characters to the upper parts of Muhaiwir Formation in this area.

Such a geological deduction has been achieved also by Al-Sayab (personal communication) based on a comparison of the fossil assemblages of the two sections where the observed fauna in the supposed Najmah formation (Wadi Horan) are an extension to those that were found in the upper parts of Muhaiwir Formation in the same area.

The dominant petrographic facies of each geochemical facies were compared with the standard microfacies SMF [11,31], and the following were deduced:
The facies of A-2a and B-1 are similar to SMF-9 representing an open platform environment of shallow waters, A-2b and B-3 are similar to SMF-16 representing a shallow restricted environment of shallow and warm waters, whereas B-2 are similar to SMF-15 representing beach shoals or tidal bars and high energy environment.

Table 6. Q-mode factor analysis for the carbonate rocks, Najmah Formation.

Factor	Eigen value	Factor % effect	Effective components	Interpretations
1	30.42	60.82	MgO, LOI, Na_2O, SrO, Na/Ca, Sr/Ca, Fe/Mn, dolomite %, depth	Intensity of dolomitization, limitation of depositional environment, closure of diagenetic system.
2	17.85	35.71	CaO, LOI, Ca/Mg, calcite %	Dolomitization, open sea depositional environment, long period of exposure.
3	1.06	2.13	IR, Al_2O_3, Fe_2O_3, MnO, SrO, Fe/Ca, Sr/Ca, Fe/Mn, depth	Direction of old shore line, shallow environment of deposition, more clastic supply.

% Total effect: 98.68
No. of samples: 50

Cluster and factor analysis of Najmah Formation   231

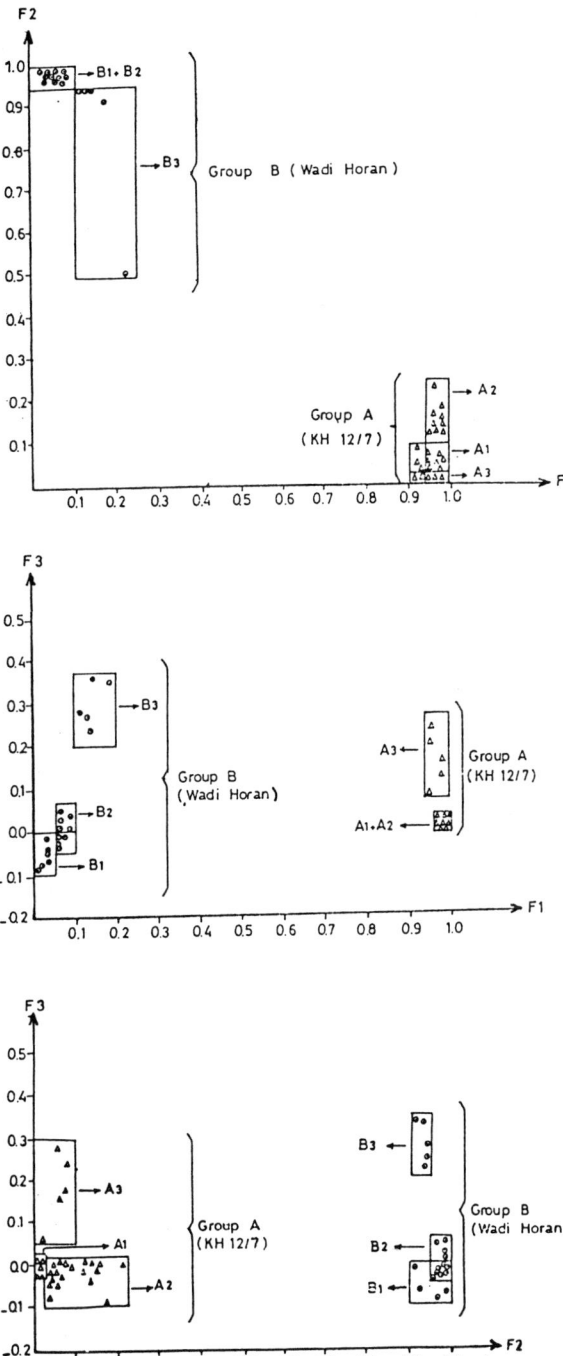

Fig. 5. Q-mode factor analysis showing the extent of affection of each geochemical facies by the three factors for the carbonate rocks, Najmar Formation.

SUMMARY AND CONCLUSIONS

1. Naimah Limeston Formation at Wadi Horan section is composed entirely of calcite, whearas in KH 12/7 dolomite is common and sometimes is dominant.
2. The I.R. content is very low indicating the domination of relatively dry and arid climate and/or relatively offshore environment during time of deposition.
3. R-mode cluster analysis revealed three major qroups of components indicating the high limestone content in Wadi Horan and an increase in the dolomite content at KH 12/7. The R-mode factor analysis gave four interpretable factors for each section. In KH 12/7 relative increase in dolomitization with depth, high salinity solutions and a limited depositional environment are suggested. In Wadi Horan, relative openning in the diagenetic system and low effect of dolomitization are suggested.
4. Q-mode cluster and factor analysis revealed eight geochemical facies, five in KH 12/7 and three in Wadi Horan. Each facies has distinguishable geochemical constituents and petrographic description and influenced differently by three deduced factors suggesting that the rocks of Wadi Horan section do not necessarily belong to Najmah Formation as it is supposed. The later section shows great resemblance to the upper parts of Muhaiwir Formation in the same area.

Acknowledgment

The authors are deeply grateful to the G.E. Geological Survey and Mining (GEOSURV) for technical assistance, access to their technical reports and financial support to M.A.Al-Hilaly and to the College of Science and the University of Baghdad for their research facilities.

REFERENCES

1. I. S. Al-Aasm and J.Veizer, Diagenetic stabilization of aragonite and Low-Mg cacite. I: trace elements in rudists. *Jour. Sed. Petrol.*, 56, 138-152 (1985).
2. W. S. Al-Hashimi and V. Skocek, Sedimentary geology of Iraq. Internal Report 1513. Geol. Surv. Miner. Invest.. Baahdad, 177p. (1984).
3. M. A. H. Al-Hilali, Geochemistry and mineralogy of Najman and Muhaiwir Formation-Jurassic System in the Western Desert-Iraq. Unpubl. Ph.D. thesis at Univ. Baghdad. 228p (1989).
4. M. Al-Mubarek, Regional geological survey of the western and Southern Desert, Iraq. Internal Report 1380. Geol. Surv. Miner. Invest., Baghdad, 636p. (1983).
5. A. S . Al_Sinjari, Stratigraphy of the Jurassic system, Rutba Area.. Unpubl. Ph.D thesis at Univ. Baghdad. 211p. (1987).
6. M.A. Bassi, S. Yousif, W. Raji, H. Odisho and L. Khalaf, Petrology, paleontology and geochemistry of Keyhole 12/7, western desert of Iraq. Internal Report 1603, Geol. Surv. Miner. Invest., Baghdad, 62p (1988).
7. T. Buday, The regional geology of Iraq, Volume 1, Stratigraphy and Paleogeography. S. E. Geological Survey and Minerals. Baghdad (1980).
8. J.C. Davis, Statistics and data analysis in geology, John Wiley and Sons Inc., New York (1973).
9. R.J. Dunham, Classification of carbonate rocks according to depositional texture. In: *Classiflcatlom of carbonate rocks*. W.E. Ham (Ed.), Amer. Assoc. Petrol. Geol. Mem. 1, 279p (1962).
10. H.V. Dunnington, Stratigraphy and stratigraphic nomenclature of Gaara-Wadi Horan region. Internal Report, Iraq. Petrol. Co. (1953).
11. E. Flugel, *Microfacies Analysls of Limestone*. SpringerVerlag. Berlin. (1982).

12. G.M. Friedman, Trace elements as possible environmental indIicators in carbonate dediments. In: *Depositional Environment in Carbonate Rocks*. G.M. Friedman (Ed). Soc. Econ. Paleont. Mineral. Spec.Publ. 14, (1969).
13. J.W. Harbough and D.F. Merriam, *Computer Applications In Stratigraphic Analysis*. John Wiley and Sons Inc., New York (1968).
14. H.A. Ireland, Insoluble residue. In: *Procedure in sedimentary Petrology*, R.E. Carver (Ed). John Wiley and Sons Inc., New York (1971).
15. S.Z. Jassim, S.A. Karim, M.A. Basi, M. Al-Mubarek and J. MunIr. Final report on the regional geological survey of Iraq. Vol. 3: Stratiaraphy. Internal Report 1447. Geol. Surv. Miner. Invest., Baqhdad (1984).
16. S.Z. Jassim, M.Y. Tamar-Agha and K. S. Al-Bassam, Excursion guid: 7th Iraqi Geol. Cong., Internal Report 1740. Geol. Surv. Miner. Invest., Baghdad (1986).
17. K.G. Joreskong, J.E. Klovan and K.A. Rayment, *Geological Factor Analysis*. Elsevier Sci. Publ., Amestrdam (1976).
18. N. Kaddouri, Triassic-Jurassic sediments in Iraq. *Jour. Geol. Soc. Iraq.*, **8**, 167-169 (1986).
19. N. Kaddouri, S'Ker Formation: a new Hauterivian-Barremian stratigraphic unit in western Iraq. *Creta. Res.*, **10**, 267-270 (1989).
20. S.A. Karim and P. Ctyroky, Stratiaraphy of eastern and southern flanks of the Ga'ara Hight. Internal Report 1185, Geol. Surv. Miner. Invest., Baghdad (1981) .
21. D.J.J. Kinsman, Interpretation of Sr^{+2} concentrations in carbonate minerals and rocks. *Jour. Sed. Petrol.*, **39**, 486-508 (1969).
22. L.S. Land and G.H. Hoops, Sodium in Carbonate Sediments and Rocks: A possible Index to the salinity of diagenetic solutions. *Jour. Sed. Petrol.*. **43**, 614-617 (1973).
23. L.S. Land, The isotopic and trace element geochemistry of dolomIte: The state of the art. In: *Concepts and Models of Dolomitization*. D.H. Zenger and J.B. Dunham (Eds), Soc. Econ. Paleont. Mineral. Spec. Publ. 28, 87-110 (1980).
24. D.N. Lumsden and R.V. Lloyd. Mn (II) partitioning between calcium and magnesium sites in studies of dolomite origin. *Geochim. Cosmochim. Acta.*, **48**, 1861-1865 (1984).
25. J. Maxwell, *Rocks and mineral analysis*. Int. Sci. Publ., 384p. (1968) .
26. A.W. Rose, H.E. Hawkes and J.S. Webb, *Geochemistry in Mineral Exploration. 2nd Edition*, Academic Press, London (1979).
27. J.T. Sanford and R.E. Mosler, Insoluble residues and geochemistry of some Llandoverian and Wenlockian rocks from Gotland. *Sveriges Geololsga Undersokulng*. Ser. C, Nr. 811, 1-31 (1985).
28. K.E. Stanton, *Rapid Methods of Trace Analysis for Geochemical Applications*. Edward Arnold Corp., London (1966).
29. R. Till, Are there geochemical criteria for differentiating reef and non-reef carbonates. *Bull. Amer. Assoc. Petrol. Geol.*, **55**, 523-530 (1971).
30. R.C. Van Bellen, H.V. Dunnington, R. Wetzel and D.M. Morton, *Lexique Stratigraphique International, V. III, Asia, Fascicule 10a, Iraq*. Center National de la Recherch Scientifique. Paris (1959).
31. J.L. Wilson, *Carbonate Facies in Geological History*, Springer-Verlag. Berlin (1975).

PLATE-1

1. Bioclastic packstone facies, representing geochemical facies A-2a. Characteristics Foram. Sp. of U. Jurassic are visilble. Sample No. S(5g7). Kh 12/7. Naimah Fm. (X33).
2. Same as above. Sample No. S(509), Kh 12/7, Naimah Fm.(X33).
3. Peloidal grainstone-packstone facles, representing geochemical facies A-2b. Sample No. S(530). Kh 12/7. Naimah Fm. (X33).
4. Dolomitized peloidal packstone facies, representing geochemical facies A-2c. Sample No. S(680), Kh 12/7, Naimah Fm.
5. Highly recrystallized and dolomitized facies. representing geochemical facies A-3, quartz grains being visible. Sample No. S(747). Kh 12/7, Naimah Fm. (X33).
6. Same as above. Sample No. S(732), KH 12/7, Naimah Fm. (X33).

PLATE-2

1. Bioclastic wackstone facies, representing geochemical facies B-1. Sample No. N1/9, Wadi Horan, Najmah Fm. (X33).
2. Same as above, a characteristic Foram. Spec. of U. Jurassic is visible. Sample No. N1/12. Wadi Horan. Naimah Fm. (X33).
3. Bioclastic oolitic packstone-wackstone facies, representing geochemical facies B-2. Sample No. N1/4, Wadi Horan, Naimah Fm. (X33).
4. Bioclastic peloidal packstone facies, representing geochemical facies B-3. Sample No. N1/5, Wadi Horan, Najmah Fm. (X33).
5. Same as above, a characteristic Foram Spec. of U. Jurrasic is visible. Sample No. Nl/7, Wadi Horan, Najmah Fm. (X33).
6. Same as above. Sample No. S(732). KH 12/7. Naimah Fm. (X33).

PLATE 1

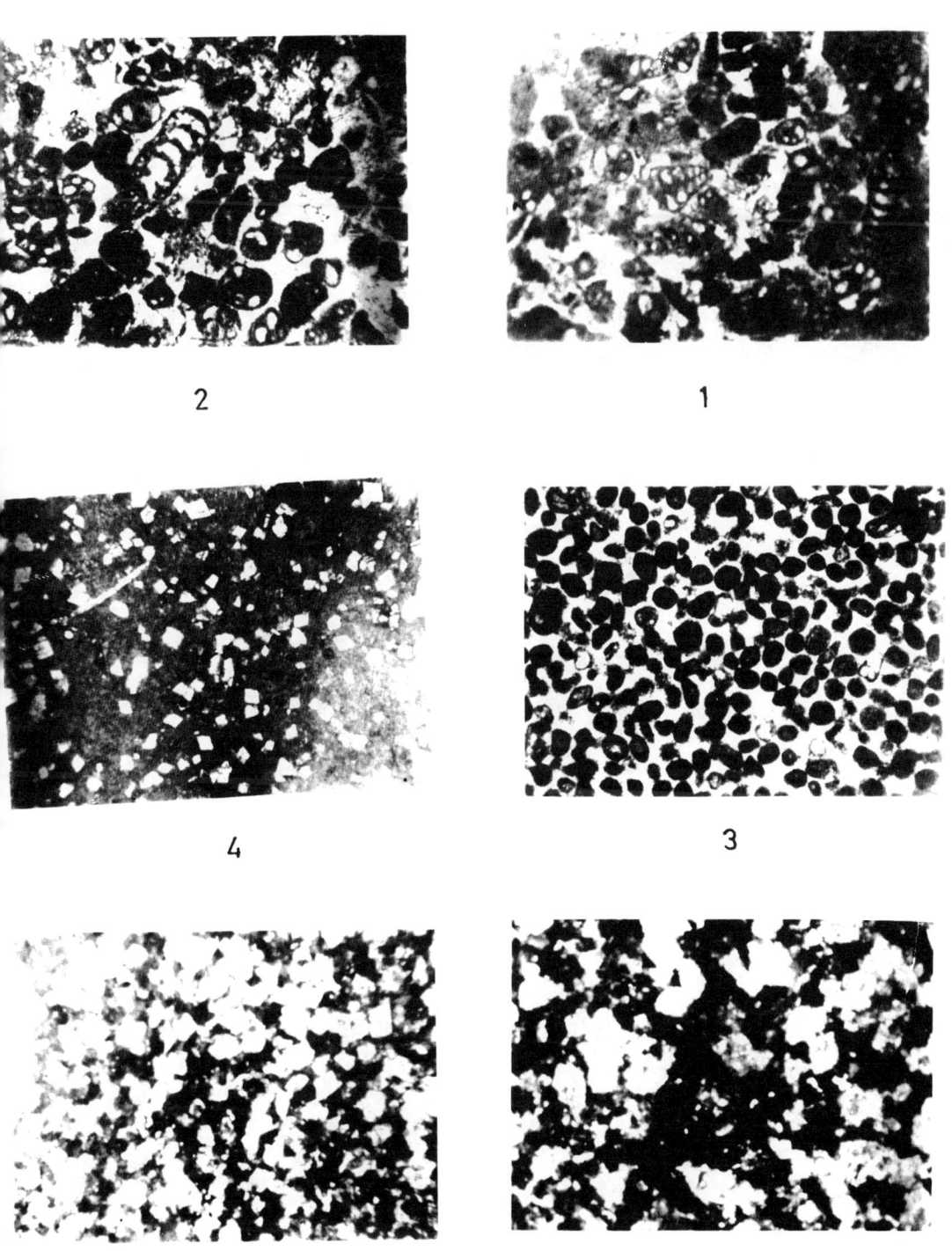

PLATE - 2 -

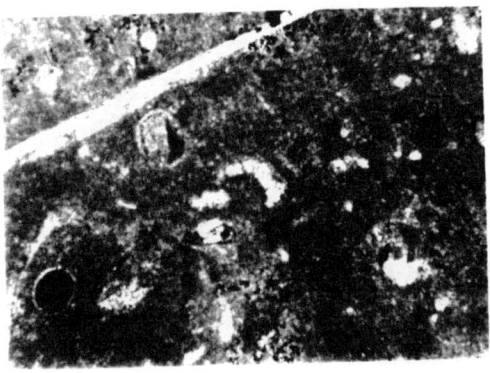

2

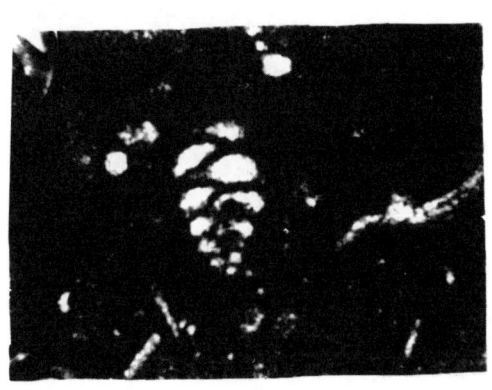

1

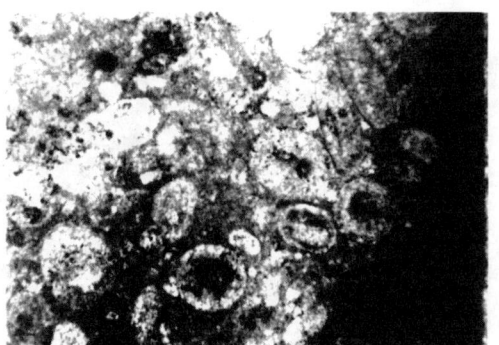

3

5

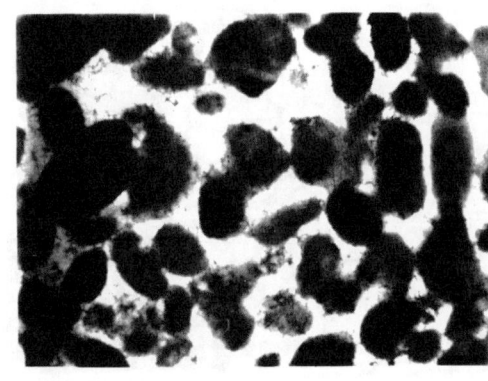

4

Shortite formation in Turkey and its geochemical properties

F. SUNER
Department of Geology, Technical University of Istanbul, 80670, Ayazaga, İstanbul, TURKEY

Abstract -- Shortite is a carbonate mineral that rarely forms in evaporative sequences in the World. In USA, this double sodium calcium carbonate ($Na_2CO_3 \cdot 2CaCO_3$) was discovered together with other soda minerals. In Turkey, Shortite was discovered within the second largest natural sodium carbonate deposits of the World in the vicinity of the Beypazari Province of Ankara. The mineral was found out as a result of arduous studies in the lower parts of the lower trona horizon, under the levels where pirrsonites ($Na_2CO_3 \cdot CaCO_3 \cdot 2H_2O$) and gaylussites ($Na_2CO_3 \cdot CaCO_3 \cdot 5H_2O$) formed. Shortite, discovered for the first time in this study, in very small sizes, was observed with large scatter. The mineral was subjected to comprehensive mineralogical, petrographical, geochemical and physicochemical investigations using DTA, SEM, XRD and other techniques. Comparison of the above results with those of other soda minerals, in view of the shortite bearing levels, reveals that shortite was formed from the low ratio of HCO_3/CO_3 and Na/Ca containing solutions within Kizilcahamam volcanism and volcano-sedimantary sequences during the Neogene that correspond to the beginning period of the formation of Beypazari Basin. The shortite bearing sequences were also found interbedded with the volcanic products during the same period.

Keywords: Shortite, Turkey, Geochemistry, Formation condition.

INTRODUCTION

Shortite, a rare forming mineral, was discovered in saline environments, with carbonate, sulfate and chlorine bearing minerals. Shortite was first found in Green River Formation [1] and in the muds of Searles Lake [2] with trona and other evaporative minerals. In Turkey, shortite ($Na_2CO_3 \cdot 2CaCO_3$) sparsely was located within the second largest natural sodium carbonate (trona) accumulations through disseminations in the Neogene sediments, in Beypazari, the province of Ankara. Since in Turkey, the mineral was first discovered in this study; its occurrence and the properties were investigated in detail. Shortite was found in the lower parts of the lower trona horizon in a very small amount and sizes. The pirrsonites and gaylussites have been observed over the shortite bearing level, in the same lithologic sequence comprising carbonate, laminated claystone, dolomite, bituminous shale, tuff and tuffite. The aim of this study, based on the experimental data, is to determine the properties of shortite and to delineate the physicochemical and geochemical conditions under which it formed.

STRATIGRAPHY

Beypazari Trona Deposits were formed in the Middle and Upper Miocene sequences as two disconnected seams each of which has contained more than eleven single trona accumulations. The Çorakli, Karadoruk and Hirka formations were the main three sequences of the basin. Çorakli formation was deposited from conglomerate, volcanic bresh, bituminous shale and coal levels in different thicknesses. This formation was overlain by Hirka formation which consisted of marl, tuff, tuffite, claystone, dolomite bearing limestone, with thicknesses up to 350 meters. Karadoruk formation has overlain this sequence and the chert bearing dolomitic limestone, was the main component of the Karadoruk formation, the thicknesses of which was about 80 meters. All the levels were conformably overlain by Pliocene sequences which consisted mainly of sandstone, limestone and gypsum bearing limestone [3-8].

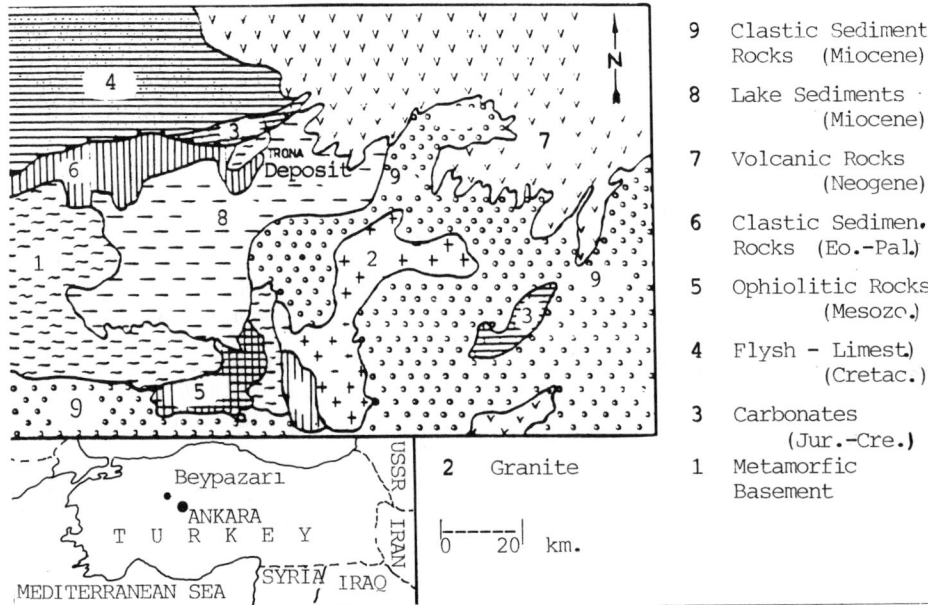

Figure 1. Geological Map of Beypazarl Basin (1)

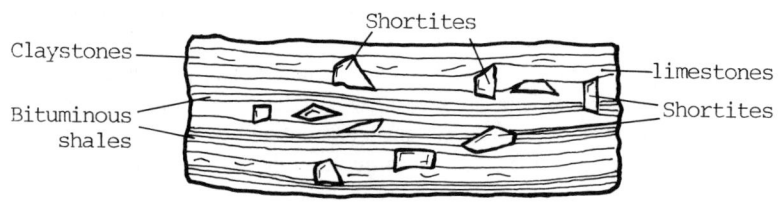

Figure 2. Schematic section of Shortite crystals

METHODS

Shortite, collected from Hirka formation, was studied systematically under binocular microscope. Upon cleaning its surface from the sedimantary impurities, the samples were prepared for futher analysis. The selected samples were investigated by XRD (X-Ray Diffraction), DTA (Differential Thermal Analysis), SEM (Scanning Electron Microscopy) methods and were analyzed by wet chemical, gravimetric, colorimetric and spectrographic

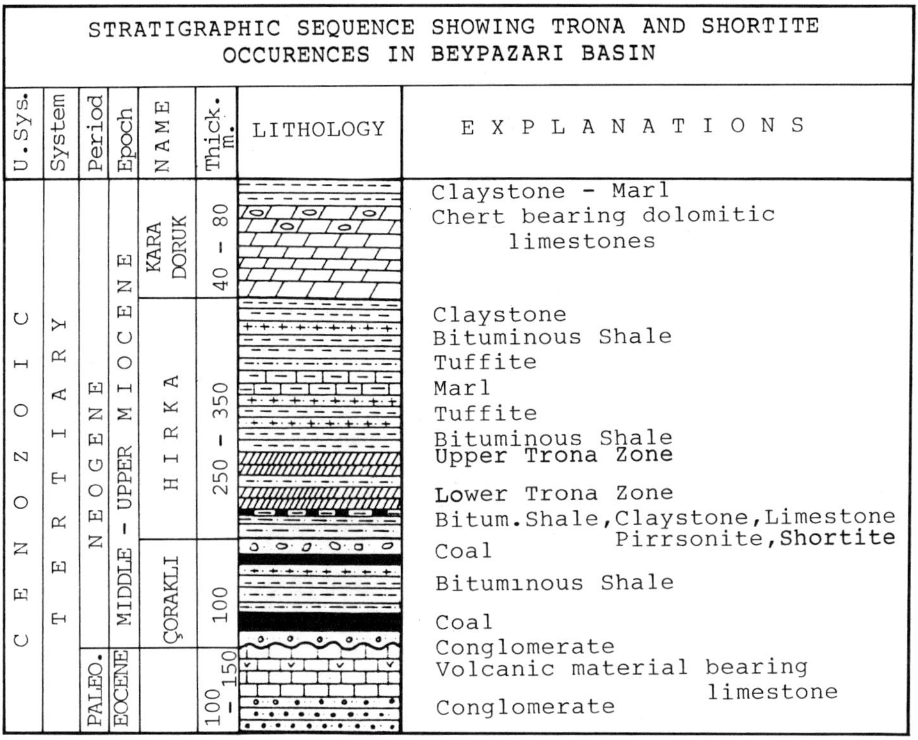

STRATIGRAPHIC SEQUENCE SHOWING TRONA AND SHORTITE OCCURENCES IN BEYPAZARI BASIN
U.Sys.
System
Period
Epoch
NAME
Thick. m.
LITHOLOGY
EXPLANATIONS
CENOZOIC
TERTIARY
NEOGENE
PALEO.
MIDDLE - UPPER MIOCENE
EOCENE
KARA DORUK
HIRKA
ÇORAKLI
40 - 80
250 - 350
100
100 - 150
Claystone - Marl
Chert bearing dolomitic limestones
Claystone
Bituminous Shale
Tuffite
Marl
Tuffite
Bituminous Shale
Upper Trona Zone
Lower Trona Zone
Bitum.Shale,Claystone,Limestone
Pirrsonite,Shortite
Coal
Bituminous Shale
Coal
Conglomerate
Volcanic material bearing limestone
Conglomerate

Figure 3. Geological stratigraphic section of Cenozoic sequence in Beypazari Basin [1].

34.96 (2.56)
32.80 (2.72)
27.23 (3.27)
24.95 (3.56)
23.24 (3.82)
17.90 (4.96)
16.05 (5.52)
14.79 (5.99)
35
30
25
20
15

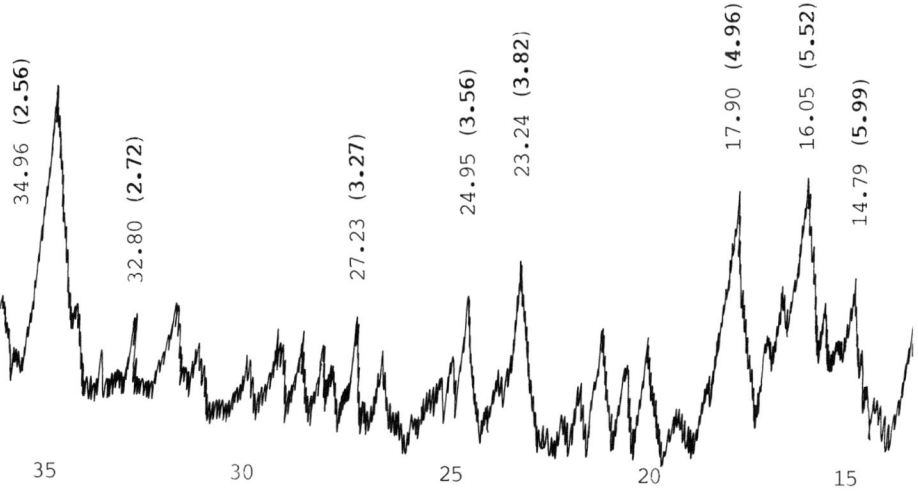

Figure 4. The XRD diffractogramme of shortite; Cu(Ni)Kα, 2θ = 1° / 1 Sec., n=1.5437; (20mA x 40kV); operating conditions.

methods. The trace element analyses of shortite were determined by means of spectorgraphic methods. Statistical evaluations have been performed for the calculation of parameters involved.

RESULTS

Shortite has been studied on hand-specimens and the physical properties were determined. The mineral exhibits many cleavages and white and grayish-white colors with waxy brightness. The measured dimension was approximately 2 mm. which is similar to that observed with Pirrsonites [9]. In host rocks and sedimentary sequences, the irregular distributions and disseminated structure of shortites are the main properties of the deposits. There is no significant differences, in optical features, between the other observed sodium and calcium carbonates and shortite. The single and crossnicol properties are very similar to those of gaylussite and partially to pirrsonite trona ($Na_2CO_3 \cdot NaHCO_3 \cdot 2H_2O$) and nahcolite ($NaHCO_3$), which has been observed in the upper stratigraphic position of the deposits. XRD methods is another technique used for investigating the mineralogical composition of shortite. The XRD results represented in Figure 4 show that a pattern similar to that observed with pirrsonite and gaylussite discovered in the basin. There are some peaks which were present both in shortite and other soda minerals. The DTA technique, one of the poverfull methods, for distingushing carbonates, was applied to shortite. The DTA curve of shortite is given in Figure 5. In order to futher define the morphology of the mineral, in addition to the polarized microscopic studies, Scanning Electron Microscopic studies were also conducted. The crystal structure of shortite using the above methods is determined to be rombic in its most beatiful form, as shown in Figure 6.

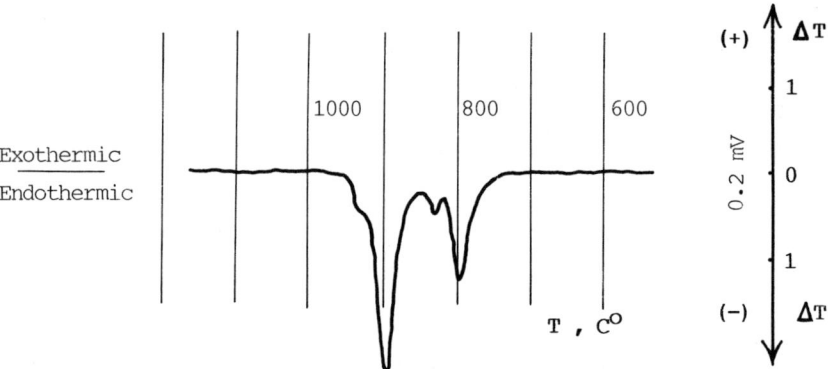

Figure 5. DTA curve of shortite. Pt-Pt/Rh thermoeleman, 10°/sec heating rate; 2.5 mm/sec recording rate, atmosphere.

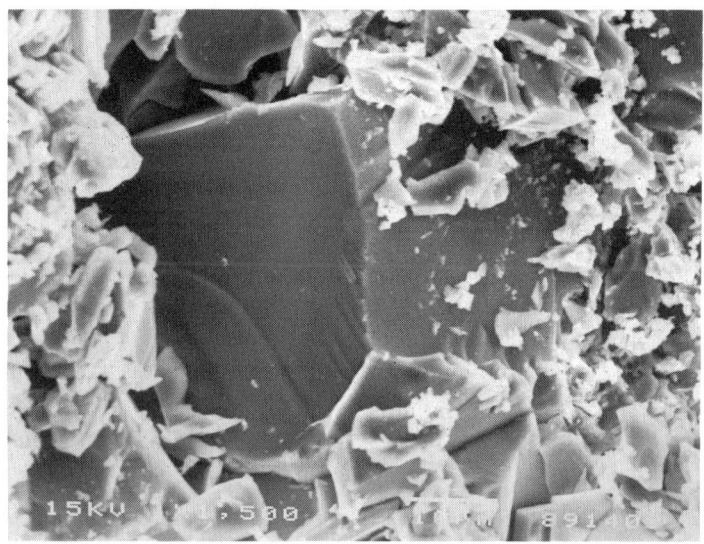

Figure 6. Electron Microphoto of Shortite

GEOCHEMICAL STUDIES

The trace and also major element contents of Shortite specimens were analyzed using different methods. The results of analyses are shown in Table 1. The statistical parameters calculated for this data are also shown in Table 1. The results of statistical evaluation reveal that all the trace elements fallow a normal-Gaussian distribution. The other geochemical studies aimed to determine the correlation coefficients of Br, F, B and Cl elements for the purpose of comparing the conditions of formations between shortite and other soda minerals. The correlation coefficients and lines of Cl versus Br, B and F are shown in Figure 7.

DISCUSSION

Shortite has been investigated in wiev of mineralogical, geochemical, petrographical properties of this mineral. Shortite and other soda minerals demonstrate similar characteristics. Therefore, it is highly difficult to find this mineral, not only because it is a rare forming mineral, but also due to difficulties encountered in its identification. It is especially difficult, to define shortite macroscopically and the evaluation of the properties of shortite is definetely necessary for correctly identifying the mineral. Mineralogical features of the evaporative minerals are approximately the same. In partlcular, there are similarities between shortite and gaylussite in view of the general features and the shape of occurences. For this reason, it was rather difficult to differentiate between the mineral groups. On the other hand, the morphologies of these groups were slightly different. The results of geochemical investigations show that there is a slight difference between the trace element contents of gaylussite and pirrsonite. This must be related to the conditions of formation, *i.e.* by taking into accout the levels where these minerals were observed. The similarities between these minerals would then be more meaningful.

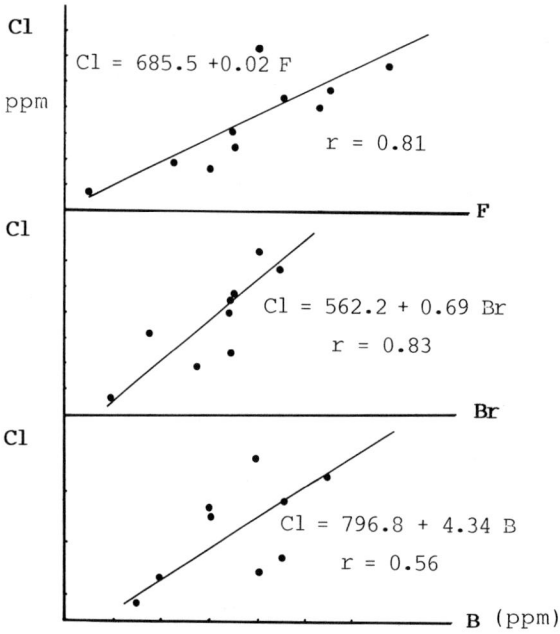

Figure 7. The relations of Cl, F, B and Br

Table 1. The statistical parameters of trace element contents in shortite.

	X (ppm)	S	S_H	M_i	M_o	S_k
K	42	3.8	4.18	41.26	39.78	0.58
Mg	4410	175.7	195.2	4380	4320	0.46
Ba	421	54.3	59.73	433	457	0.66 (−)
Rb	34.5	4.50	4.95	33.6	31.8	0.60
Cs	336	20.56	22.66	332.5	325.5	0.51
Si	172.4	24.70	27.44	175.4	181.4	0.36
Cl	878.4	8.6	9.5	878	877.2	0.12
P	86	22.50	24.95	82.7	76.1	0.44
Al	185	10.70	11.77	183.4	180.2	0.45
Ti	48	9.6	10.55	46.2	42.6	0.56
Li	3	1.2	1.32	2.7	2.1	0.75
F	8970	1340	1474.6	9060	9240	0.20 (−)
I	3225	930	1203	3360	3630	0.43 (−)
SO4	3950	791.3	870.4	3780	3440	0.64
B	18.7	1.07	1.18	18.75	18.85	0.14 (−)
Br	452	7.81	8.59	453.3	4.56	0.51 (−)

X:Mean ; S:Standart Deviation ; S_H:Real Standart Deviation
M_i:Median ; M_o:Mod ; S_k:Skewness

Furthermore, the high correlation coefficients between the volatile element contents in these coefficients among the other observed soda minerals have supported the formation to occur the identical conditions.

Physicochemical properties of Shortite are also important to understand the formation and transformation mechanism of shortite. Similar to other carbonate and evaporative minerals, shortite can be transformed easily because of the changes in activity of CO_3, activity Of H_2O, molar concentration of CO_2 and temperature. This mineral, according to ion and Eh-pH levels, can form from nahcolite, pirrsonite and gaylusite. The presence of other salt-type minerals is also important from the point of deposition and transformations of the mineral. Shortite may form from the solutions which contain (Na) and (Ca) ions in low temperature and high activity of CO_3. As a result of an increase in activity of HCO_3, Na % and activity of H_2O, shortite may transform to other soda minerals as shown in equations such as

$$2NaHCO_3 + 2CaCO_3 \longrightarrow Na_2CO_3 \cdot 2CaCO_3 + CO_2 + H_2O \quad (1)$$
$$Na_2CO_3 \cdot 10H_2O + 2CaCO_3 \longrightarrow Na_2CO_3 \cdot 2CaCO_3 + 10H_2O \quad (2)$$
$$Na_2CO_3 \cdot CaCO_3 \cdot 2H_2O + CaCO_3 \longrightarrow Na_2CO_3 \cdot 2CaCO_3 + 2H_2O \quad (3)$$
$$Na_2CO_3 \cdot CaCO_3 \cdot 5H_2O + CaCO_3 \longrightarrow Na_2CO_3 \cdot 2CaCO_3 + 5H_2O \quad (4)$$

CONCLUSION

In Beypazari, some evaporative minerals have been discovered, in which shortite constitutes an important one. This mineral have been observed randomly and scatteraly in sedimentary sequences which were mainly tuff / tuffites, claystone, and bituminous shale at lower parts of the Hirka formation. Based on the properties observed, the salient features of this study can be summarized as fallows :

1) Shortite ($Na_2CO_3 \cdot 2CaCO_3$) has been located in the lower parts of pirrsonite and gaylussite bearing levels, in a very small amount. Therefore, this mineral was most probably formed from the low Na / Ca ratio bearing cooler solutions, at the beginning period of deposition.
2) The microscopic studies reveal that shortite has an equal crystal dimension and shows no transformation marks. In the level where shortite was formed, there were no other soda minerals such as nahcolite, pirrsonite and trona. It may be therefore assumed that shortite was deposited primarily from the solutions.
3) The physicochemical parameters, which were responsible for the deposition of shortite, such as low Na / Ca ratio, low temperature, high activity of CO_2, had been subjected to rapid change such that the shortite formation had not been continued for a long time. The increased Na ion concentration, temperature and activity of H_2O were also responsible for the ceasing of formation. The volcanic products and sedimentary materials protected formed minerals from tranformations.
4) As a result of increase in temperature, Na / Ca ratio and activity of H_2O, pirrsonite and gaylussite had been deposited.
5) The trace element contents of shortite and pirrsonite were similar with no significant differences. For this reason, the solutions which were responsible for shortite and pirrsonite were from the same origin. Especially, there is strong relationships among the volatile matters such as F, Cl, Br and B both in shortite and pirrsonite, which supports the mechanism proposed.

REFERENCES

1. W. Bradley and H. P. Eugster. Geochemistry and Paleolimnology of Trona Deposits and Associated Autogenic Minerals of Green River Formation of Wyoming, *U. S. Geol. Surv. Prof. Paper,* **69. B**, 1-71, (1969)
2. H. P. Eugster and G. I. Smith, Mineral Equilibra in Searles Lake Evaporites, California, *Jour. Petrol.*, **6**, 48-74, (1965)
3. F. Suner, Beypazari Trona Deposits, Ph. D. Thesis, Fen Bilim. En. İTÜ, (1989)
4. F. Suner, Beypazari Tronalarinda Ca, Mg, Rb ve K Dagilimi, *İTÜ Bull.* **47**, 10-14, (1989)
5. F. Suner, Diferansiyel Termal Analiz Yönteminin Soda Minerallerine Uygulanisi, *İTÜ Bull.*, **48**, 1-5, (1990)
6. Y. Bürküt and F. Suner, Mineralogical and Geochemical Properties of Turkish Pirrsonites, *Geosound Spec. Iss.,* **20**, 191-200, (1990)
7. C. Helcaci, U. İnci, F. Yagmurlu and H. Yilmaz, Geologic Framework of The Beypazari District and Neogene Trona Deposits, TR, *Doga. Bil. Dergi.,* **13,** 245-256, (1989)
8. F. Suner, Beypazari Nakolitlerinde Ca, Mg, Rb ve K Dagilimi, *Doga. Bil. Dergi.,* **14**, 67-76, (1991)
9. F. Suner, Beypazari Trona Deposits, *Abstract. First South. Asia. Geol. Cong.* 71, *(1992)*

Mineral assemblages and formation of the Kestelek and Sultançayiri borate deposits

Cahit HELVACI
Dokuz Eylül Üniversitesi, Mühendislik Fakültesi Jeoloji Mühendisligi Bölümü, 35100 Bornova-İZMİR, TURKEY

ABSTRACT--The Kestelek borate deposit occurs within Neogene playa lake sediments, and is interbedded with clay, limestone and tuffs. The volcanosedimentary sequence in the Kestelek region consists of, from bottom to top, basement conglomerate, volcanic rocks, the tuff-claystone-limestone, borate zone, limestone, and conglomerate-sandstone-limestone units. In the Sultançayiri borate deposit, borates are interbedded with gypsum, claystone, limestone and tuffs, and the sequence in this region consists of, in ascending order, basement conglomerate, tuff, limestone, the borate zone and a gypsum zone.

Colemanite, ulexite and probertite predominate in the Kelestelek deposit, and hydroboracite occurs rarely, Calcite, quartz, zeolites, smectites, illite and chlorite are the associated minerals in this deposit. In the Sultançayiri deposit, pandermite predominates, but other borates include colemanite and howlite. Gypsum is abundant and calcite, zeolites, smectites, illite and chlorite are the other associated minerals in the Sultançayiri deposit. Borate minerals occur in many different forms; massive, fibrous, as thin layers interbedded with clay, as nodules, and in euhedral crystals.

The colemanite and pandermite nodules occurring respectively in the Kestelek and Sultançayiri deposit formed directly from solutions within the unconsolidated sediments below the sediment/water interface, and have continued to grow as the sediments have been compacted in the borate zones. Because these minerals are readily dissolved, secondary colemanite, occurring as transparent and euhedral crystals, is found in the cavities of nodules, in cracks and in vughs. Due to the effect of carbon dioxide-rich surface waters, all borates at or near the surface may be easily weathered, and completely replaced by calcite. Probertite, which forms in the same chemical environment as ulexite in the Kestelek deposit, indicates a period of greater evaporation within the playa lake. Howlite in the Sultançayiri deposit has apparently grown in the clays alternating with thin pandermite and colemanite bands, its growth having coincided with a period of increasing Si concentration. As a result of diagenesis in places small howlite nodules are embedded in unconsolidated pandermite and colemanite masses.

Hydrothermal solutions, thermal springs and tuffs associated with local volcanic activity are thought to be the source of the borates. The initial solutions from which the borates crystallized in the Kestelek deposit are deduced to have been very poor in chloride, low in sulfate and to have had abundant boron and calcium with subordinate sodium. On the other hand, in the Sultançayiri deposit, the initial solutions carried abundant boron, calcium and sulfate.

INTRODUCTION

The known borate deposits of Turkey occur in Western Anatolia, south of the Marmara Sea within an area measuring roughly 300 km east-west by 150 km north-south. The main borate districts are Bigadic, Kestelek, Sultançayiri, Emet and Kirka [1-5]. This area contains the largest borate reserves in the world [6-8]. The Kestelek and Sultançayiri borate deposits are situated in the northern part of the area (Fig. 1).

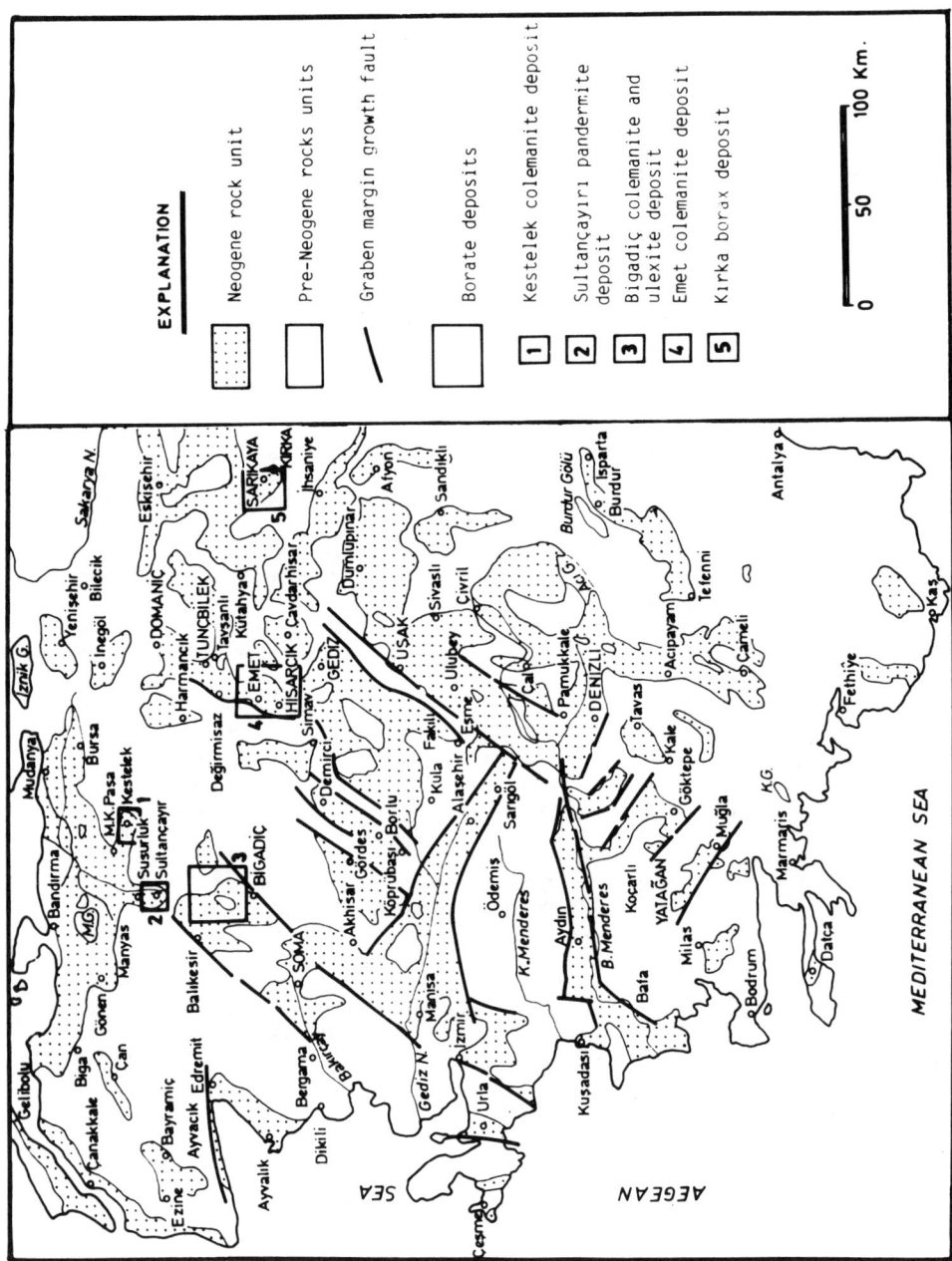

Fig. 1. Distribution of Neogene basins and borate deposits in Western Anatolia, Turkey.

The Kestelek deposit was discovered accidentally during a survey of lignite deposits for the Mineral Research and Exploration Institute of Turkey (MTA) in 1954. The only mineral recorded was colemanite, a very common calcium borate known in the other Turkish borate deposits [9]. Reserves here are about 10 million metric tons with a high ore grade.

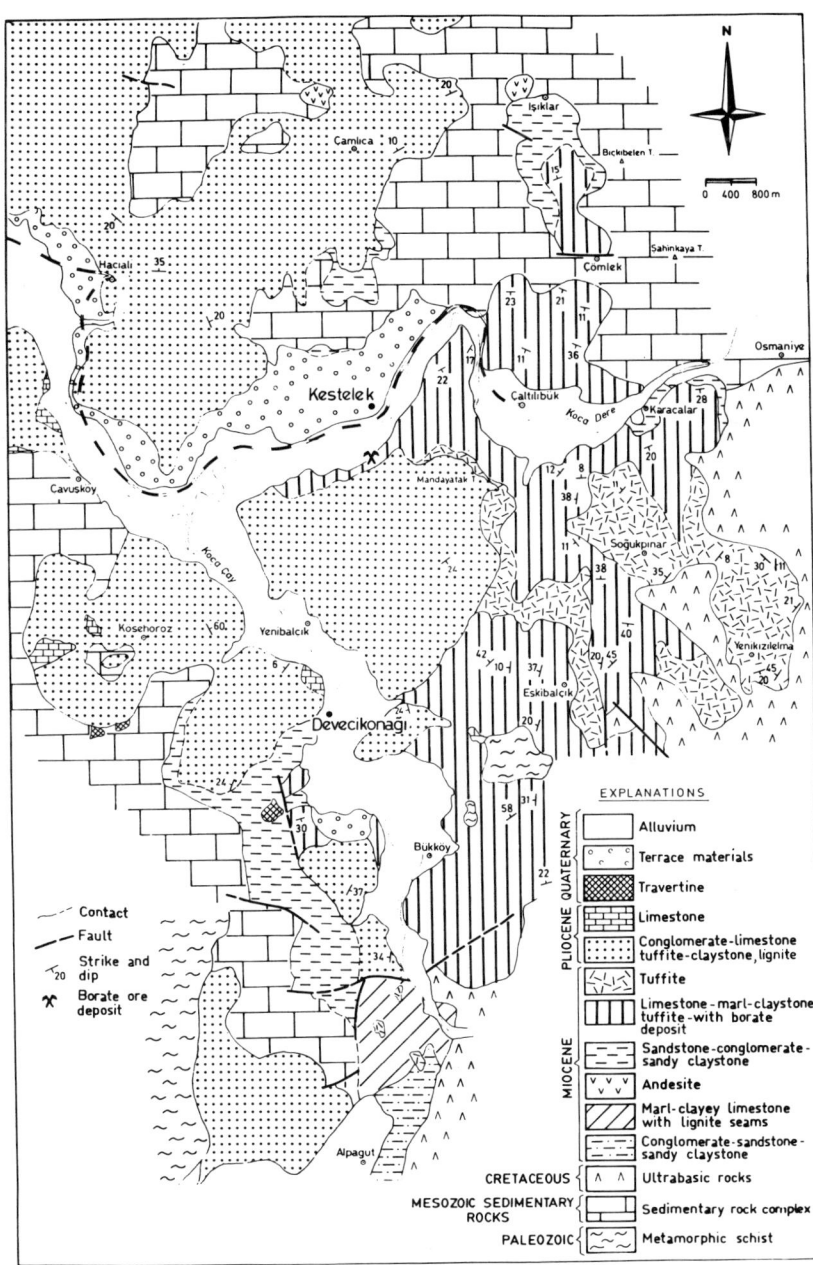

Fig. 2. Geological map of the Kestelek borate basin.

The Sultançayiri boratiferous gypsum basin contains the oldest workings among the known borate deposits of Western Anatolia, and the pandermite was mined from this deposit by open cast and underground mining methods. It is estimated that the reserves of the Sultançayiri borate deposit are approximately 1 million metric tons. However, much of these reserves was mined out by French and English companies between the years 1865 and 1955 [10].

The Kestelek and Sultançayiri borate deposits were first visited by the writer in 1979, and visited subsequently for several years for the purpose of making detailed maps and observations and for collecting samples, both from the open pits and underground workings.

The aim of the present paper is to report geological, mineralogical and geochemical aspects of the deposits, and to discuss the genesis of the borate minerals occurring in the Kestelek and Sultançayiri borate deposits.

GEOLOGICAL SETTING

The Turkish borate deposits formed during the Miocene in lacustrine environments in closed basins with abnormal salinity and alkalinity during periods of volcanic activity which commenced in early Tertiary time and continued at least until the beginning of the Quaternary [5,8,11,12]. The pre-Neogene basement of the basins comprises Paleozoic and Mesozoic rocks, partly belonging to the Menderes massif (Fig. 1). All these basins have been filled with a series of volcanic materials, such as volcanic muds, tuffaceous rocks, and flows. Ion-rich fluids have also been circulated into these basins [5,8,11].

Although the lithologies of the borate deposits show some differences from one to another, they are generally represented by Ca, Na, Sr, and Mg borates, and are interbedded with conglomerates, sandstone, tuff, tuffite, claystone, marl and limestone; and are usually enveloped by or grade into, limestones or claystones. Sandstones and conglomerates occur near the base of each deposit. Sediments in the borate basins show clear evidence of cyclicity and exhibit both lateral and vertical facies changes [3,5,8,9].

Fig. 3. Photograph of clay, marl, limestone and tuff alternating with borates at the Kestelek open-cast mine.

Volcanic rocks in the vicinity of the playa lakes in which the borate deposits formed a re extensive. The volcanic rocks comprise a calc-alkaline series of flows ranging from acidic to basic, and pyroclastic rocks which are interbedded with the sediments.

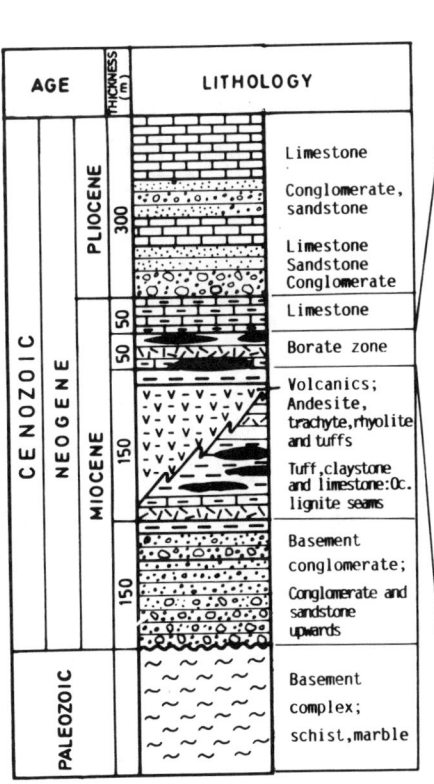
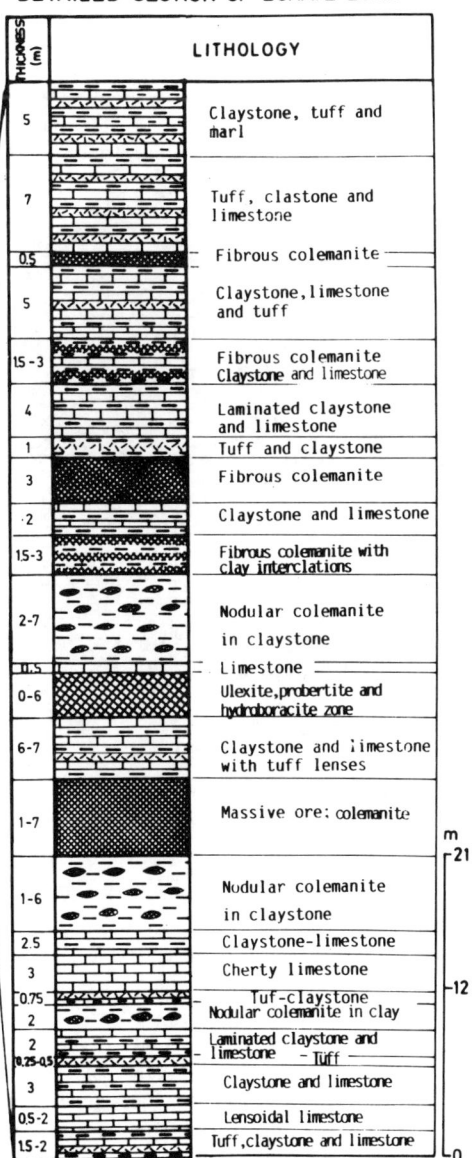

Fig. 4. Generalized columnar section of the Kestelek area.

Kestelek Deposit

The Neogene sediments which contain the borates in the Kestelek region rest unconformably on the Paleozoic and Mesozoic basement complex (Fig. 2). The Neogene sequence consists of, from bottom to top: basement conglomerate and sandstones; claystone with lignite seams, marl, limestone, and tuff; agglomerates and volcanic rocks; the borate zone consisting of clay, marl, limestone, tuff and borates (Fig. 3); limestones with thin clay and chert bands (Fig. 4). These sediments were deposited during a tectonically stable period accompanied by extensive volcanic activity. During this period, the volcanic activity gradually increased and produced tuff, tuffite and agglomerate, and andesitic, trachytic and rhyolitic volcanic rocks which are interbedded with sediments (Figs. 2 and 4). K/Ar age dating of one tuff sample taken from the borate zone yields an age of 17.4 m.y. This sequence is capped by loosely cemented Pliocene conglomerate, sandstone and limestone [9,13] (Fig. 4).

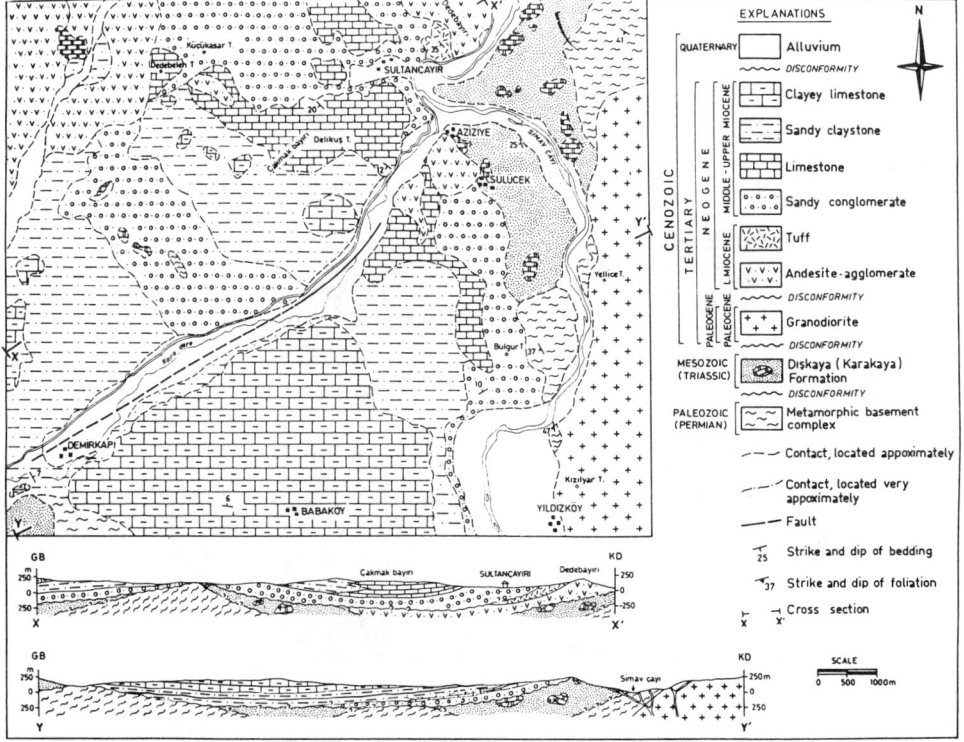

Fig. 5. Geological map of the Sultançayiri boratiferous gypsum basin.

Sultançayiri Deposit

The Miocene volcano-sedimentary rocks units which contain borates in the Sultançayiri region rest unconformably on the Paleozoic metamorphic basement complex, Mesozoic Diskaya (Karakaya) Formation, and a Paleogene granodiorite which intruded the basement rock units (Fig. 5). The Miocene sequence consists of, in ascending order: andesite and agglomerate; tuffs; basement sandy conglomerate; limestone; sandy claystones containing boratiferous gypsum, bedded gypsum and tuff; and clayey limestone (Fig. 6).

The borates are interbedded with gypsum, claystone, limestone and tuffs within the borate zone of the sandy claystone unit. Borate beds which are predominantly pandermite (priceite) typically alternate with gypsum in the lower section of the boratiferous gypsum of the sandy claystone unit (Fig. 6). Boratiferous gypsum beds, varying from 10 to 15 meters in total thickness, occur within the sandy claystone unit (Fig. 7). Pandermite and randomly distributed howlite nodules are common within these gypsum horizons (Fig. 8). K/Ar age dating of one tuff sample taken from the tuff unit yields an age of 20.01 m.y.

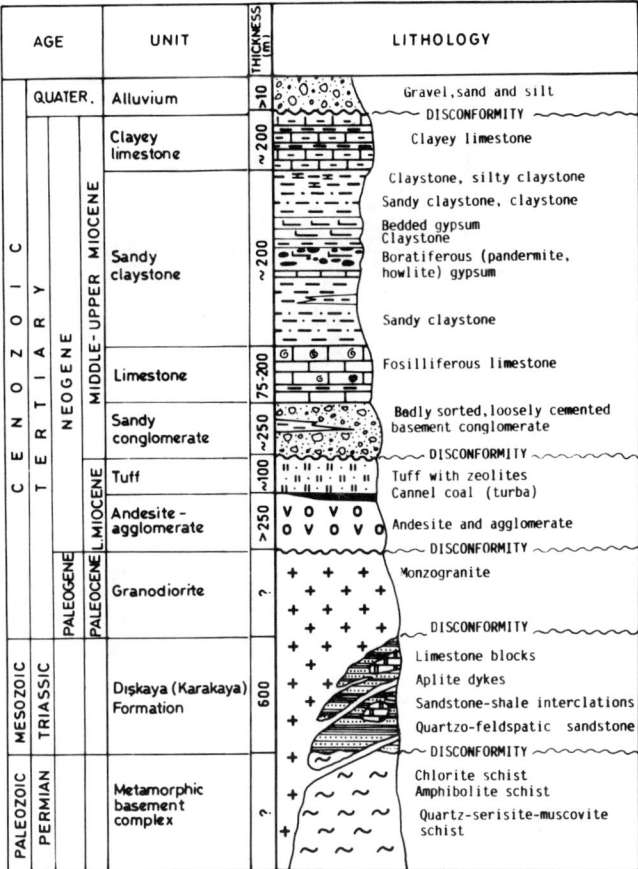

Fig. 6. Generalized columnar section of the Sultançayiri basin.

SAMPLE PREPARATION AND ANALYTICAL METHODS

All collected borate and non-borate minerals are bulk mine samples. Detrital material and clay fractions have been eliminated from the borate samples. Borate and non-borate minerals were determined by Phillips direct-recording X-ray diffractometer analyses by standard powder and oriented-sample techniques.

Si, Fe, Al, Ca, Mg, Sr, As, S, B and L in the borate, non-borate and rock samples were determined using a 1272 Beckman atomic absorption spectrophotometer. Na and K were determined by flame photometry. All samples were run in duplicate.

Fig. 7. Bedded gypsum within the sandy claystone unit, Sultançayiri deposit.

Fig. 8. Pandermite (priceite) nodules within the gypsum beds of the sandy claystone unit, Sultançayiri deposit.

MINERALOGY

Borate minerals occur interbedded with claystones and tuffs as nodules, often with discoidal shape, or in masses, and as thin layers of fibrous and euhedral crystals. Borate minerals also occur in fibrous aggregates, massive forms, and in euhedral crystals in vugh fillings.

Kestelek Deposit
Colemanite, ulexite and probertite predominate, with hydroboracite occurring only rarely in the Kestelek deposit [13,14] (Table 1). Secondary colemanite occurs in transparent, euhedral

crystals in the cavities of nodules, in cracks and in vughs. The presence of probertite, which forms in the same chemical environment as ulexite in the Kestelek deposit, indicates a period of higher temperature within the playa lake.

Calcite, quartz, zeolites, smectite, illite and chlorite a re the associated minerals. Clinoptilolite and heulandite are the only known authigenic silicate minerals occurring within the interbedded tuffs. The tuffs consist of oligoclase, quartz, biotite and smectite.

Table 1. Borate minerals recorded from Kestelek and Sultançayiri borate deposits.

Mineral Name	Oxide Formula	Kestelek Deposit	Sultançayiri Deposit
Colemanite	$Ca_2B_6O_{11} \cdot 5H_2O$	+	+
Pandermite(priceite)	$Ca_4B_{10}O_{19} \cdot 7H_2O$	-	+
Ulexite	$NaCaB_5O_9 \cdot 8H_2O$	+	-
Probertite	$NaCaB_5O_9 \cdot 5H_2O$	+	-
Hydroboracite	$CaMgB_6O_{11} \cdot 6H_2O$	+	-
Howlite	$Ca_4Si_2B_{10}O_{23}\ 5H_2O$	-	+

Sultançayiri Deposit

Pandermite (priceite) predominates, but other borate minerals include colemanite and howlite (Table 1). Howlite, which has apparently grown in the clays alternating with thin pardermite (priceite) and colemanite bands, indicates a period of increasing silica concentration. As a result of diagenetic events, some small howlite nodules are embedded in the pandermite (priceite) and colemanite masses and in the gypsum beds [13,14].

Gypsum is abundant, and calcite, zeolites, smectites, illite and chlorite are the other associated minerals in this deposit. Gypsum occurs in independent masses, thin bedded layers, or as intergrowths with pandermite. Boron-bearing K-feldspar, clinoptilolite and opal-CT are the authigenic silicates. Volcanogenic high sanidine, anorthoclase, quartz, and clay minerals such as illite and smectite are also present in the associated tuffs and claystones.

Borate minerals occurring in the Kestelek and Sultançayiri deposits are described and discussed in detail below.

Colemanite ($Ca_2B_6O_{11} \cdot 5H_2O$)

Colemanite is by far the most common mineral in the Kestelek deposit, whereas it occurs rarely in the Sultançayiri deposit. It coexists with ulexite, probertite and hydroboracite in the Kestelek deposit, and pandermite, howlite and gypsum minerals in the Sultançayiri deposit. It occurs in many different forms ranging from minute stellate crystal clusters in clay, to ovoid nodules up to 50 cm in diameter, and in continuous layers. Among the commoner habits are nodular forms with radiating structures (Fig. 9); massive granular colemanite; disseminated crystals, often stellate, in a clay matrix; fibrous layers surrounding nodules (Fig. 9); thin layers interbedded with clay, sometimes brecciated; vugh fillings (Fig. 10); and transparent and euhedral crystals (Fig. 11).

Nodules are by far the most common form of colemanite, but these nodules occur in a large variety of shapes and sizes. There is a tendency for the smaller nodules to be spherical or ellipsoidal (Fig. 9), and the larger ones to be ovoidal. Some, irrespective of size, contain vughs; which sometimes contain a liquid; other have a core of granular colemanite which is coarsely crystalline and second-generation in origin (Fig. 9). This secondary colemanite formation is mainly represented by transparent and euhedral crystals and is often present in the cavities of nodules (Fig. 11), in cracks and in vughs (Fig. 10). These euhedral crystals from the

Kestelek deposit are probably the best known crystals from the Turkish and other borate deposits of the world.

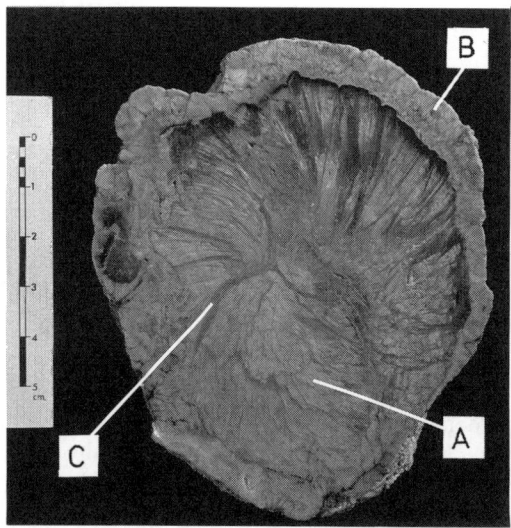

Fig. 9. Section of a colemanite nodule with three generations of colemanite: A, massive colemanite crystals; B, fibrous colemanite crystals; and C, coarsely crystalline colemanite in cracks (possibly septarian)

Fig. 10. Euhedral colemanite crystals on massive or granular colemanites, and in cracks and vughs, Kestelek open-cast mine.

Fig. 11. Euhedral colemanite crystals up to 5 cm in length in the vugh of a colemanite nodule, Kestelek open-cast mine.

Careful inspection reveals that these nodules grew in successive stages, each layer being separated by a thin, discontinuous veneer of caly (Fig. 9). Later generations of colemanite crystals radiate from separate centers of nucleation on the original nodule. It is often difficult to identify all stages of nodule growth, but judging from the presence of included clay, it is clear that these nodules formed within the clays and tuffs below the sediment/water interface, and probably continued to grow as the sediments were compacted in these deposits (Fig. 9).

Due to the effect of carbon dioxide-rich surface and underground waters, colemanites at or near the surface may be easily weathered, and completely replaced by calcite.

Pandermite (=priceite) ($Ca_4B_{10}O_{19} \cdot 7H_2O$)

Pandermite predominates at the Sultançayiri deposit. It also occurs rarely at the Bigadic deposits [8]. It was named after the town Bandirma (NW Anatolia, Turkey) from whence it was exported. Later it was discovered that pandermite is identical to priceite, described from Oregon, USA [15,16]. The identity of that pandermite is identical to priceite has been established by chemical and optical studies. Pandermite is found as nodules and masses up to a ton in weight, and as lenses interbedded with gypsum and clay in the lower part of the boratiferous gypsum zone in the Sultançayiri deposit. It is white, compact with fine-grained crystals, and sometimes resembles magnesite and chalky limestone (Fig. 12). It is associated with colemanite, howlite and gypsum and alters calcite under surface or near surface conditions (Fig. 13).

Ulexite ($NaCaB_5O_9 \cdot 8H_2O$)

Ulexite occurs with following textures in the Kestelek deposit: massive; cauliflower-like nodules; fibrous; cone; rosette; and columnar (Fig. 14). Locally, very thin, fibrous ulexite crystals growing on top of the massive and cauliflower-like ulexite nodules occur, and cauliflower-like nodules composed of randomly oriented crystals, 1-5 cm long, form independent layers up to a few decimeters in thickness. Thin veins of secondary diagenetic ulexite commonly fill the small cracks in the ulexite, clay and tuff mixture (Fig. 14).

Fig. 12. Compact, white and powdery occurrence of pandermite (priceite), Sultançayiri deposit.

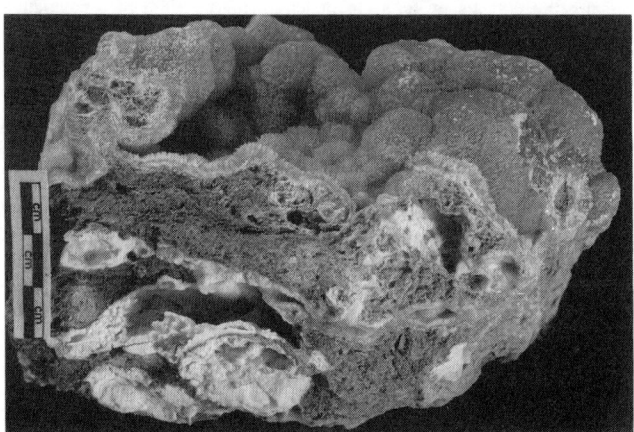

Fig. 13. Altered pandermite nodules and secondary calcite formation after pandermite, Sultançayiri deposit.

Ulexite is generally associated with colemanite, probertite and hydroboracite in the Kestelek deposit, but no alteration to colemanite and probertite, or from any other mineral has been observed. It is usually white and very soft. The purest forms of ulexite are white, but many are gray due to the fact that the nodules grow in clay and mud. Thus, like colemanite and other borates occurring in this deposit, ulexite nodules appear to have developed within and not on the sediments.

Probertite (NaCaB$_5$O$_9 \cdot$5H$_2$O)

Probertite has a restricted distribution and occurs only in the underground mine in the Kestelek deposit. It coexists with ulexite and is present in the deeper parts of the calcium-sodium borate zone.

Probertite is light yellow to white, and occurs as radiating or fibrous crystals (Fig. 15). The size of the crystals ranges between 5 mm and 5 cm. Crystals show generally radiating structures, and clay fills the spaces between crystals. Probertite coexists with ulexite, colemanite and hydroboracite. This mineral which forms in the same chemical environment as ulexite in the Kestelek deposit, indicates period of greater evaporation within the playa lake.

Fig. 14 Massive cauliflower-like ulexite nodules with silky appearance and radiating crystals, and diagenetic ulexite veins cutting ulexiteclay-tuff mixture.

Hydroboracite $(CaMgB_6O_{11} \cdot 6H_2O)$

Hydroboracite is found in all the major borate deposits of Turkey [14] and occurs sporadically at different horizons and particularly in clay layers in the Kestelek deposit. It forms small clusters and nodules in which radiating needle-shaped crystals, 0.5-5 cm, are randomly oriented. Radiating needle-shaped crystals of hydroboracite are mutually intersecting and groups of them have a conical appearance (Fig. 16).

Fig. 15. Coarsely radiating probertite crystals, Kestelek underground mine.

Fig. 16. Radiating, mutually intersecting crystals of hydroboracite and groups of them having a conical appearance, Kestelek underground mine.

Locally, hydroboracite forms thin layers within the clay interbeds. In thin section, the needle-shaped crystals of hydroboracite have a fibrous texture. This mineral is usually white, but in some cases light gray or yellowish. It is clearly associated with colemanite and ulexite in the Kestelek deposit.

Howlite ($Ca_4Si_2B_{10}O_{23} \cdot 5H_2O$)

Howlite has a very restricted distribution and has been found only in the Sultançayiri deposit and a few mine pits of the Bigadic deposits. It occurs as small, compact, nodular white masses, internally dense and structureless, resembling unglazed porcelain (that is, chalk-like and earthy) (Fig. 17). Howlite is associated with colemanite, pandermite and gypsum in the Sultançayiri deposit.

Fig. 17. Howlite nodules of different sizes, Sultançayiri deposit.

Fig. 18. Howlite nodule embedded in a pandermite mass, Sultançayiri deposit.

Fig. 19. Howlite nodules of different sizes embedded in a pandermite gypsum mixture.

Howlite in the Sultançayiri deposit apparently has grown in the clays alternating with thin pandermite and colemanite bands, its growth having coincided with a period of increasing Si concentration. As a result of diagenesis, locally small howlite nodules have grown, embedded in unconsolidated pandermite (Fig. 18), and in pardermite-gypsum mixtures (Fig. 19).

Non-borates

A number of non-borate minerals associated with borates occur in the borate zones and interbedded claystones and tuffs of these deposit. Generally, borate minerals are associated with calcite, dolomite, gypsum, zeolites and clay minerals. Calcite, dolomite and quartz are common in the Kestelek deposit, whereas gypsum is widespread in the Sultançayiri deposit. Calcite also occurs, however, in surface outcrops and adjacent to faults as a result of modern weathering. Therefore, much of the calcite may be of very recent origin, and occurs as an alteration product of borate minerals at or near the surface (Fig. 13).

Gypsum occurs with massive, laminated, crystalline and bedded textures in the Sultançayiri deposit (Fig. 7). The probable distribution of gypsum beds in the Sultançayiri deposit has been determined and 300 million tons of reserves are estimated. There gypsum beds have a mean SO_3 content of 34 %, and furthermore, the properties of the gypsum from these beds are suitable for the cement industry.

The occurrence of the clay minerals, predominantly smectite and less frequently illite, is ubiquitous in both deposits. Clinoptilolite and heulandite in the Kestelek deposit, and boron-bearing K-feldspar, clinoptilolite and opal-CT in the Sultançayiri deposit are the authigenic silicate minerals occurring in the associated tuff intercalation.

ORIGIN AND DIAGENESIS OF THE DEPOSIT

Borate minerals occur in many different forms, such as nodular, disseminated euhedral crystals, thin layers ,and as vein and vugh fillings. Nodular forms, massive-granular, thin layers and disseminated crystals in a clay and/or tuff matrix have been considered as early minerals, having formed penecontemporaneously with the clastic sediments [5,12,17]. Vein and vugh fillings of borate minerals are believed to be late diagenetic products, and similar observations have been made for the Kramer, Sijes, Loma Blanca and Emet deposits by different authors [5,12,17-20]. From field and textural evidence it is clear that all early-formed borate minerals, such as colemanite, pandermite, ulexite, probertite, hydroboracite and howlite, were deposited penecontemporaneously with the unconsolidated sediments. It is likely that they were formed within the clays and tuffs below the sediment/water interface and probably continued to grow as the sediments were compacted.

The Kestelek deposit is characterized by high Ca (colemanite), very low Na, and very low chloride and sulphate, whereas the Sultançayiri deposit is characterized by abundant Ca, B and sulphate relative to the other Turkish borate deposits [3,5,8,12].

Table 2. Partial chemical analysis of colemanite ores and clays from the Kestelek borate deposit.

Oxide %	Colemanite ore samples								Clay Sample
	1	2	3	4	5	6	7	8	9
B_2O_3	40.81	39.92	37.77	44.12	45.05	40.05	43.19	39.88	2.10
Fe_2O_3	0.70	0.23	0.74	0.3~	0.34	0.43	0.44	0.68	4.48
SiO_2	2.97	1.83	8.03	1.90	1.30	3.43	3.70	6.13	17.27
Al_2O_3	0.65	0.70	0.84	0.84	0.55	0.65	0.60	0.80	26.50
CaO	36.58	32.90	29.40	35.00	30.10	30.80	33.60	28.00	6.65
MgO	0.65	0.56	0.47	0.45	0.65	0.60	0.70	0.60	9.76
Na_2O	0.80	0.40	0.05	0.60	0.33	0.94	0.40	0.33	1.27
K_2O	0.02	0.02	0.07	0.01	0.01	0.03	0.03	0.04	0.60
Li_2O	0.03	0.04	0.54	0.05	0.06	0.07	0.07	0.06	0.10
SrO	0.43	0.40	0.06	0.35	0.37	0.34	0.38	-0.51	0.06
SO_3	0.00	0.00	0.00	0.00	0.00	0.00	0.00	0.00	3.43
As	0.0005	0.0005	0.004	0.0006	0.0005	0.0008	0.0007	0.0007	0.013

The chemical analyses of the colemanite, and other borate and gypsum samples from the Kestelek and Sultançayiri deposits indicate that there are several differences between these two deposits beyond their different mineral assemblages (Tables 2 and 3). The main difference is the sulphate content of the deposits; overall the Sultançayiri deposit is characterized by a very high sulphate content (Table 3).

Table 3. Chemical analyses of borate and boratiferous gypsum samples from the Sultançayiri borate deposit.

Oxide %	Pandermite	Howlite	Boratiferous gypsums					
	1	2	3	4	5	6	7	8
B_2O_3	46.49	40.69	7.65	3.31	Trace	Trace	Trace	Trace
SO_3	0.57	1.49	40.69	38.69	51.47	34.94	42.31	43.62
SiO_2	0.21	15.56	2.20	3.34	1.61	0.20	0.80	1.40
Al_2O_3	0-04	0.08	0 37	0 30	0 41	0.05	0.34	0.36
Fe_2O_3	0.04	0.05	0.23	0.32	0.25	0.39	0.24	0.24
CaO	34.50	29.98	35.75	30.56	36.20	39.86	33.56	32.22
MgO	0.12	0.19	1.28	1.63	1.99	0.30	0.03	0.33
K_2O	0.05	0.10	Trace	0.12	Trace	0.04	0.04	0.02
Na_2O	0.80	0.92	0.63	0.07	0.62	0.05	0.04	0.06
Li_2O	0.03	0.02	0.03	0.01	0.02	0.03	0.02	0.01
LOI	18.09	11.48	10.76	19.29	7.43	22.39	22.19	21.80
Total	100.94	100.56	99.89	97.64	100.00	98.25	99.57	100.06

Table 4. Chemical analysis of the Yildiz thermal spring, the Sultançayiri area, which deposits thenardite, nahcolite and quartz at present.

CATIONS (mg/l)		ANIONS (mg/l)	
Na	330	Cl	66
K	17	F	8.1
Mg	0.60	SO_4	211
Ca	21	CO_3	0.1
Li	0.80	HCO_3	630
B	1.21		
As	0.95		
Sn	1.36		

Some of the major oxide distribution such as B_2O_3, SO_3, CaO and SiO_2 along the vertical section of the boratiferous gypsum, which is part of the sandy claystone unit in the Sultançayiri deposit is shown on Fig. 20. Positive correlation between B_2O_3 versus CaO and SO_3 versus CaO is clearly seen on Fig. 20.

Borate minerals in the Kestelek deposit, like the tuffs and clays in the borate zone, are characterized by very low concentrations of As, Sr and S (Table 2), relative to minerals from the Emet deposits [5], the Kirka deposit [3], and the Western U.S. borate deposits [18]. These elements were chemically precipitated from brines, mainly as sulphates, which may have been partly derived from thermal springs located adjacent to the borate deposits. One thermal spring in the Sultançayiri deposit, the Yildiz thermal spring, issuing from the metamorphic basement complex with no evidence of buried borates nearby, contains 0.96 mg/l boron, and deposits major thenardite, nahcolite and quarts at present (Table 4).

Major-oxide compositions of clays and tuffs interbedded with the borate deposits are given in Tables 2 and 5. Generally, the particle size of the tuffaceous material is coarser at Sultançayiri than in the Kestelek area. Na_2O values are higher in the Kestelek tuffs which contain oligoclase as a major constituent. The highest SiO_2 concentration was detected in the

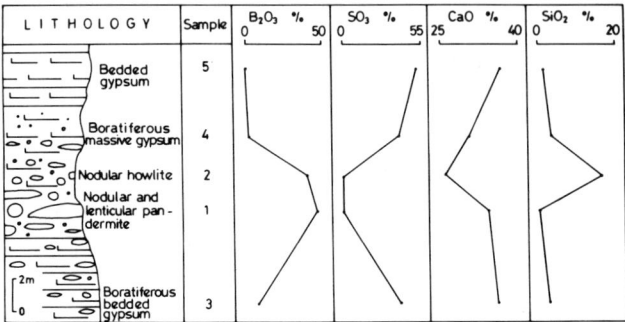

Fig. 20. Vertical variation of major oxides of borate and boratiferous gypsum in the Sultançayiri deposit.

tuffaceous rocks from Sultançayiri, where both opal-CT and quartz are the major constituents. High CaO and MgO contents depend on the abundance of dolomite and calcite of some tuffaceous samples (Table 5).

Where authigenic K-feldspar is present (the Sultançayiri deposit), a pH greater than 9 can exist basinward, in the final stage of lake evolution. The pH values of the lake margins may be lower (7< pH <8). At the Kestelek area the presence of only clinoptilolite and opal-CT seems to indicate lower pH values than the Sultançayiri deposit, namely pHs between 8 and 9 [21]. This points to the conclusion that a standard pH value is not necessary to precipitate borate salts but, rather, a range of values, higher than 8 is sufficient. Ataman and Baysal [22] measured clay minerals from the western Anatolian borate deposits found that there are differences between the pHs of the Emet, Bigadic and Kirka deposits. The formation and paragenetic association of authigenic silicates, such as clinoptilolite, analcite and K-feldspar, requires pH values ranging from >8 through >10. This pH range in borate-bearing basins shows that the precipitation of borates starts at a pH around nine and continues at higher alkalinities.

There is much geological and mineralogical evidence to show that secondary alteration and diagenetic-mineral formation played an important role in the modification of mineralogical and chemical distributions within these deposits. Diagenetic borate minerals with euhedral

Table 5. Chemical analyses of tuffaceous rocks from the Kestelek and Sultançayiri borate districts.

Oxide %	Kestelek		Sultançayiri		
	1	2	3	4	5
SiO_2	55.00	61.81	73.00	73.50	73.50
Al_2O_3	16.90	11.96	11.60	11.50	11.60
TiO_2	0.54	0.22	0.04	0.03	0.03
FeO	0.22	0.20	0.00	0.00	0.00
Fe_2O_3	2.33	2.29	0.70	0.65	0.63
MnO	0.02	0.07	0.05	0.05	0.04
MgO	3.23	3.71	1.16	0.32	0.22
CaO	4.90	5.61	1.30	1.20	1.30
Na_2O	3.57	3.09	2.35	2.50	2.30
K_2O	2.90	2.20	4.75	4.45	4.65
LOI	10.47	8.15	5.20	5.30	5.30
Total	100.08	99.30	100.15	99.50	99.57

crystals, authigenic boron-bearing K-feldspar, zeolites (clinoptilolite and heulandite), and chlorite are good examples of this process. Much of the volcanic ash and glassy matrix of the tuffs is converted to authigenic minerals, such as K-feldspar, zeolites and clay minerals during diagenesis.

The geochemical association of boron, arsenic and sulphur suggests that the initial brines were always fed by an abundance of calcium and boron in the Kestelek deposits, whereas calcium, boron and sulphur are associated in the Sultançayiri deposit. Arsenic was present only in trace amounts. Surface streams may have carried Ca, Mg and Sr in solution into the basins as a result of weathering of rocks exposed in the adjacent area, but the major source is considered to be leaching of underlying basement metamorphic rocks by thermal springs and surface streams. Volcanic rocks, tuffs and clays interbedded with the borates are the most likely sources for Ca, Mg, Na, Sr, B, As and S. It seems likely that thermal springs were important sources of B, As, S and probably Ca and Na when the borates formed, as similar observations have been made for many other deposits elsewhere [5,17,18,20,23]. The association of boron-bearing hydrothermal solutions and thermal springs with volcanic rocks is known elsewhere in the world [23-26].

In the Kestelek and Sultançayiri areas, the extensive volcanic rocks associations and tuff intercalations with the borates indicate that much of the sediment was derived from a volcanic terrain. Hydrothermal solutions, thermal springs and tuffs associated with local volcanic activity are thought to have been the source of the borates. The initial solutions crystallizing the borates in the Kestelek deposit are deduced to have been very poor in chloride, low in sulphate and to have had abundant boron and calcium with subordinate sodium. On the other hand, in the Sultançayiri deposit, the initial solutions carried abundant boron, calcium and sulphate.

ACKNOWLEDGMENTS

A special acknowledgment goes to Steven K. Mittwede, who provided constructive criticism of the manuscript. Field work was supported by the Etibank State Company in Ankara and the local mineral production division in Kestelek, and I thank the management and technical staff for their assistance. I also thank Mualla Gurle for considerable drafting assistance, Erol Sanll for photographing assistance and Meral Akdere for typing assistance.

REFERENCES

1. Meixner, H., Mineralogische Beobachtungen an Colemanit, Inyoit, Meyerhofferit, Tertschit und Ulexit aus neuem Turkischen Boratlargestatten. *Neidelb. Beitr. Miner. Petrogr.* **3**, 445-455 (1953).
2. Baysal, O., Mineralogic and genetic studies of the Sarikaya (Kirka) borate deposits. Ph.D. Thesis, Hacettepe University, Turkey, 157p (1972) (in Turkish)
3. İnan, K., Dunham, A.C. and Esson, J., Mineralogy, chemistry and origin of Kirka borate deposit, Eskisehir Province, Turkey. *Trans. Inst. Min. Metall., Sect. B*, **82**, 114-123 (1973).
4. Helvaci, C., Geology, mineralogy and geochemistry of the borate deposits and associated rocks at the Emet Valley, Turkey. Ph.D. Thesis, University of Nottingham, England, 338 p (1977).
5. Helvaci, C., Occurrence of rare borate minerals: Veatchite-A, tunellite, teruggite and cahnite in the Emet borate deposits, Turkey*Mineral Deposita* **19**, 217-226 (1984).
6. Lyday, P.A. (1982) Boron: *Minerals Yearbook 1980*, U.S. Bureau of Mines, 1: p 3
7. Kistler, R.B. and Smith, W.C., Boron and borates. In: S.J.Lefond (ed.) *AIME: Industrial Minerals and Rocks*, New York, 533-560, (1983).

8. Helvaci, C., Mineralogy and genesis of Bigadic borate deposits, western Turkey. 28 th. Int. Geol. Congress, Washington D.C., Book of abstracts, 2, 49-50, (1989).
9. Özpeker, I., Western Anatiolian borate deposits and their genetic studies. Ph.D. Thesis, Technical University of İstanbul, Turkey, (1969).
10. Travis, N.J. and Cocks, E.J., *The tincal trail, a history of borax*. Harrap Limited, London, 311 p, (1984).
11. Helvaci, C., Stratigraphic and structural evolution of the Emet borate deposits, western Anatolia, Turkey. *Dokul Eylül Univ., İzmir, Turkey, Research Paper MM/JEO-86*, AR 008, 28 p, (1986).
12. Helvaci, C. and Firman, R.J., Geological setting and mineralogy of Emet borate deposits, Turkey. *Inst. Mining Metallurgy Trans., Sect. B.*, **85**, 142-152, (1976).
13. Helvaci, C., Mineral assemblages and formation of the Kestelek and Sultançayiri borate deposits, western Turkey. 29 th International Geological Congress, Kyoto, Japan, Abstracts, 2, p 327, (1992).
14. Helvaci, C., A review of the mineralogy of the Turkish borate deposits. *Mercian geology*, **6**, 257-270, (1978)
15. Palacne, C., Berman, H. and Frandel, C., *Dana's System of Mineralogy*, John Willey, New York, 7 th ed., 2, p 320-389, (1951).
16. Meixner, H., Borate deposits of Turkey. *M.T.A. Institute Bull.Ankara*, No. 125, 1-2, (1965).
17. Alonso, R.N., Helvaci, C., Sureda, R.J. and Viramonte, J.G., A new Tertiary borax deposit in the Andes. *Mineral. Deposita*, **23**, 299-305, (1988).
18. Bowser, C.J., Geochemistry and petrology of the sodium borates in the non-marine evaporitic environment. Ph.D. Thesis, University of California, Los Angeles, 282 p, (1965).
19. Barker, C.E. and Barker, J.M., A re-evalution of the origin and diagenesis of borate deposits, Death Valley region, California. In: *Borates: Economic Geology and Production*. Society of Mining Engineers of the American Institute of Mining, Metallurgical and Petroleum Engineers, Inc., pp 101-135m (1985).
20. Alonso, R.N., Occurrence, stratigraphic position and genesis of the borate deposits in Puna, Argentina. Ph.D.Thesis. Universidad Nacional de Salta, Argentina, 196 p, (1986) (in Spanish).
21. Surdam, R.C. and Parker, B.R., Authigenic aluminosilicate minerals in the tuffaceous rocks of the Green River Formation, Wyoming. *Geol.Soc. Ame. Bull.*, **83**, 689-700, (1972).
22. Ataman, G. and Baysal, O., Clay mineralogy of Turkish borate deposits. *Chem. Geol.*, **22**, 233-247, (1978).
23. Muessig, S., Recent South American Borate Deposits. In: *Second Symp. On Salt* (Clereland Ohio: Northern Ohie Geological Society), 1, 151-159, (1966).
24. White, D.E., Thermal waters of volcanic origin. *Geol. Soc. Amer. Bull.*, **68**, 1637-1658 (1957)
25. White, D.E., Hem, J.D. and Waring, G.A., Chemical composition of subsurface waters. In: *Data of Geochemistry, 6 th Ed.* U.S. Geol. Survey, Prof. Paper, 440F, 67p, (1963).
26. Kitano, Y., Geochemistry of calcareous deposits found in hot springs. *J. Earth Sci., Nagoya Univ.* **11**, 68-100, (1963).

Hydrogeochemical controls on the formation of primary dolomite in some ephemeral lakes in the Coorong region of South Australia

R. AHMAD[1] and P. B. HOSTETLER[2]

[1] *Department of Geology, The Australian National University, Canberra, ACT 0200, AUSTRALIA.*
[2] *School of Earth Sciences, Macquarie University, North Ryde, NSW 2109, AUSTRALIA.*

Abstract-- Active precipitation of protodolomites/dolomites and some other carbonate minerals has been taking place in a number of lakes in the Coorong region of South Australia during the Holocene. Detailed sedimentological and hydrogeochemical investigations of these lakes have been carried out over a number of years. Newly formed protodolomites/dolomites are fine grained (< 1 μn), disordered or partly ordered to ordered, containing 41 to 53 mole% $MgCO_3$ in their crystal lattices. Their physicochemcial properties vary among the lakes as well as within individual lake basins. The hydrogeochemistry of the surface as well as the subsurface waters varies widely among the lakes. Some, at least, of the protodolomites/dolomites are primary, and have formed by direct chemical precipitation from surface waters as a result of evaporative concentration, and at elevated pH caused by plant photosynthesis and CO_2-degassing. pH, salinity (TDS), and the Mg/Ca ratio of the lake waters are the most important factors that control the physicochemical properties of the protodolomites/dolomites as well as the type of the carbonate mineral assemblages in the lakes. Active precipitation of protodolomite/dolomite in the lakes takes place within the Mg/Ca ratio range of 10 to 50 and the salinity (TDS) range of 10000 ppm to 250000 ppm in the surface waters of the lakes. During the late stage of evaporation, at salinites above 250000 ppm, the Mg/Ca ratio in the lake waters/brines grow exponentially probably due to the inhibition in the precipitation of protodolomite/dolomite. Higher pH (8 - 10.5) and lower salinity (10500 ppm - 52900 ppm) tend to favor the precipitation of dolomite - magnesite assemblage, whereas relatively lower pH (7.3 - 9) and higher salinity (27500 ppm - 400300 ppm) favor the precipitation of dolomite - hydromagnesite or hydromagnesite - dolomite assemblages. Lower salinity and higher pH give better stoichiometry/ordering to the dolomites. Lower salinity and longer duration of crystallization favor the growth of relatively larger size and better developed rhombic shape for the crystals of dolomites in the lakes.

Keywords: Coorong, South Australia, Holocene, ephemeral lakes, carbonate facies/microfacies, primary dolomite, protodolomite, fine grained, dolomite ordering, hydrogeochemical controls, temperature, evaporation, pH, density, Mg/Ca ratio, salinity, plant photosynthesis, CO_2-degassing.

INTRODUCTION

The Coorong region of South Australia encompasses a number of small, shallow, ephemeral lakes (Fig. 1). Active precipitation of primary protodolomite/dolomite and other carbonate minerals from the surface waters of these lakes has occurred during the Holocene, and continues today [1]. Dolomite descriptions are common in the geological record [2,3] and dolomite is currently forming in modern lakes [4], sabkhas and shallow marine environments [5], yet the physicochemical and hydrogeochemical controls are not well understood. Several researchers [6-8] have synthesized protodolomite/dolomite like materials in the laboratory but no one has yet been able to make true dolomite. A number of models for the formation of

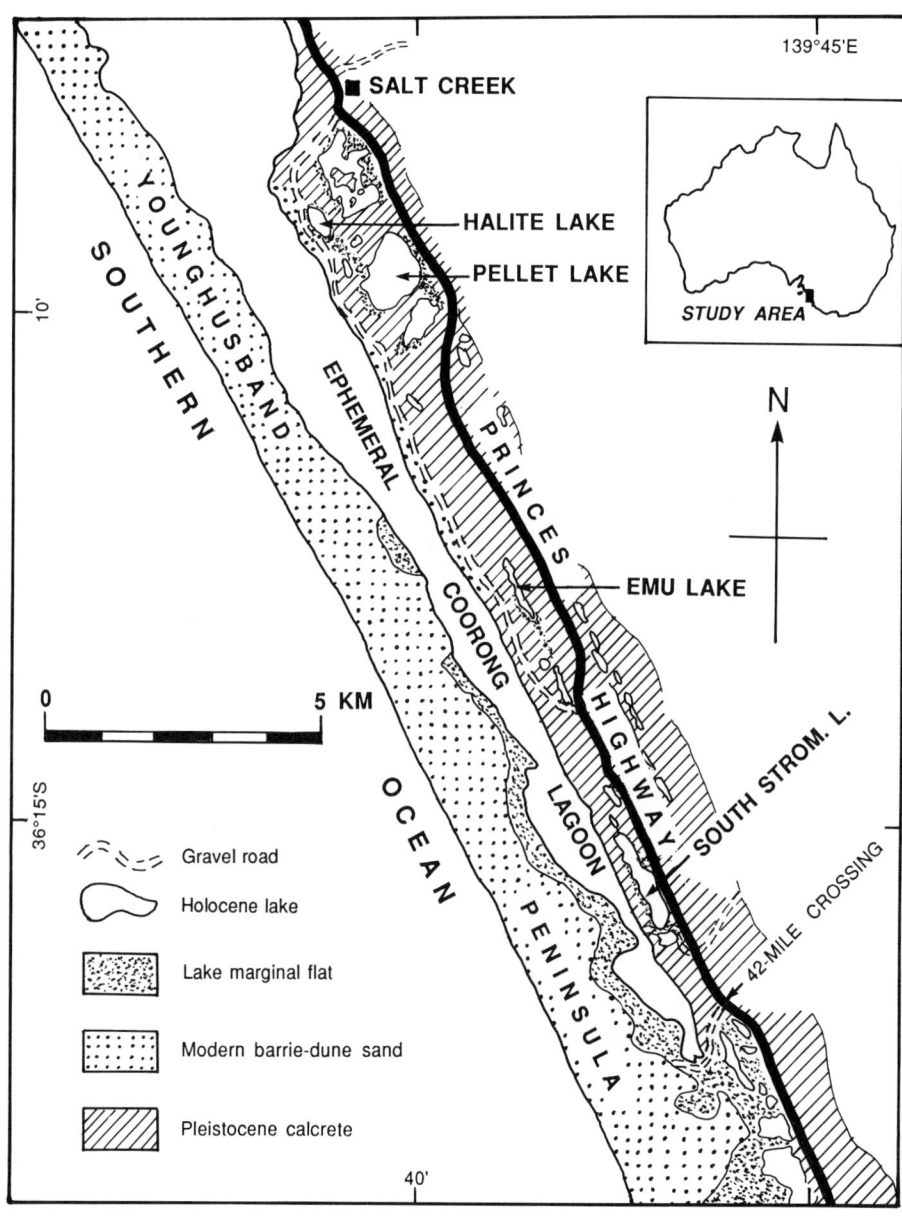

Figure 1. Map showing the locations of Halite, Emu, South Stromatolite, and Pellet Lakes.

dolomite have been put forward by previous workers, but none of them adequately explain the origin of this mineral [9].

Although the sedimentology in many lakes precipitating dolomites around the world [4] have been well studied, detailed investigations of their hydrogeochemistry have not generally been done [10,11]. In the Coorong region, the sedimentology of dolomite bearing Holocene carbonate sediments in several lakes has been studied by previous workers [12-21]. For modelling the processes of protodolomite/dolomite formation in the Coorong lakes, which differ greatly from one another with respect to sedimentology, carbonate mineralogy, geochemistry, and hydrogeochemistry [1], a thorough investigation of the hydrogeochemistry, among other types of studies is necessary.

The purpose of the present paper is (1) to present the results of the hydrogeochemical investigations of the surface and the subsurface waters of Halite, Emu, South Stromatolite, and Pellet Lakes of the Coorong region, and (2) to show how the hydrogeochemistry of the lakes control the formation of protodolomite/dolomite and other carbonate minerals in them. The location of the lakes is shown in Figure 1.

GEOLOGICAL EVOLUTION OF THE LAKES

The Coorong region is characterized by a series of Pleistocene, parallel barrier beach-dune ridges that have been progressively stranded across the coastal platform as a result of falling sea-level [22]. The interdunal corridors of these beach-dune ridges represent Pleistocene lagoons. The present coast line in the region is marked by a modern barrier beach-dune ridge known as the Younghusband Peninsula. The Coorong Lagoon lies on its landward side. The ephemeral lakes under investigation lie in the interdunal corridors of the Pleistocene beach-dune ridges that are closest to the Coorong Lagoon (Fig. 1). The low-lying portions of the Pleistocene interdunal corridors in the vicinity of the ephemeral lakes were inundated when the Late Holocene sea-level rise had been stabilized relative to the Coorong region about 6500 years ago, as shown by radiocarbon studies [17,23]. Prior to the incursion of seawater some of these low-lying areas seem to have been inundated by fresh swamp water during the early phase of the Late Holocene sea-level rise at about 10000 years ago [1].

At the peak of the Holocene transgression, the sea-level stabilized at 1 m to 2 m higher than that of today, as indicated by the extent of stranded lagoon and marine deposits [16,24]. This was the time during which low-lying areas of the Pleistocene topography marginal to the Coorong Lagoon were inundated. Here, small coastal lagoons formed from the successive flooding of topographic depressions, giving rise to the lakes (Fig. 1) [17,18], which became the sites of Holocene carbonate sedimentation. As sea-level stabilized during the Late Holocene, the modern barrier dune-ridge of the Younghusband Peninsula began to form as an offshore line of barrier islands.

Subsequent accretion of dominantly shelf derived skeletal sands then caused barrier building and lagoon restriction, until ultimately the present continuous barrier evolved, burying most of its Pleistocene counterparts [17]. Along with the formation of a continuous modern barrier dune-ridge system, the main Coorong Lagoon was slowly sealed from direct connection to the open sea. As the degree of restriction progressively increased, portions of the lagoon were cut off by sediment accretion from direct access to the sea to form the shallow lakes of the study area, which became the sites of Holocene carbonate sedimentation. At this time, the lakes trapped sea/lagoon water as ground water. The lakes are ephemeral in nature. They are inundated in winter by the rain water through surface run-off and subsurface seepage from the local lake catchment, and dry up in summer due to evaporation. Saline

Table 1.
Summary of the important physicochemical and mineralogical properties of the protodolomites/dolomites, and the surface carbonate sediments of the lakes. A = aragonite, D = dolomite, HM = hydromagnesite, HMC = high-Mg calcite, LMC = low-Mg calcite, M = magnesite, PD = protodolomite, CF = carbonate fraction of the bulk sediment, TCC = total carbonate content, and TOC = total organic carbon. The dolomite content has been recalculated as the weight percentage of total carbonate content in the stratigraphically uppermost samples of boreholes, surface sediment samples, and the surface films.

	Halite Lake	Emu Lake	S. Strom. L.	Pellet Lake
Dolomite content (Wt. % of TCC)	7	84	Trace	77
Carbonate mineral association	A > D > HM > LMC	D > M > A > HM	A > HM > LMC > D	D > HM > A > LMC
Mole% MgCO3 in dolo./protodol.	41 - 45	48 - 51	42	46 - 53
Dolomite/protodolomite ordering	Disord. - partly ord.	Partly ord. - ordered	Disordered	Disord. - partly ord.
Dolo./protodol. crystal size	<2μm	<2μm	<2μm	<3μm
Dolo./protodol. crystal shape	Globular	Globular	Globular	Imperfect rhombs, globular
Total carb. content (TCC, Wt.%)	57	83	73	90
Total organic carbon (TOC, Wt.%)	0.22	1.4	0.21	0.86
Sr in Dolo./protodol./CF (ppm)	11842	10585	10959	8750
(Sr/Ca)X1000	71	92	55	63

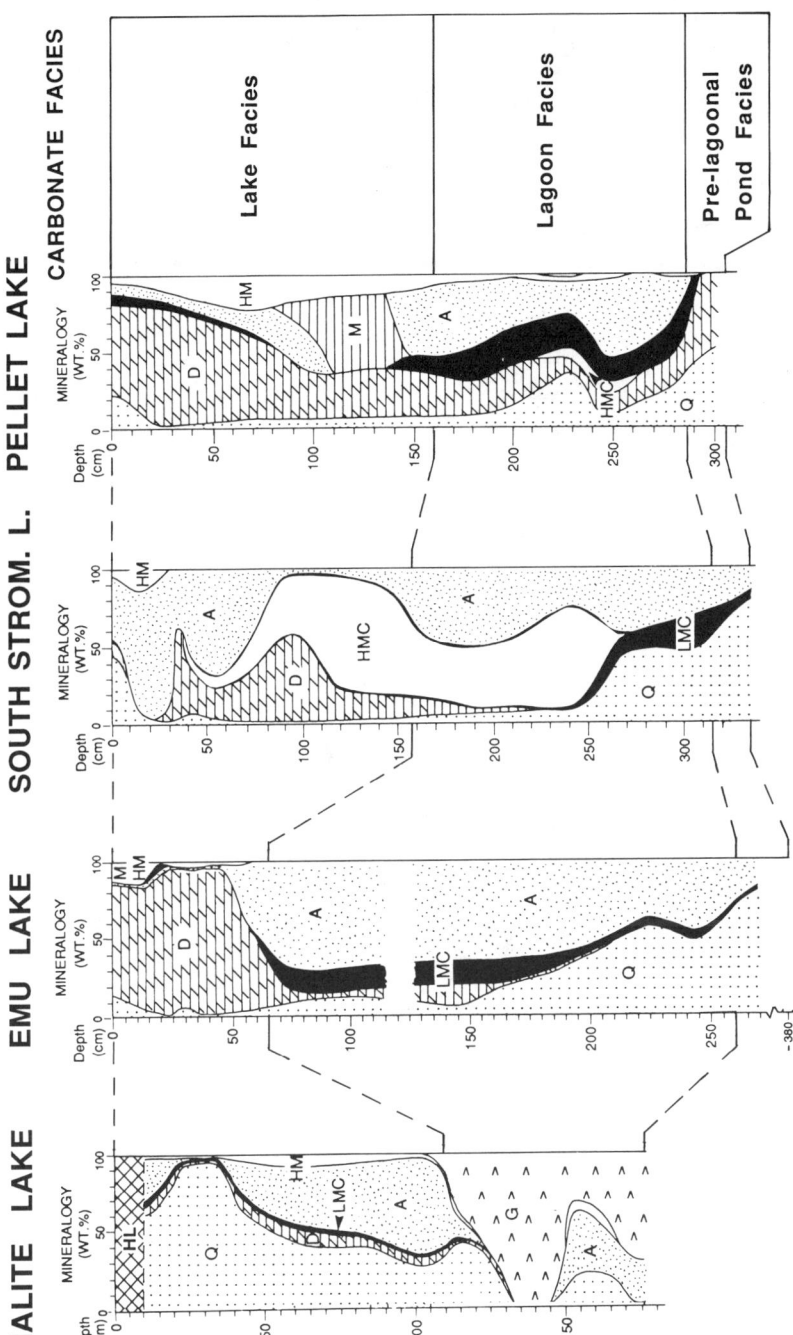

Figure 2. Carbonate mineralogy and correlation of the carbonate facies in the Holocene stratigraphic sequences of Halite, Emu, South Stromatolite, and Pellet Lakes. A = aragonite, D = dolomite, G = gypsum, HL = halite, HM = hydromagnesite, HMC = high-Mg calcite, LMC = low-Mg calcite, M = magnesite, and Q = quartz.

Figure 3. Scanning Electron photomicrographs showing the dolomite crystal sizes and morphologies. A. Abundant globular crystals of dolomite in Emu Lake surface sediments, and B. Abundant imperfectly developed rhombic crystals of dolomite in Pellet Lake surface sediments. Arrow points to a better developed rhombic crystal of dolomite.

ground waters prevail at the shallow subsurface. Among these lakes, Halite and Pellet Lakes retain surface water for a longer duration than those of Emu and South Stromatolite Lakes during the yearly cycle.

METHODS

Piezometers were set at different sites starting from the lake centre to the marginal flats (Fig.1). The bottoms of the piezometers were set at different depth levels below the surface (e.g. Fig.4). Surface and subsurface water samples were collected from the piezometers at different times during each yearly cycle over a period of several years. Temperature and pH of the collected water samples were measured on-site. The density of the waters was measured in the laboratory. Major cations such as Na, K, Ca, Mg, and Fe (total iron as Fe^{3+}) were determined by AAS. Cl and SO4 were determined by Ion Chromatography, and TCO_2 was determined by using Ion Selective Electrode. The salinity was calculated as parts per million (ppm) total dissolved solids (TDS). Hydrogeochemical properties such as the temperature, density, pH, the cations and anions, and their ratios for the surface as well as the subsurface water samples were plotted against stratigraphic depth in separate diagrams for each individual lake, and were discussed and evaluated by Ahmad [1]. Such detailed descriptions are beyond the scope of this paper. In this report, the important physicochemical properties of the surface and the subsurface waters of the lakes are summarized in Table 2, and the data for Pellet Lake only are plotted against stratigraphic depth in Figure 4. Triangular plots (Fig.5) are also made in order to show the compositions of waters/brines of the lakes in the system Ca-Mg-Na-K-TCO_2(HCO_3+CO_3)-SO_4-Cl [10].

The bulk mineralogy was determined semiquantitatively by X-ray diffraction (XRD) of a large number of samples taken at intervals of 5 cm to 7 cm from core samples obtained from the central locations in the lakes, washed (free of halite salts) and dried, using the techniques described by Ahmad [1]. The carbonate facies/microfacies boundaries within the Holocene stratigraphic section in each of the lakes were defined by cluster analysis techniques using a large number of physicochemical variables, and aided by the visual/microscopic observations of the fossils [1].

CARBONATE MINERALOGY

The carbonate mineralogy varies widely between the surface and the subsurface in each lake as well as between the Holocene stratigraphic sequences of the lakes. Figure 2 shows the distribution of the carbonate mineralogy, and the correlation of the carbonate facies among the lakes. Important characteristics of the surface carbonate sediments, particularly of the protodolomite/dolomite in the lakes, are summarized in Table 1.

Generally three carbonate facies exist in the Holocene stratigraphic sections of the lakes. These are, in ascending order, the pre-lagoonal pond, lagoon, and the lake facies. However, no evidence of deposition of the pre-lagoonal pond facies has been found in Halite Lake (Fig. 2). The pre-lagoonal pond facies is nonfossiliferous, and has 50% or less total carbonate content (TCC), comprising low-Mg calcite (LMC) and/or very fine grained (< 1 μm) protodolomite/disordered dolomite. The lagoon facies is characterized by restricted to open marine fossils, and contains up to 90% TCC, including abundant aragonite, LMC and high-Mg calcite (HMC), some protodolomite/disordered dolomite, trace amounts of hydromagnesite (found only in Halite and Pellet Lakes), occasional large amounts of gypsum (e.g. Halite Lake), and a high content of organic-rich, sapropelic materials (TOC).

The lake facies is characterized by the general absence of restricted to open marine fossils and the presence of very high TCC (up to 98%), comprising highly abundant protodolomite/dolomite (Fig. 3), and some aragonite, HMC, LMC, hydromagnesite, and magnesite. The dolomites are disordered to partly ordered, and occasionally ordered (e.g. those of Emu Lake). The mole% $MgCO_3$ in the dolomites of the lake facies generally varies from 41% to 53%. Detailed descriptions of the facies as well as the microfacies contained in them are given by the separate paper [1].

DISCUSSION

The high rate of evaporation of the lake waters leads to the precipitation of protodolomite/dolomite, other carbonate minerals and halite at the air/water interface, thereby producing thin films at the lake surface. Collection of the surface films and their analyses by XRD after washing and drying (halite free) show the presence of highly abundant, disordered to partly ordered dolomites, thereby giving a clear proof of the primary origin of the dolomites as a result of direct chemical precipitation from the lake waters.

The criteria generally used by the sedimentologists to characterize primary dolomite are: low absolute age (generally ^{14}C-age of only a few hundred years, for example, the primary dolomites of the Deep Spring Lake in California with ^{14}C-age of 500 - 1000 years, [25]), small crystal size (< 1 μm - 5 μm) generally without well developed rhombic shape, poor ordering (disordered to partly ordered), nonstoichiometry (Ca-rich) (e.g. dolomites of many Holocene lakes around the world, [4]), and significantly
differing (> 2‰) $\delta^{13}C$ and $\delta^{18}O$ values with coexisting calcite/aragonite [3]. The dolomite obtained from the surface of Emu Lake gives a ^{14}C-age of 590 ± 220 years [23]. The ages of surface dolomites from other lakes in the region, determined by previous workers, are less than 600 years [26], and 300 ± 250 years [27].
The protodolomites/dolomites of the lakes under investigation are commonly Ca-rich (41 - 48 mole% $MgCO_3$; rarely 50 - 53 mole% $MgCO_3$), generally disordered to partly ordered, and have very fine (< 1 μm - 4 μm) crystal size (Fig. 3). Stable isotopic investigations of the dolomites obtained from Pellet, Halite, and the South Stromatolite Lakes [19] showed that their $\delta^{13}C$ and $\delta^{18}O$ values differ at least by 2‰ from those of the coexisting LMC and/or HMC. The above attributes give further testimony for the primary origin of the dolomite.

The hydrogeochemical conditions under which the protodolomites/dolomites are formed in the lakes vary from one another although there are certain properties which are very much similar in all of them. Triangular plots of the hydrogeochemical data of the surface water samples of the lakes in the system Ca-(Na+K)-Mg (Fig. 5) show that Halite, Emu, and the South Stromatolite Lakes form a group that lies along the (Na+K)-Ca edge, and close to the (Na+K)-pole, whereas most of the samples from the Pellet Lake plot along the (Na+K)-Mg edge and close to the (Na+K)-pole. However, a few of the Pellet Lake samples show significant spread along the (Na+K)-Mg edge because of higher concentration of Mg, whereas a few of the Halite Lake samples show a similar spread along the (Na+K)-Ca edge because of the higher concentration of Ca. Most of the samples from these lakes are more enriched in (Na+K) than seawater as a result of evporative concentration as well as for continuous input from aerosol and seaspray [1].

Triangular plots of the surface water samples from the lakes in the system $Cl-SO_4-TCO_2$ (Fig. 5) show that most of the samples form a cluster adjacent to the $Cl-SO_4$ edge and

Table 2.
Summary of the important hydrogeochemical data obtained from the lakes. The concentrations of elements are in parts per million (ppm). SW = surface water, SSW = subsurface water, TDS = total dissolved solids. The classification of the water/brine is according to Hardie and Eugster (1970).

		Halite Lake	Emu Lake	S. Strom. L.	Pellet Lake
Temperature (T°C),	SW	10±2 - 26±3	9±2 - 23±3	9±2 - 25±3	9±2 - 26±3
pH-	SW	7.3 - 8.5	8 - 10.5	8.5 - 9	7.3 - 9
	SSW	6.7 - 7.9	7.2 - 8.2	7.1 - 8	6.9 - 8.6
Mg-	SW	3150 - 43625	200 - 4050	625 - 3850	612 - 44500
	SSW	6625 - 41360	440 - 2888	1225 - 5500	475 - 15760
Ca-	SW	97 - 261	25 - 91	35 - 101	20 - 145
	SSW	357 - 565	70 - 443	199 - 462	20 - 248
Na-	SW	53000 - 111000	3250 - 53000	9000 - 44750	16000 - 106670
	SSW	27333 - 91750	3500 - 32250	16000 - 35000	5000 - 104000
K-	SW	1200 - 9750	75 - 1390	225 - 1140	250 - 8500
	SSW	2800 - 15133	175 - 717	400 - 1350	100 - 1200
Cl-	SW	117799 - 212516	5808 - 30118	14902 - 47147	27044 - 193762
	SSW	94323 - 208984	8016 - 32880	25195 - 62163	9323 - 107183
SO_4-	SW	9518 - 86109	509 - 2505	2604 - 6007	1268 - 14064
	SSW	18881 - 46928	500 - 6522	2218 - 8977	752 - 14128
TCO_2-	SW	119 - 491	57 - 1202	233 - 827	155 - 759
	SSW	286 - 13751	110 - 1201	293 - 1206	83 - 1254
Mg/Ca-	SW	19 - 254	4.5 - 37	9.6 - 64	8.5 - 297
	SSW	17 - 311	2 - 7.5	4.9 - 17	2.5 - 98
Cl/Ca-	SW	647 - 1430	232 - 861	327 - 1048	520 - 1112
	SSW	53 - 557	35 - 156	100 - 168	68 - 2897
Salinity (ppm TDS)-	SW	72756 - 365442	10584 - 52924	27533 - 82784	52062 - 400337
Type of water/brine-	SW	Cl-type	Cl-type	Cl-type	Cl-type

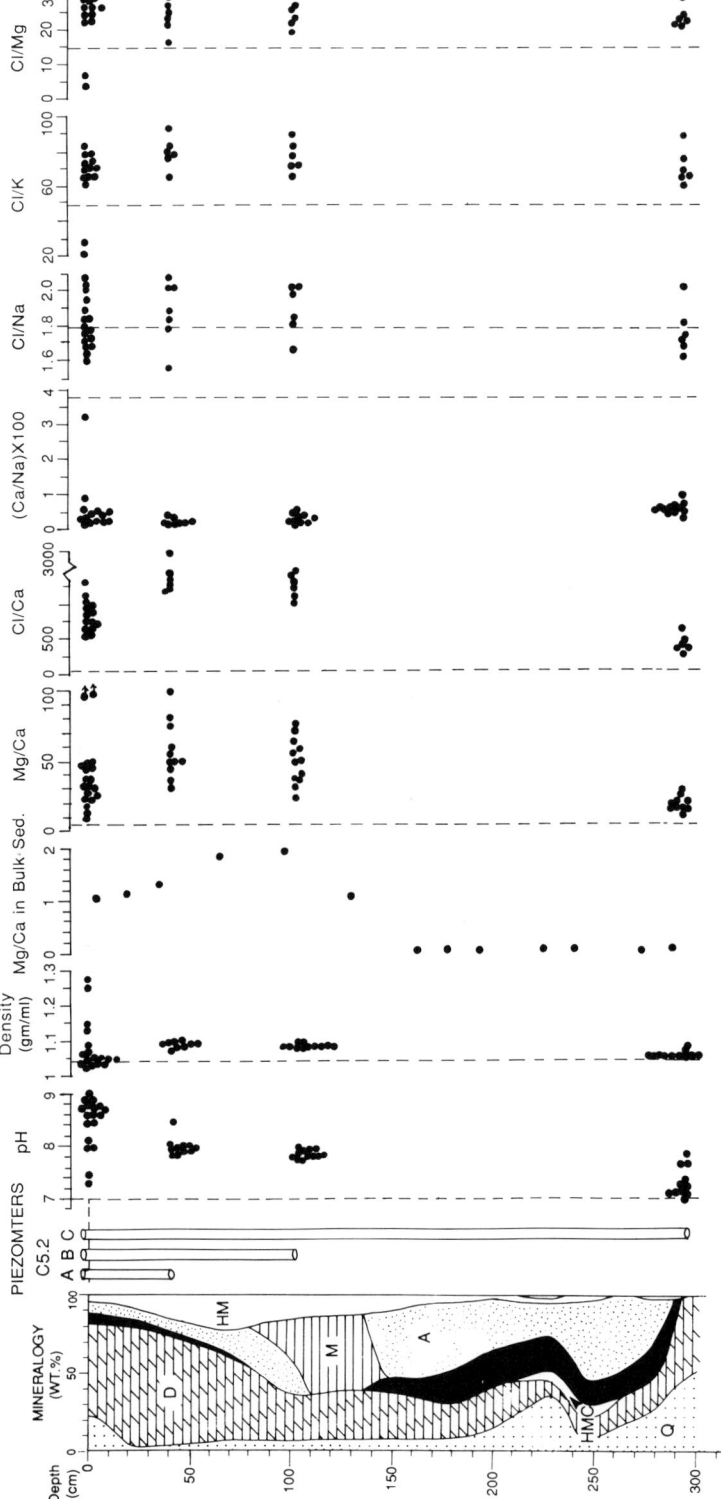

Figure 4. Plots of important hydrogeochemical data and Mg/Ca ratios in the bulk carbonate sediments against depth in the Holocene stratigraphic section in Pellet Lake. The dashed lines represent the average seawater values (Holser, 1979). The mineralogy in the Holocene stratigraphic section and the depth levels of the lake centre piezometers C5.2A - C5.2C are also shown. A = aragonite, D = dolomite, HM = hydromagnesite, HMC = high-Mg calcite, LMC = low-Mg calcite, M = magnesite, and Q = quartz.

near the Cl- pole. These surface waters are thus of Cl-type and are similar to those from other reported regions [6,10,28]. However, the surface water samples of the Halite and the Pellet Lakes spread considerably along the Cl-SO4 edge and towards the SO4- pole. The Pellet Lake waters are relatively richer in SO4 than those of Halite, Emu, and the South Stromatolite Lakes.

Among the important factors that control the precipitation of carbonate minerals, such as the temperature, solution composition and concentration, rate of crystallization, and the presence and concentration of certain organic compounds, the more important ones for the formation of dolomite are, apparently, the Mg/Ca ratio and the salinity [29]. The temperatures of the surface waters of the lakes are very much similar to each other (Table 2). The average temperature of the surface waters lies at $9 \pm 2°C$ in winter and $25 \pm 3°C$ in summer. The precipitation of dolomite in these lakes takes place partly as a result of CO_2-degassing and plant photosynthesis and partly as a result of evaporative concentration of participating ions in the lake water [1]. CO_2-degassing and plant photosynthesis yield progressively higher pH values and carbonate/bicarbonate ratios. All of these factors account for primary protodolomite/dolomite formation under surface condition.

In the surface waters of the lakes, the concentration of Mg generally lies above the average seawater value of Holser [30] (Fig. 6). In the subsurface waters, the concentration of Mg is even higher than that of the surface water. On the other hand, the concentration of Ca in surface waters varies widely, but is usually lower than that of average seawater. The lower abundance of Ca in the surface waters is caused by the removal of this element by the precipitation of dolomite and other carbonate minerals listed in Table 1. Similarly, depletion of Mg is caused by the precipitation of HMC, protodolomite/dolomite, and/or hydromagnesite. In subsurface waters, Mg lies close to or at, a slightly higher level than that of the average seawater. Plots of the surface water for the lakes (Fig. 6) reveal that the concentration of Ca lies below the seawater value for all of them, whereas Mg varies from less than the seawater to orders of magnitude higher than this at a late stage of evaporation.

At a late evaporatic stage, the concentrations of Mg in surface waters of Halite and the Pellet Lakes are relatively much higher than those of Emu and the South Stromatolite Lakes. The Mg/Ca ratio in the surface waters of all the lakes in question lie above the average seawater Mg/Ca ratio (3.18) and become orders of magnitude higher during the very late stage of evaporation. During the very late stages of evaporation, the Mg/Ca ratios in the surface waters of the Halite and the Pellet Lakes are much higher than those of the Emu and the South Stromatolite Lakes. This suggests that, at the higher stages of evaporation, the formation of dolomite and/or hydromagnesite is relatively more pronounced in Emu and South Stromatolite Lakes than those of Halite and Pellet Lakes. The presence of highly abundant dolomite (up to 84% of the TCC) in the surface sediments of Emu Lake and the high abundance of hydromagnesite in South Stromatolite Lake (Table 1) support the above phenomenon. With respect to the Mg/Ca ratio in the surface waters, the lakes are similar to other dolomite forming lakes around the world, having an Mg/Ca ratio greater than five [4]. In fact, the Mg/Ca ratio for all plotted surface lake water is ≥ 10.

With respect to salinity (Fig. 6), most of the samples of Emu Lake and two samples of South Stromatolite Lake are lower than that of the average seawater, whereas one of Emu Lake, the majority of South Stromatolite Lake, most of Pellet Lake, and all of Halite Lake samples are higher than that of seawater. It is also important to note that, within the range of 10000 ppm to 250000 ppm TDS, the Mg/Ca ratio in the surface waters of the lakes shows a very gentle, linear increase from 10 to 50. As the TDS in the waters exceeds 250000 ppm during the late stages of evaporation, the Mg/Ca ratio increases exponentially up to about 300 (e.g. Pellet Lake) as the TDS attains a value of 400000 ppm. This observation suggests that

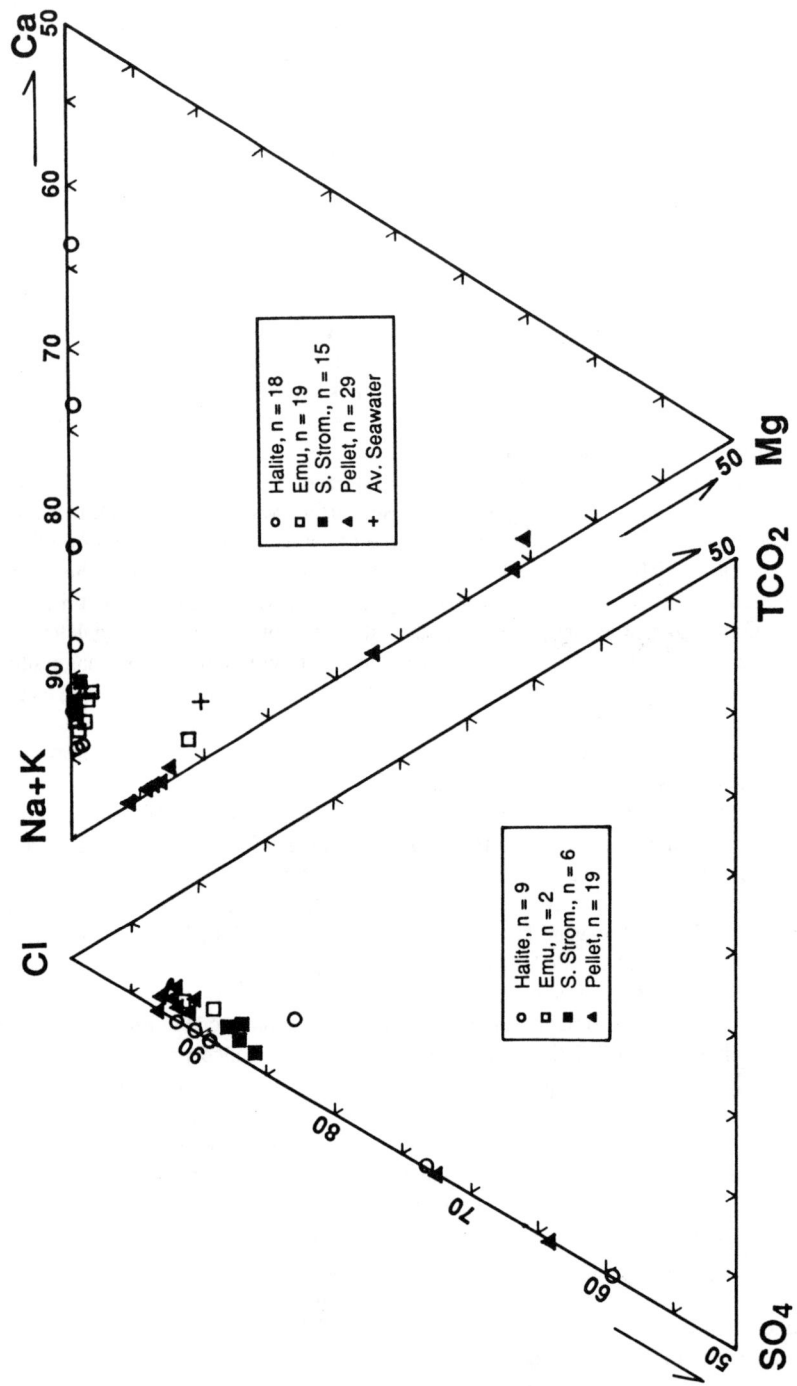

Figure 5. Triangular plots showing compositions of the surface waters/brines of the lakes in the system Ca-Mg-Na-K-TCO$_2$-SO$_4$-Cl. Plot of the average seawater composition is based on the data of Holser (1979). For clarity, not all points are shown.

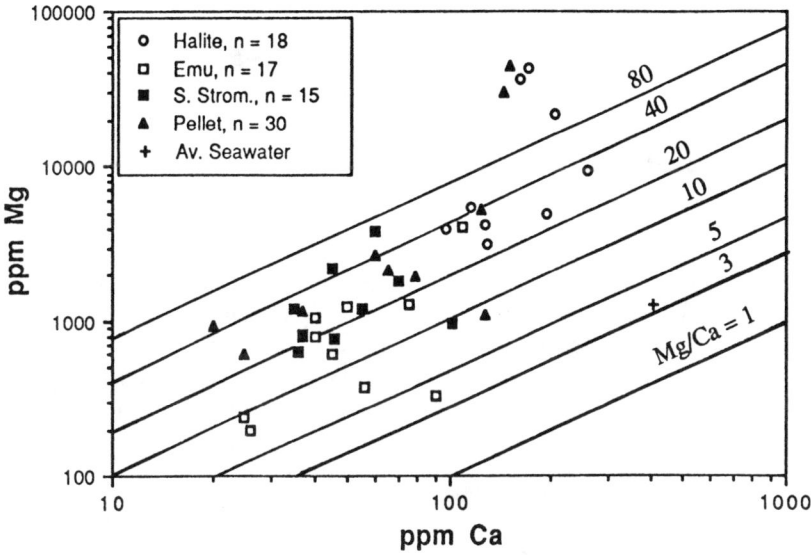
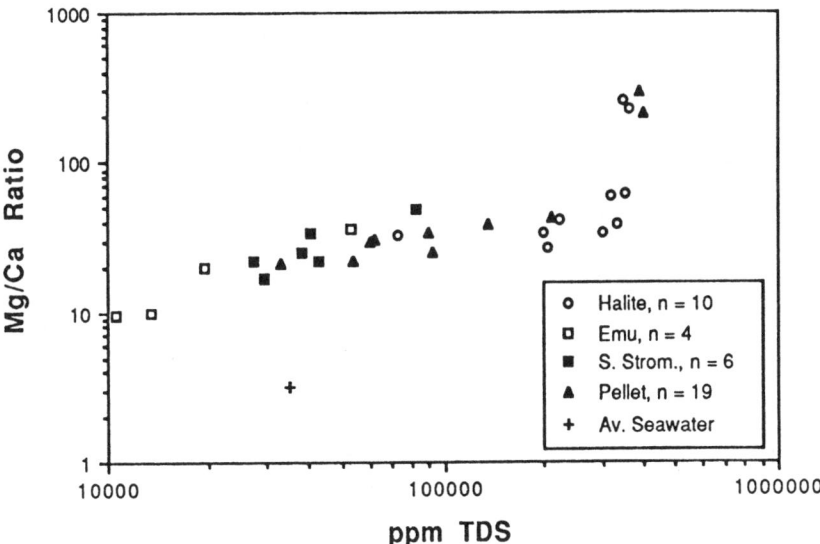

Figure 6. Mg versus Ca, and Mg/Ca ratio versus Salinity (ppm TDS) plots of the surface water/brine samples of the lakes. The average seawater composition (Holser, 1979) is plotted for comparison. For clarity, not all points are shown.

the very active precipitation of protodolomite/dolomite in the surface waters of the lakes takes place within the Mg/Ca ratio range of 10 to 50 and the salinity (TDS) range of 10000 ppm to 250000 ppm. With a salinity rise above 250000 ppm, the precipitation of dolomite is inhibited. However, other Ca-bearing carbonate minerals could, perhaps, form at such high salinity values.

It is apparent from Figure 4 that, in Pellet Lake, the density of the surface water varies widely from less than that of average seawater to remarkably higher than this due to the effects of dilution by the rain water in winter followed by high evaporative concentration in summer. At the subsurface, the density gradually decreases downward, but stays almost stable during the yearly cycle. Mg/Ca ratios show significant variations in the surface as well as in the shallow subsurface waters during the yearly cycle because of their participation in the precipitation of protodolomite/dolomite and other carbonate minerals. However, this ratio change is much less pronounced in the subsurface waters of the lowermost part of the Holocene stratigraphic sequence, perhaps indicating that the formation of protodolomite/dolomite and/or other carbonate minerals is pronounce at that subsurface level. The Mg/Ca ratios in the bulk sediments of the Holocene stratigraphic section show good correlation with those in the coexisting subsurface waters, thereby indicating that this ratio has an important control on the precipitation of protodolomite/dolomite and/or hydromagnesite, magnesite etc.

The vertical profiles of the Mg/Ca ratios in the waters of Halite, Emu, and the South Stromatolite Lakes are similar to that of the Pellet Lake as shown in Figure 4 [1]. In the surface and the subsurface waters of Pellet Lake, the Cl/Ca, and the Cl/Mg ratios remain significantly higher than those of average seawater because of the incorporation of Ca and Mg ions into the crystal lattices of protodolomite/dolomite and other carbonate minerals, thereby decreasing their abundances in the water. For the same reason, the Ca/Na ratio in the waters of this lake remain significantly lower than that of average seawater.

Previous researchers [15,18-21] have all suggested that the lakes under investigation are fed by the regional ground water, which supplies the necessary ions, particularly the Mg ions, that are required for the precipitation of dolomite and other magnesium carbonate minerals. However, detailed sedimentological and hydrogeochemical studies of the lakes by Ahmad [1] revealed that the lower part of the Holocene carbonate sequences in the lakes are enriched in organic-rich sapropelic materials that unconformably overlie an almost impermeable Pleistocene calcrete layer of several tens of centimetres to a few meters in thickness. This calcrete layer, combined with the overlying sapropelic layer, form 'seals' at the bottom of the lakes, making each of them an individual closed/semiclosed basin. These basins are fed by the meteoric water through surface run-off as well as by the input of shallow ground water through seepage from the local catchment of each lake. Based on the monitoring of aerosol and seaspray fall-outs in the catchments of these lakes, Ahmad [1] also indicated that the supply of ions necessary for the precipitation of protodolomite/dolomite and other carbonate minerals in the lakes comes from these sources. Other sources of the ions, especially Ca, are the dissolution of the aragonitic and calcitic shell materials and lagoon waters that were trapped initially during the early phase of evolution of the lakes as already described.

Based on the isotopic investigation of the dolomites from the lakes in question, the presence of two types of dolomites are indicated [20,21]: type A - relatively coarser grained, slightly heavier in its oxygen isotopic signature, and is typically 3-6 ‰ heavier in carbon than type B, and occurs in the central areas of the lakes. Type B dolomite is generally fine grained (< 0.5 μm), Ca-rich, isotopically lighter, and occurs along the lake margins. These authors believe that the isotopically lighter, type B dolomites are found in the zone of mixing of the lake water with the regional ground water along the edges of the lakes, and the heavier, type A is formed in the lake centers from the evaporation of the lake water. However, detailed microfacies analyses of the surface carbonate sediments in the lakes [1] reveal that fine grained, poorly ordered to disordered protodolomite/dolomite that corresponds to the type A dolomite [20,21] also forms in the marginal flats around Halite, Pellet, and other lakes, wherein no sign of outcropping regional ground water exists. The regional ground water and the lake water mixing zone model of Rosen et al. [20] and Warren [21], therefore, cannot provide an adequate explanation for the formation of fine grained protodolomite/dolomite that occurs in the marginal flats.

Very fine grained dolomite without definite crystal shape occurs in high abundance throughout the central areas of Emu Lake (Fig. 3) as well as in the marginal flat around it. The mixing zone model [20,21] that is restricted only to lake marginal areas is also inadequate to explain the formation of very fine grained dolomites that occur in the entire surface of this lake. We believe that the protodolomites/dolomites in the marginal flat as well as within the central areas of the lakes are formed as a result of direct chemical precipitation from the lake waters, and the waters that becomes stranded in the marginal flats in closed basin systems. The protodolomite/dolomites in the marginal flats and in the narrower areas of the lake basins have a relatively shorter duration for crystal growth than those that form in the lake centers, thus presumably allowing more time for the lake center crystals to grow larger. Relatively lower salinity and higher pH in the surface water of Emu Lake are possible factors that give rise to better stoichiometry (48 51 mole% $MgCO_3$ in the dolomite lattice) and better ordering (partly ordered to ordered) to these dolomites relative to those of Halite, South Stromatolite, and Pellet Lakes (Table 1).

CONCLUSIONS

1. The protodolomites/dolomites of the Halite, Emu, South Stromatolite, and Pellet Lakes are primary and have formed as a result of direct chemical precipitation from lake waters.
2. Elevated pH caused by plant photosynthesis and CO_2-degassing, accompanied by increased salinity as a result of evaporation, and higher Mg/Ca ratio (2 10), bring about the precipitation of protodolomite/dolomite from the lake waters.
3. Although Mg/Ca ratios in the surface as well as in the subsurface waters remain greater than that of average seawater, the higher pH (7.3 - 10.5) in the surface waters relative to that of their subsurface counterparts make the lake surfaces the major sites for the precipitation of protodolomite/dolomite as a result of high supersaturation.
4. The most active precipitation of protodolomite/dolomite from lake surface waters takes place within the Mg/Ca ratio range of 10 to 50 and the salinity (TDS) range of 10000 ppm to 250000 ppm. During the late stage of evaporation, at salinities above 250000 ppm, the Mg/Ca ratio in the surface waters of the lakes grows exponentially, probably because of an inhibition in the precipitation of protodolomite/dolomite.
5. Higher pH (8 - 10.5) and lower salinity (10500 ppm - 52000 ppm) tends to favour the precipitation of a protodolomite/dolomite - magnesite assemblage, whereas relatively lower pH (7.3 - 9) and higher salinity (27500 ppm - 400300 ppm) favors the precipitation of dolomite - hydromagnesite or hydromagnesite - dolomite assemblages in the lakes.
6. Lower salinity and higher pH of the surface waters likely account for the better stoichiometry and ordering to the dolomites of Emu Lake compared to those of Halite, South Stromatolite, and Pellet Lakes.

7. Low salinity and longer duration of crystallization favor the growth of larger and better developed rhombic crystal shape for the dolomites in the lakes.

Acknowledgements. This study was carried out as a part of the Ph. D. research project of the senior author that was completed at Macquarie University. Jack Davies, Steven O'Riley, Barry Batts, and Lucinda Coote provided valuable assistance in the field. Pham Di Hung helped in the water analyses. Some of the piezometers were emplaced by Dave Lock and Chris Von der Borch. The research was supported by a Macquarie University Postgraduate Research Award to the senior author.

REFERENCES

1. R. Ahmad. *Genesis and diagenesis of the Holocene dolomitic carbonate sediments in the Coorong region of South Australia,* Ph. D. Thesis, Macquarie University, North Ryde, NSW 2109, Australia, 748p. (1991) (unpub.).
2. D.H. Zenger and J.B. Dunham. Concepts and models of dolomitization. In: *Concepts and Models of Dolomitization.* D.H. Zenger, J.B. Dunham and R.L. Ethington (Eds.) SEPM Spec. Pub. No. 28, pp. 1-9, Tulsa, Oklahoma (1980).
3. M.E. Tucker and V.P. Wright. *Carbonate Sedimentology.* Blackwell Sci. Pub. (1990).
4. W.M. Last. Lacustrine dolomite - an overview of modern, Holocene and Pleistocene occurrences, *Earth Sci. Rev.* **27**, 221-263 (1990).
5. R.G.C. Bathurst. *Carbonate Sediments and their Diagenesis.* 2nd Ed. Elsevier, Amsterdam (1971)
6. D.L. Graf and J.R. Goldsmith. Some hydrothermal syntheses of dolomite and protodolomite, *Jour. Geol.* **64**, 173-186 (1956).
7. A.M. Gaines. Protodolomite synthesis at 100°C and atmospheric pressure, *Sci.* **IX3**, 518-520 (1974).
8. S. Ohde and Y. Kitano. Synthesis of protodolomite from aqueous solution at norrnal temperature and pressure, *Geochem. Jour.* **12**, 115-119 (1978).
9. L.A. Hardie. Dolomitization: A critical view of current views, *Jour. Sed. Petrol.* **57-1**, 1966-183 (1987).
10. L.A. Hardie and L.P. Eugster. The evolution of closed basin brines, *Mineral. Soc. Am. Spec. Pap.* **3** 273-290 (1970).
11. M. Magaritz and J.E. Luzier. Water-rock interactions and sea water-fresh water mixing effects in the coastal dune aquifer, Coos Bay, Oregon, *Geochim. Cosmochim. Acta.* **49**, 2515-2525 (1985).
12. D. Mawson. South Australian algal limestone in the process of formation, *Quart. Jour. Geol. Soc. London. Lxxxv, pt. 4,* 613-623 (1929).
13. A.L. Alderman and H.C.W. Skinner. Dolomite sedimentation in the south-east of South Australia, *Am. Jour. Sci.* **255**, 561-567 (1957).
14. C.C. Von der Borch. The distribution and preliminary geochemistry of modern sediments of the Coorong area, South Australia, *Geochim. Cosmochim. Acta.* **29**, 781-799 (1965).
15. C.C. Von der Borch, D.E. Lock and D. Schwebel. Ground-water formation of dolomite in the Coorong region of South Australia, *Geol.* **3**, 283-285 (1975).
16. C.C. Von der Borch. Stratigraphy of stromatolite occurrences in carbonate lakes of the Coorong lagoon area, South Australia. In: *Stromatolites.* M.R. Walter (Ed.), pp. 413-420, Chap. 8.3. Elsevier, Amsterdam (1976a).
17. C.C. Von der Borch. Stratigraphy and formation of Holocene dolomitic carbonate deposits of the Coorong area, South Australia, *Jour. Sed. Petrol.* **46**, 956-966 (1966b).
18. D.E. Lock. *Ground water controls on dolomite formation in the Coorong Region of South Australia and its ancient analogues,* Ph. D. Theis, Flinders University of South Australia, 275p. (1982) (unpub.).
19. R. W. Botz and C. C. Von der Borch. Stable isotope study of carbonate sediments from the Coorong Area, South Australia, *Sedimentol.* **31**, 837-849 (1984).
20. M.R. Rosen, D.E. Miser, M.A. Starcher and J.K. Warren. Formation of dolomite in the Coorong region, South Australia, *Geochim. Cosmochim. Acta.* **53**, 661-669 (1989).
21. J.K. Warren. Sedimentology and mineralogy of dolomitic Coorong Lakes, Sou~h Australia, *Jour. Sed. Petrol.* **60-6**, 743-756 (1990).

22. P.J. Cook, J.B. Colwell, J.B. Firman, J.M. Lindsay, D. Schwebel and C.C. Von der Borch. The Late Cainozoic sequence of southeast South Australia and Pleistocene sea level changes, *Bur. Mineral Res. Jour. Aust. Geol. Geophys.* **2**, 81-88 (1977).
23. R. Ahmad and J. Head. Radiocarbon ages of the Holocene lacustrine carbonate sediments from the Coorong region of South Australia: Implications for determination of the Great Australian Arid Period and sedimentation rates in the lakes. In: *Geol. Soc. Aust. Abstracts* No. 27, p.1, 7th Internatl. Conf. on Geochron. Cosmochron. and isotope Geol., Canberra, Australia, 24 - 29 Sept. (1990).
24. R.G. Brown. *Sedimentation in the Coorong Lagoon, South Australia,* Ph. D. Theis, University of Adelaide, Australia, 224p. (1965) (unpub.).
25. M.N.A. Peterson, G.S. Bies and R.A. Berner. Radiocarbon studies of Recent dolomite from Deep Spring Lake, California, *Jour. Geophys. Res.* **68**, 6493-6505 (1963).
26. H.C.W. Skinner, B.J. Skinner and M. Rubin. Age and accumulation rate of dolomite bearing carbonate sediments in South Australia, *Sci.* **139**, 335-336 (1963).
27. C.C. Von der Borch, *M.* Rubin and B.J. Skinner. Modern dolomite from South Australia, *Am. Jour. Sci.* **262**, 1116-1118 (1964).
28. A. Lerman. Chemical equilibria and the evolution of chloride brines, *Mineral. Soc. Am. Spec. Pap. 3*, 291-306 (1970).
29. R.L. Folk and L.S. Land. Mg/Ca ratio and salinity: two controls over crystallization of dolomite, *Am. Assoc. Petrol. Geol. Bull.* **59**, 60-68 (1975).
30. W.T. Holser. Bromide geochemistry of some non-marine salt deposits in the Southern Great Basin, *Mineral. Soc. Am. Spec. Pap. 3*, 307-319 (1970).

Biomineralization of the mirabilite deposits by the exemplification of the Barkol Lake

W. DONGYAN, L. ZHENMIN, D. XIAOLING and X. SHAOKANG
Geological Institute for Chemical Minerals, Ministry of Chemical Industry, Zhuzhou, China

Abstract-- With respect to genesis of mirabilite deposits, classical geologists interpret them as resulting from precipitation from pure solutions, but we should modify it because organisms do play a important role in causing mirabilite deposits, as in the Barkol.

BRIEF INTRODUCTION TO THE BARKOL LAKE

The Barkol Lake, situated in the eastern Xinjiang, is a large brine basin which is labeled hydrochemically by sodium sulfate subtype of sulfate type. The other hydrochemical parameters include salt content of 215 to 338 g/l, pH of 7.2 to 7.81, δD and $\delta^{18}O$ of -23.7 to +2.4 ‰ and -11.13 to +4.53 ‰ respectively. A combination of microscopic examination and SEM studying indicates that there are numerous faecal pellets of brine shrimps and halophilic bacteris enclosed within several meters of mirabilite and thenardite sediments beneath the surface waters.

BIOCOENOSIUM OF THE SURFACE BRINES

Living things observed visibly or microscopically consist mostly of bacteria alga, protozoa, brineshrimp, ephydra, and so on. Relations of these halophilic organisms to the salinity of the brines is better illustrated by Fig. 1. One can note that spherical halophilic bacteris, bacillar halophilic bacteris, brine shrimp and halophilic alga can be subjected a higher salinity than the rest.

BIOREMAINS AND BIOTRACES CONTAINED IN MIRABILITE SEQUENCES

The author has found some fossilized faecals and ova of the brine shrimps in the mirabilite and thenardite layers as well as vast quantities of halophilic bacteria and algae associated with gypsum, bassanite, halite, mirabilite and thenardite minerals. These findings have not yet been reported at home and abroad.

DISCOVERY AND GEOLOGICAL SIGNIFICANCE OF FOSSIL FAECAL PELLETS

Fossilized faecal pellets and ova of the brine shrimps are noticed to be accommodated within or between the mirabilite minerals as a kind of semieuhedral-bioclastic texture. This texture is

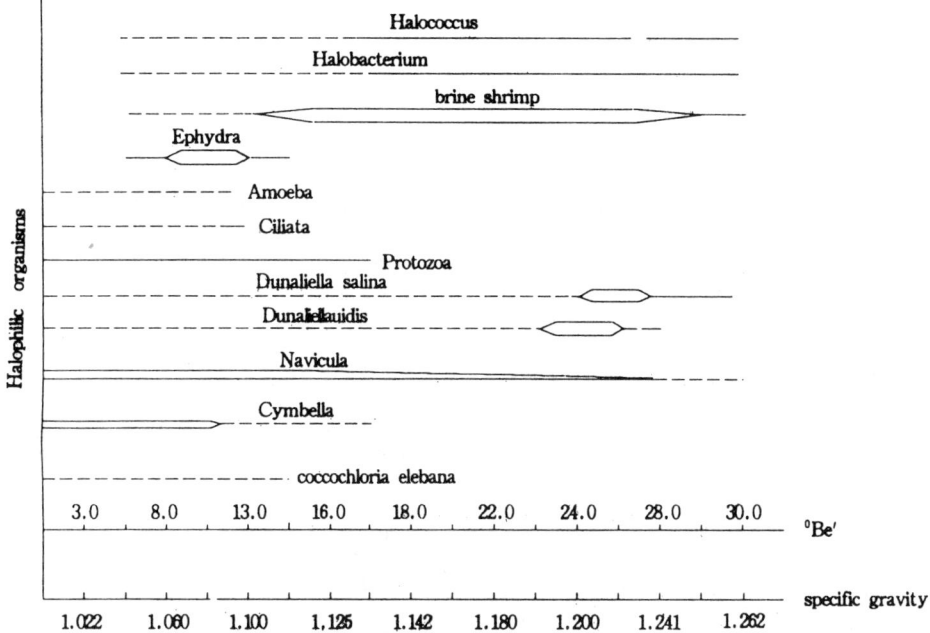

Fig. 1. Correlation of halophilic organisms vs. salinity in the barkol Lake

fairly rare in chemical precipitation of pure solutions, completely differing from those generally noted in any evaporite. A great deal of thin section identification reveals that the faecal pellets were subjected to the salt formation-diagenesis, mostly replaced by mirabilite and thenardite. The unreplaced pellets are dominated in composition by aragonite, calcite, and gypsum with less illite, chlorite, quartz, and feldspar.

Extensive incorporation of the pellets into the mirabilite-rich sediments clearly argues that not only were considerable amounts of brine shrimps flourishing during depositions of mirabilite and thenardite phases also the other microorganisms, as their foods, halophilic bacteria, halophilic algae and protozoa, on which they lived. Their higher reproductivities and physicochemical effects played an important role in accumulation, concentration, and precipitation of sodium sulfate.

FOSSILS OF BACTERIA AND ALGAE HOSTED IN SALT MINERALS

SEM examination of gypsum, bassanite, thenardite, and halite from the Barkol's salt-bearing sequences exhibits wide existence of fossils of spherical and bacillar bacteria and, in some cases, of algae. These microfossils distinguish apparently from one another in morphology, size and others, as detailed by Table 1. Their identification in the salt minerals undoubtedly suggests that the depositions of salt minerals were nearly synchronous with growing of numerous, shape-varying bacteria and algae which resided in the same brines.

By counting the number of the bacteria contained in the salt minerals based on the microphotography by SEM, it is estimated that they approached 7×10^{17} per cubic meter of brine during gypsum deposition, 3×10^{18} during bassanite deposition, and 2×10^{18} during thenardite deposition. Except for the number differences, bacteria from different salt minerals are very distinctive in morphology. This is properly attributed to fluctuations in the composition and concentration of the original brines which was capable of significantly influencing the growths of these microorganisms. Another fact exposed by SEM analyses is, some salt minerals are characterized by so-called, biosupermicroscopic textures, important indicators of biomineralization.

Table 1 Characters of fossils of halophilic organisms

Minerals	Characters of bacteria			Other organisms
	Shape	Size	Others	
Gypsum	Spherical	Maximum 2.6 μm, subordinate 1 to 1.4 μm, minmum 0.2 μm	Smooth edges	Unknown algae as elliptical, circular, bacillar, so on
Bassanite	Bacillar	Maximum 1.2x0.8 μm, Small 0.6 to 0.8 μm x 0.4 to 0.5 μm	Bodies in fixed directions	
Thenardite	Spherical to bacillar	Maximum 0.88 to 0.66 μm x 0.77 μm, subordinate 0.68 x 0.55 μm	Bodies in fixed directions	
Halite	Bacillar	Maximum 2 x 1.5 μm, Mean 0.7 x 0.5 μm, Small 0.5 x 0.3 μm		Unknown algae as branchlike, elliptical, bacillar, and so on

ECOLOGICAL BALANCE AGAINST MINERALIZATION IN THE BARKOL LAKE

As described above, in the Barkol's modern brines a large amount of halophilic organisms are living and a great many of fossils are noted in ancient mirabilite and thenardite layers. The present is the key to the past. By comparison of these with the geologic past, one can shed light on the biomechanism for the formation of mirabilite deposits through researching on relationship between ecological balance and salt mineralization. Where did the mineralizing materials of the Barkol come from. This answer is given by the petrochemical analyses that the weathered and eroded rocks of mountains surrounding the lake supplied sufficient sulfide, alkalimetal and alkali earth metal. The sulfide was oxidized by microorganisms into sulfate. But the metals were enriched more and more under biochemical functions of halophilic organisms, at last reaching a higher concentration.

In terms of relationship between biologic community and inorganic environment, the Barkol itself represents a typical ecological system which, according to features and functions, can be divided into autotrophic, consuming and decomposing biologic colonies (Fig. 2). Among these, a simple food chain is formed from planktonic algae to protozoon to brine shrimp. While some members of the chain are dead and sunk down, they are quickly decomposed by bacteria into chemical elements or primary compounds which in turn become the nutritious materials necessary for the planktonic algae. At the same time, the photosynthesis by the algae prepares sufficient oxygen for the bacteria during decomposition.

The protozoa and brine shrimps take up the algae for living. Once this occurs, it means that a new nutritious cycle begins again. Such a go-round-and-begin-again process may make

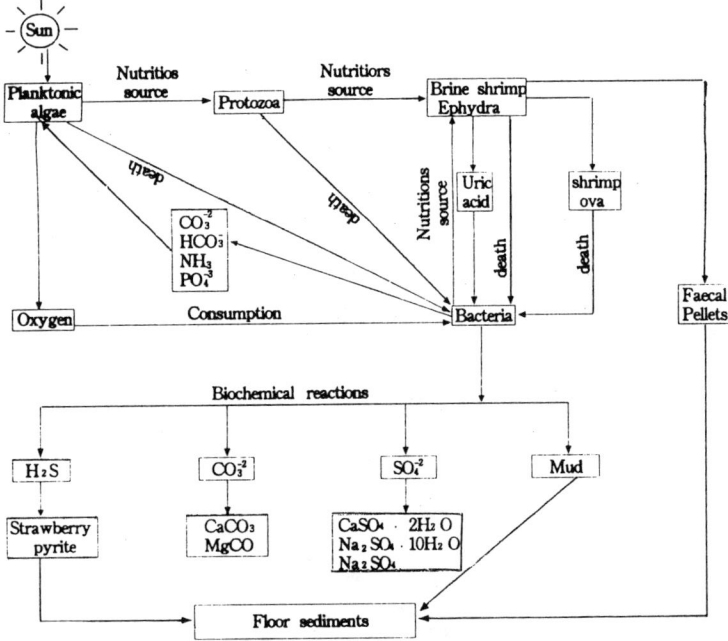

Fig. 2. Correlation of ecological equilibrium against mineralization in the Barkol Lake

the whole ecological system keep in relative equilibrium, but this is only relatively, transiently stable and, very often, unstable due to the flux of mass or energy or both into or out of the ecological system. The frequent alternations of stability and unstability lead inevitably to the enrichment of the brine in ore-forming materials such as SO_4^{2-}, alkali metal and so on. Na-sulfates common in the brines owe their enrichments just to the alternation.

The sulfur incorporated into the lake was derived mainly from the metallic sulfides of the mountains round the lake and less from the sulfates of the Tertiary rocks. It is known that sulfur is crucial matter for growth of living things, as has been supported by existence of sulfur in the form of organic sulfur in cysteine, cystine, methionine and amino acids. As bacteria decompose animal and plant remains, sulfur is always released out as H2S. This is then oxidized by inorganic sulfur fed bacteria into sulfur and sulfate following the below equation

$$CO_2 + H_2S + O_2 + H_2O \xrightarrow{\text{Bacteria}} CH_2O + SO_4^{2-} + 2H^+$$

Biogeochemical circulation of sulfur demonstrate that the reduced sulfur in the lake is going to be oxidized anyway into sulfates, a process for which bacteria, needless to say, are directly responsible. Similarly, alkali metal and alkali earth metal transported into the lake are controlled in accumulation more by bacteria's activities than anything else. Researches on organic chemistry of organisms point out that such elements as Na, K, Ca, and Mg have a large percentage of concentration in their cell tissues. For example, Na appears as cation out of the cells against K as cation within the cells. Millions of brine organisms have been capable of enriching ample amounts of mineralizing elements by taking up of alkali metals. This is followed by halophilic organisms which cause variations in pH and Eh values of the brines, when they drain metabolically ova, faecal materials and ureal solutions into the lake. As a result, the entire environment of the lake changes. This change plays an important part in enhancing the precipitation of sodium sulfates. Further studies state that the sodium sulfate-rich brines tend to lead to death of a part of brine shrimps who can not adapt themselves to the increasing salinity. From the dead shrimps some red halophilic bacteria capture considerable protein for subsequent decomposition, thereby quickly and thrivingly grow. Results of the quickness and thrivingness are elevations in the lake water temperature, evaporation and salinity. In this way, conditions would be created extremely favorable for precipitation of mirabilite and thenardite. Simultaneously, the brine shrimps are a cleaner of the lake water through controlling the growth of the algae by swallowing them and the other microorganisms, even mineral debris. Meanwhile, the purity of mirabilite and its kind increase to a larger level.

Alternative variations from balance to unbalance or reversibly in the ecological system of the upper brines are coincident with repeated depositions of salt minerals on the lake floor. In the course of the coincidence, halophilic organisms, especially halophilic microorganisms, display a striking role.

CONCLUSION

Sufficient amounts of halophilic organisms and their remains found in the Barkol's mirabilite deposit are completely identical to those of the modern biologic community residing in its surface brines.

Investigation on how far organisms contributed to the mirabilite mineralization in the Barkol demonstrates that genetic mechanism of this type of ore deposit is explained rather by biochemical deposition than by purely chemical deposition which, though, is approved by classical geologists.

Acknowledgment The author is indebted to Prof. Yuan Jianqi for fruitful direction and Chang Bingxi for reviewing the manuscript and giving opinions about it, also to Researchers Wang Dazhen, Dai Aiyun and Senior Engineer Xu Baozheng for direction and help in examination of the brine organism.

Halogenic basins: facial and paleotectonic models

G. A. BELENITSKAYA

All-Union Scientific Research Geological Institute, Department of Lithology, St.Petersburg, 199026, Russia

ABSTRACT -- Halogen (salt, sulfate) complexes are everywhere facially and cyclically associated with a stable set of non-halogen formations: carboniferous, dolomitic, bioherm, stromatolitic, red bed. All of them form regularly constructed parageneses - halogen-bearing associations (HA), localized around halogenesis depocentres. Two geological models of "facial profile" and "riftogenic structure", proposed by authors, reflect main features of structure, location and formation of these associations. The model of "facial profile" represents generalized scheme of a single HA macrocicle in different facial settings. It demonstrates principles of association structure, typomorphic features of every its member, spatial-time interrelations between them, facial settings of formation. The model of "riftogenic structure" is based on a common feature of all geodynamical types of halogenesis basins: on their subordination to continental riftogenic systems s.l. The model uses regular character of evolution of sedimentational halogenesis settings during tectonic cycles of development of riftogenic systems and outlines type vertical successions (series) of halogenesis basins and different types of halogen-bearing associations.

Key words: halogenesis, evaporites, facial analysis, paleotectonic analysis, riftogenic structure.

INTRODUCTION

Characteristical features of structure, facial and formational relations, paleotectonic position were revealed during compilation of the map of halogen (evaporite) formations of the territory of the USSR, scale 1:10,000,000, in VSEGEI [2]. It was established that halogen (salt, sulfate) complexes are everywhere cyclically and facially related with a number of non-halogen formations: highly carboniferous (black shale), dolomitic, bioherm, stromatolitic, red beds. Various links of these relations were discussed in literature [2, 4, 8, 9, 10, 12, 13]. All these complexes form a single naturally constructed halogen-bearing association (HA). Main features of composition, structure, facial and paleotectonic distribution and formation of these associations is discussed in the paper.

FACIAL MODELS

Two major indices define HA structure: cyclicity and lateral zonation (every cycle and association on the whole). Both indices are to different extent discussed in literature [2, 4, 5, 7, 12]. Cyclicity of 3-4 orders is traced. Cycles of the first order (macrocycles) have a different thickness in formations of different types: from several hundreds of meters, sometimes more than 1000m in potassium-bearing formations, to several tens of meters in sulphate-calcium formations. Cycles of lower orders complicate them: thickness of the second

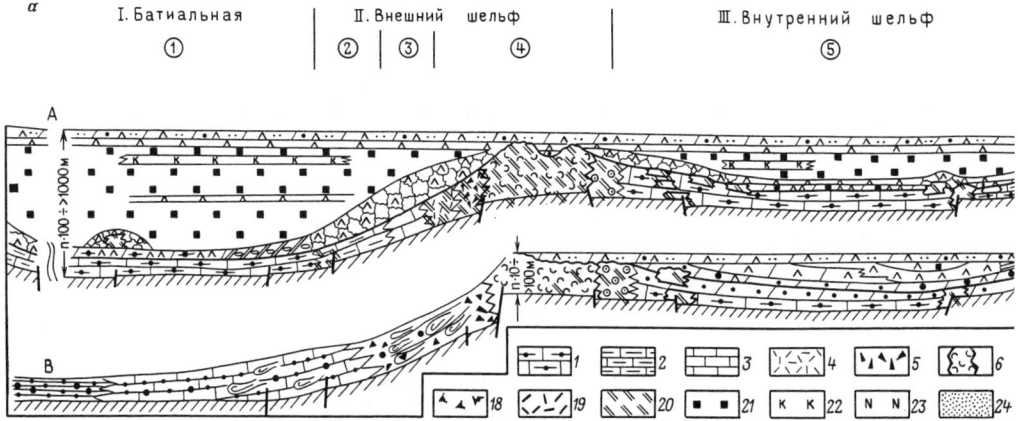

Figure 1. Facies-paleogeographic models of halogen-bearing associations: **a**, single sedimentary macrocycles; **b**, polycyclic formations. Main paleogeographic settings of halogenic sedimentation: **I-IV**, marine (I - bathyal, II - outer shelf, III - inner shelf, IV - coastal sabkha-lagoonal), **V-VII** - continental (V - lowland-lacustrine, VI - piedmont, VII - mountain-lacustrine); **A**, with depression basins of halogenesis; **B**, with relatively shallow-water basins; **1-13**, facies zones (in the circle, subaqueous; in the square, subaerial).**1-10**, carbonate formations: **1**, dark clayey-carbonate deposits with elevated carbon content, sometimes metal content; **2**, layered clayey limestones; **3**, layered ballstones; **4**, detrital limestones; **5**, coarse clastic train of organic structure; **6**, core of organic structure; **7**, calcarenite-oolitic limestones, dolomites; **8**, bioherm masses, bioherms, biostrome layers; **9**, stromatolites; **10**, dolomites; **11**, silty-dolomitic rocks; **12-20**, different types of anhydrites and anhydrite-gypsiferous rocks: **12**, massive microcristalline anhydrites, often glassy, nodule-like (anhydrite bars, "pillows"); **13,14**, indistinctly inequilayered massive and sheeted anhydrites and gypsa (13) and dolomite-anhydritic rocks (14); **15**, dark thin-bedded anhydrites, dolomite-anhydrites with a markedly elevated carbon content, sometimes metal content; **16**, inequilayered silty-dolomite-anhydritic rocks; **17**, avalanche-landslide chaotic formations with anhydrite cement; **18**, gypsa and anhydrites of sabkhas; **19**, dispersed inclusions of anhydrite, gypsum, glauberite, halite in red beds; **20**, metasomatic and filling anhydrite; **21**, rock salt; **22**, potash salts; **23**, lacustrine halogenic deposits, often of a sulphate-sodium, sometimes soda type; **24**, terrigenous deposits; **25**, elevated carbon content, sometimes metal presence (a, insignificant; b, high); **26**, red beds; **27-29**, flysh, flyshoid deposits, essentially carbonate, often with elevated silicon content, carbon content; **27**, coarce clastic; **28**, turbidite; **29**, predominantly pelitic; **30**, boundary between the lower and upper units of the cycle (isochronous surface); **31**, a break, partial erosion; **32**, substrate; **33**, disturbance zones; **34**, facies transitions; **35**, deposits of cover horizons of macrocycles and rocks of background sedimentation, separating macrocycles (a, essentially carbonate; b, essentially terrigenous red beds).

order is from parts of meter to several meters, thickness of the third order is from parts of centimeter to several centimeters, sometimes to 30-40 cm. The latter corresponds to the socalled annual cycles. Usually cyclicity of two-three orders is distinctly manifested.

Macrocycles with thickness from several hundreds of meters to more than 1000m are characteristical only for potassium-bearing formations (East Siberian basin, €1-2 and West European basin, P2ž - 4-5 macrocycles; North Kaspian basin, P1 - one macrocycle and others). Single macrocycles of similar scale sometimes occur in formations of sodium chloride (halite) type, but as a rule with features of potassium content (North Caucasus basin, J3). It is possible that the presence of macrocycles of such scale is an indirect indication to probable presence of potassium in them.

Structure and composition of macrocycles, reflecting major features of structure and composition of formations as a whole, are of principle significance. Major features of their

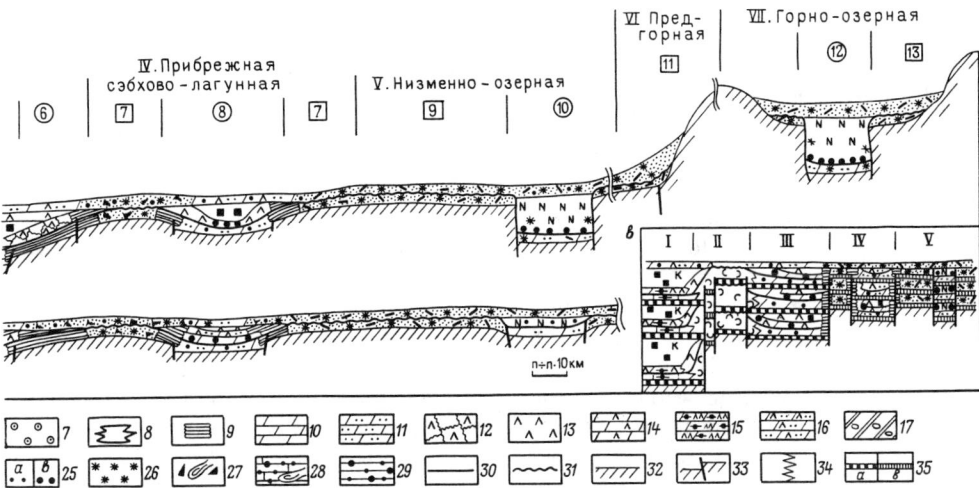

Figure 1. Continued.

construction for marine settings were revelated by M.M. Grachevsky and J. Wilson. Generalized scheme of single macrocycle structure in different paleogeographical settings is given in Figure 1.

Among marine settings, which are fundamental for halogenesis, three major settings are distinguished: the first of them is bathyal (I) with limiting outer open shelf (II); the second is a setting of inner shelf (III); and the third is a coastal sabkha-lagoon setting (IV). Three settings are distinguished among continental ones: lowland-lacustrine (V), piedmont (VI) and mountain-lacustrine (VII). In every setting (except bathyal) two variants of halogenesis are realized: with arising of tectono-sedimentational depression, complicating relief ("A"), or without it ("B"), i.e. with kettle or shallow-water (plane) basins of halogenesis correspondingly. It is important that all maximums of halogenesis are associated with variant "A".

Every cycle is represented by two major members: upper, considerably halogen, and lower, non-halogen. Cycles are of zonal structure with conjugated facial transitions in composition both of halogen and non-halogen elements.

In the first setting subhalogen complex in depression zone is represented by relatively thin clay-carbonate deposits with somewhat higher bituminosity. Organogenic structures (reefs, reefoids, bioherm masses) forming barrier frames frequently occur in board zones. A horizon of dark highly carboniferous (black shales, up to combustible shales) thin-laminated rocks of domanicoid appearance with ranging siliceous-clay-dolomite-anhydrite composition occurs at the boundary with overlapping halogen components. In many cases only this horizon is a recurrent element of the cycle. The upper element of the cycle is formed by salt-bearing, frequently potassium-bearing, strata, a gigantic lense of salts with maximum thickness over depression deposits (sometimes more than 500-1000m). Anhydrites are mainly subordinate and represented by a series of characteristical morphological and structural-textural varieties, demonstrated in Figure 1.

In the region of the inner shelf in the variant "A" (depression-shelf), when the shelf is complicated by one or more depressions, the profile on the whole resembles the profile of bathyal pre-reef zone, especially on large scales, depth and contrast range of depressions. Characteristical features of both subhalogen (bioherm, carboniferous) and halogen formations of the cycle and their ratio are also similar (but no so distinct). In the variant "B" ("shelf lagoon", after J. Wilson), in subhalogen parts of the section dolomitic and limestone-dolomitic beds with organogenic constructions of simpler structure predominate: biostrome beds, bioherms, isolated or chain-like, sometimes bioherm masses. Carboniferous formations are concentrated in interreef zones and bottom depressions, but are not very distinct. Carbon contents are frequently manifested as thinly dispersed dolomite pigmentation. Upper element of the cycle is represented by dolomite-anhydrites, sometimes with subordinate amount of salts, and rather thin (up to several tens of meters). Transitions between members of the cycle are gradual.

The variant "B" is more common for coastal sabkha-lagoon region. Subdivision of the cycle into lower and upper elements is indistinct. Stromatolite beds of dolomitic composition, often with higher bituminosity and sulfate content, are wide-spread. Their interbedding and interlacing with sulphate-bearing, sometimes saliferous formations; predominating in upper parts of the cycle, is characteristical. A set of textural and structural varieties of anhydrites and gypsums of "sabkha" appearance is original. Cycles are rather thin (up to first tens of meters). In subaerial (sabkha proper) conditions high sulfate content combines with red (parti-colored) rocks. When the region is complicated by a depression, the section to some extent comes closer to the section of the inner shelf.

In all zones in the roof of macrocycles a horizon of mixed anhydrite-dolomite-silty composition occurs, frequently with high carbon content, though lesser significant than subhalogen horizon. In marginal (coastal) zones it is replaced by red beds.

Thus, distinctly traced all over marine past of the profile is stable facial conjugation between halogen and non-halogen formations (paragenerations): bioherm, stromatolitic, carboniferous, dolomitic and in subaerial areas red bed. All of them are localized around halogen depocentre. From variant "A" to variant "B" and from region I to region IV zonal change of structure and composition of every its members, decrease of scale, distinction and differentiation took place from maximum in bathyal and depression-shelf settings, where all of them are the most extensive, distinctive and almost isolated, to thin interbedding with partial interlacing in shallow-water-shelf and sabkha-lagoon settings.

Paragenerations under discussion, conjugated in one cycle are mainly asynchronous between each other. Usually they alternate in the following succession: bioherm -> carboniferous and stromatolitic -> halogen, frequently with partial overlapping. In differentiated areas of the profile, major phase of formation of every of them corresponds to a pause (or slowing down) in development of others. Complexes of sabkha-lagoon zones are an exclusion, because isolation both in time and space is minimum and all these formations can be generated synchronously with every type of sedimentation in marine settings there.

Continental conditions are characterized by lacustrine halogen deposits, frequently of sodium sulfate or sodium carbonate types. It is likely that in plane lakes individual beds are generated and in kettle (depression) lakes-thick halogen lenses are formed. In many cases they contain beds of highly carboniferous rocks (sometimes combustible shales) occurring in the basement, rarer in the roof of halogen elements of cycles, and in policyclic formations they form interbedding. Halogen complexes contain a lot of terrigenous materials; they are framed and sometimes overlapped by red beds, gradually passing seawards into sabkha-lagoon deposits. Rather large amounts (to tens of per cent) of carbonates (frequently dolomite), anhydrite, gypsum, salts, carboniferous materials occur in the form of interlayers, lenses, inclusions in red beds. Thus, in continental settings halogenesis epicenters are accompanied by the same set of characteristical formations (except reefogenic). Lateral conjugation between marine and continental halogen-bearing associations deserve consideration.

Admixture composition in halogen members of cycles of all types is usually close to composition of their non-halogen members proper.

In successions of concrete HA from one to 4-5 microcycles are represented. From one to three zones are usually manifested in every of them. HA zonation reflects zonation of single cycles, complicated by their repetitions, sometimes by the presence of dividing interbeds consisting of rocks of background sedimentation (carbonate, terrigenous) and transgressing or regressive displacements between cycles, especially noticeable in marginal parts of formations.

Thus, following statements can be drawn up. (1) Major characteristical features of HA reflect that of composition and structures of macrocycles, complicated by their repetitions. (2) Zonal structure of macrocycles is characterized by conjugated facial transitions in structure both of halogen and non-halogen members. (3) In both members of cycles all elements that form them are associated with definite facial settings and in every of them they are characterized by a set of typomorphic features and interrelations. As a result, every setting has its own set of paleoindicating features. (4) As to time of formation, cycle elements are most frequently close but asynchronous.

PALEOTECTONIC MODELS

Geodynamic and paleodynamic settings of galogenesis basins generation consist of two groups. I. Divergent, riftogenic and epirift settings: (1) midland rifts, aulacogens, overrift troughs and shift depressions within rift systems; (2) intercontinental rifts; (3) basins of passive margins. II. Subduction and collision convergent settings: (4) secondary rifts within subduction belts; (5) superimposed and residual marginal troughs in zones of collisions in such systems as microcontinent-continent, continent-continent; (6) midland depressions (intermountain troughs) in the same settings of collisions; (7) collision rifts.

According to predominating type of background strains, the above mentioned structures can be also divided into two types, which are not completely coincide with two above mentioned groups of settings. First group includes riftogenic structures of different times and different scales, associate with common tension and destruction of continental crust (extensions, displacements, full rupture and block or microblock breaking off); the group includes rift structures of different generations: primary, secondary (subduction) and collision. The second group includes structures, associated with strains of general construction, arising due to convergence and collisions of ruptured continental blocks (however, at the background of these regime, troughs, intermediately controlling halogenesis, are also subordinated to structures of extension and warping, though of the lower rank). Analysis of factual material demonstrates that structures of the second orogenic group usually inherit position of riftogenic structures and develop over them. It enables to consider orogenic formations as evolving link of riftogenic systems or as a stage of their ontogenic cycles, following the riftogenic stage proper. Similar approach to continental riftogenic systems is used in works of E.E. Milanovskiy, V.A. Sokolov and other researchers. In such broad understanding of riftogenic systems, their selective association with halogenesis basin passes into a rank of general pattern. Different manifestations of this association were discussed in a series of works [1-3, 6, 7, 11, etc.). Disclosing of its universal character enables to propose a single, relatively simple model of "riftogenic structure" for paleotectonic and paleogeodinamic analysis of halogenesis basins. The model includes following interrelated statements. (1) All large halogenesis basins are spatially subordinated to riftogenic structures s.l.-RS (to rifts proper, rift-like structures and their tectonic derivatives, including orogenic). (2) Halogenesis impulses correspond in time to any places of tectonic activity as structures under discussion - both of extension/warping and compression/inversion. (3) RS evolution during tectonic cycles controls regular alternation of landscape-sedimentation halogenesis settings, determinating successive change of HA types and, as consequence, formation of their monotypic vertical series. (4) Common style of tectonic development of RS different geodynamic types is similar, that determines comparability of HA series and their combination.

Let us consider some of the above mentioned statements in detail.

Completeness of disclosing (destruction of crust), position within continental blocks relative to their margins, stage character of development, etc. are important characteristics of RS, controlling HA types and scales. A series of structures can be outlined from completeness of continental crust destruction: I - small graben, embrio-rift (partial crust destruction) -> II - rifts proper, aulacogens (crust is sharply degenerated, right up to appearance of areas of oceanic crust) -> intercontinental rifts (full break of continental crust in axial zone of rift). Halogen formations are controlled by paleostructures of all three groups with maximum in structures transitional from the second group to the third. In rift structures of the oceanic type HA are not established. According to position within continental blocks, it is convenient to subdivide RS into midland and marginal-continental.

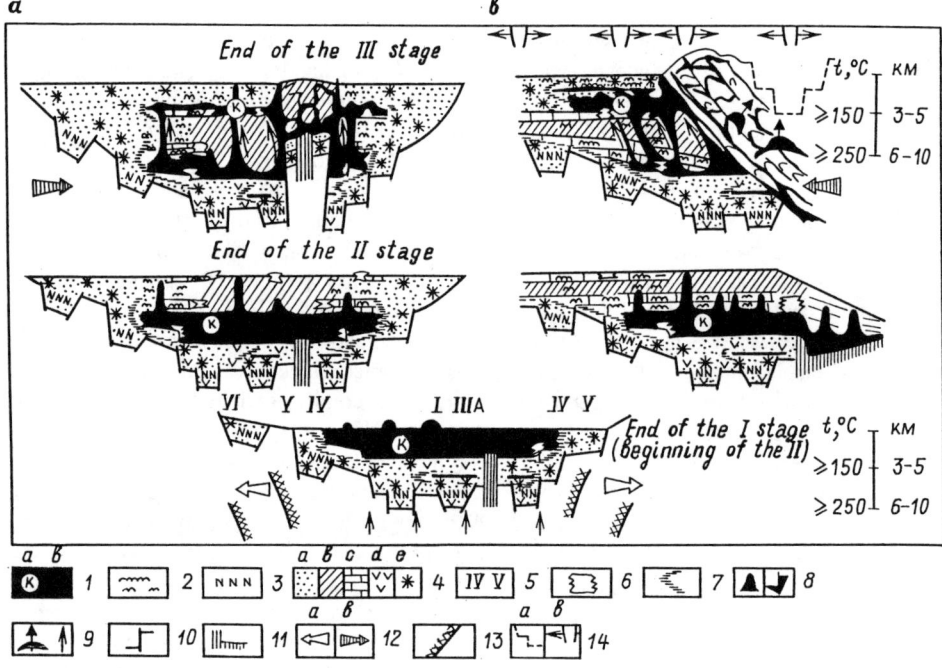

Figure 2. Scheme showing halogenesis evolution within intracontinental (**a**) and marginal (**b**) riftogenic structures for one tectonic cycle. **1-3**, types of halogenic complexes: **1**, salt-bearing (a, potassium-bearing; b, halititic-chloride-sodium); **2**, sulphate-calcium; **3**, sulphate-sodium, carbonate-sodium; **4**, predominant type of deposits (a, terrigenous; b, terrigenous and carbonate in different proportions; c, carbonate; d, volcanogenic; e, red beds); **5**, main paleogeographic setting (see in Figure 1); **6**, the most extensive reefogenic complexes; **7**, approximate displacement of the coastal zone; **8**, salt in a disturbed occurrence (a, in diapirs; b, in zones of tectonic faults); **9**, directions of salt, brine, hydrocarbon flow; **10**, tectonic faults; **11**, oceanic crust; **12**, prevailing type of stresses (a, extension; b, compression); **13**, riftogenic structures of preceding cycles; **14**, incipient riftogenic structures of the next generation (a) and different versions of their emplacement (b). See I-VI in Figure 1.

Three macrostages are observed in RS development: pre-rift activization, rift and post (inter)-rift. Halogen formations are closely associated with the second macrostage and to a lesser extent with the third macrostage. Within the limits of the major rift macrostage, three stages are distinguished in every cycle (E.E. Milanovskiy, B.A. Sokolov) (Figure 2): I - rift proper (initial or regeneration stage); II - syneclise or subsidence; III - orogen or inversion. The rift stage is correspondingly subdivided into substages: I_1 - continental rift, I_2 - marine rift, I_3 - microoceanic rift. RS development can be interrupted at any stage and substage with the result that in natural settings quite definite structural equivalent corresponds to every of them.

At the first stage, midland and marginal-continental structures are similarly developed: rift parts of aulacogens, taphrogens, grabens, including that at the basement of future passive margins, and so on. Some of them are obviously of shift nature. At the second stage, deep epirift troughs, superimpose on the above mentioned structures arise; overrift subsidences correspond them in internal parts of blocks and overrift areas of pericratonic regions and passive margins correspond them along periphery. At the third stage, small inversion uplifts and conjugated troughs (like foredeep) are generated over some of internal structures, and foredeep proper and residual inherited troughs are generated over marginal structures. The most extensive subsidences at all stages are typical for zones of triple rift conjugation. At relatively third passive post-rift macrostage, overrift troughs, which are lesser extended that at the rift macrostage, form during impulses of high activity.

Dynamics of halogenesis development is clearly conjugated with stage character of RS evolution (Figure 2). Definite facial type of setting and, conformably definite HA type, correspond to each stage. At beginning of the first stage, halogenesis is concentrated in mountain, rarer in lowland-plane settings (VI, V in Figure 1), in lakes, sea-lakes frequently of depression type. Association with volcanogenic and redbed formations is characteristical. Typical paleorepresentatives are following: Miocene sodium-sulphate formations of Tien-Shan intermountain area, Eocene sodium-carbonate formation of the Green River basin. Modern representatives are following: sodium-sulphate and sodium-carbonate deposits in the southwest regions of USA and rift zone of East Africa. At the substage of the marine rift halogenesis takes place in marine midland basins of gulf-like type. Formations are halitite, dimensions are limited. Association with volcanogenic and redbed formations is still characteristical and it is weak with reefogenic formations. The largest manifestations of halogenesis within the limits of midland microoceanic basins, in axial zones of which oceanic crust begins to form, are associated with microoceanic substage. Thick, usually potassium-bearing formations are cyclically conjugated with reefogenic formations, frequently with volcanogic complexes. Observed HA series of the rift stage has common transgressive direction with increase of carbonate material in associations, dimensions and completeness of halogenesis.

In some of the most studied basins HA are established correlatable with both, marine and microoceanic, substages (in Dnieper-Pripyat basin, D_2ef and D_3; in North Sea-Germany basin, P_1 and $P_2\check{z}$; in Red Sea basin, N_1^1 and N_1^3, etc). Formations of the oceanic substage are mainly revealed at basement of sections in majority of non-deformated or weakly deformated complexes of passive margins: along coasts of young oceans, Atlantic, Indian; within the limits of folded belts in composition of relic passive paleomargins of Mesozoic and Paleozoic Tethys oceans (T_3-J_1(?) in Mediterranean and Mexican basins, V(?)-$\mathcal{C}$, Mesopotamian basin, etc.). In aulacogens and aulacogen-like structures, HA are established in many Phanerozoic and, as relics, in some Precambrian structures.

At the second stage HA are related with shallow water-shelf and, to a lesser extent, with lagoon-sabkha settings (IIIB, IVB). Policyclic sulphate-dolomite complexes, entering the composition of thick shelf carbonate bodies of "carbon platform" type are characteristical. In composition of bathyal complexes halogen structures there, as a rule, are lacking. Association with volcanism is not marked.

The third stage is characterized by regressive (reverse in comparison with stage I) succession changes in time of settings and halogenesis types: from marine (and microoceanic), frequently deep (I, IIIA) to continental (V, VI), from thick potassium-bearing formations, fixing the second (upper) maximum of halogenesis at the beginning of the stage to weakly sulphate-bearing redbed predominantly terrigenous at it end. Potassium-bearing strata are characterized by pronounced reefogenic formations raising barrier complexes and carbonate platforms of preceding stage in their frames.

At the third macrostage of RS development (post- or interrift), HA are generated in structures of overrift syneclises and troughs. They are thin and usually do not contain chloride salts (sulphate-calcium type predominates).

Thus, to there stages of active development three series of HA correspond: transgressive, inundation and regressive. Every of them includes definite sets of HA in determined succession. Two maxima of halogenesis correspond to the end of the first (I) and the beginning of the third (III) stages. The above mentioned types of geodynamic settings, controlling generation of halogen basins, are correlated with one of these series (in their full or reduced variant) and thus can be manifested through them. Settings of destructional group are characterized by HA series of transgressive type, orogenic settings, by that of regressive type and passive margins, by HA series of inundational type.

This is a generalized scheme, reflecting basic direction of HA evolution in ontogenic cycles of RS and in different geodynamic settings. In concrete sections these series are represented with different fullness and in different combinations. At the first approach, fullness and dimensions of manifestation both of separate HA and their combinations in series are in dependence on scales of RS, controlling them, their relative activity in the given cycle, at the given stage and substage and also on dimensions of continental blocks, subjected to destruction. The most complete and extensive series and their sets with fully manifested halogenesis are characteristical of settings of two groups: (1) aulacogens and (2) triad of vertically conjugated settings, such as marginal-cratonic rifts-passive margins-marginal troughs. Maxima of all mentioned indices are typical of marginal-cratonic zone of riftogenic split intersections, which usually correspond to points of intersection of two mentioned groups (triradiate conjugation, etc.). Such points control sections the most repleted with halogen complexes and also the largest HA of the world (North Kaspian, East Siberian, North Sea-Germany basins, the Gulf of Mexico, ets.).

It should be noted that active manifestation of every next stage or subsequent cycle of RS development (and regeneration) is entailed with probability and even inevitability of destruction of HA salt bodies of previous cycles and stages (with parallel partial redeposition of salt to higher statigraphical levels)[1]. Therefore, salinity levels and extents really occurring in sections of paleoriftogenic structures are far from being fully reflected primary sedimentational situation. The presence of the above mentioned near-halogen parageneses (i.e. HA), including a whole series of well-preserved formations, gives the possibility, orienting to these formations, to reconstruct more fully picture of the past and to test the given model even in the case of salt destruction.

CONCLUSION

Thus, halogen complexes and a series of paragenerations cyclical associated with them (highly carboniferous, dolomitic, bioherm, stromatolitic, redbed) form stable determined constructed systems -halogen associations. Major features of their composition, structure and formation are concentrated in two models: "facial profile" and "riftogenic structure". The first demonstrates principles of association structure, facial settings of their formation, typomorphic features of different facial varieties and character of lateral interrelations between them. The latter reveals dynamics of replacement of HA facial types during tectonic cycles and features of construction of their vertical series in different geodynamic settings.

REFERENCES

1. G.A. Belenitskaya. Some paleotectonic disposition patterns of salt-bearing strata. In: *New data on geology, geochemistry, underground waters and mineral resources of salt-bearing strata*. A.L. Yanshin and G.A. Merzliakov (Eds). pp. 176-187. Nauka, Novosibirsk (1982).
2. G.A. Belenitskaya and N.M. Zadorozhnaya (Eds). *Reefogenic and sulphate-bearing Phanerozoic formation of the USSR*. Nedra, Moscow (1990). (in Russian).
3. N.M. Dzhinoridze, S.D. Gemp, A.F. Gorbov and V.I. Raevskiy. *Dispositionand criterion patterns of potassium salts prospecting of the USSR*. KIMS, Tbilisi (1980). (in Russian).
4. M.M. Grachevskiy, Yu.M. Berlin, I.T. Dubovskiy and G.F. Ulmishek. *Correlation of strata of different facies in oil and gas prospecting*. Nedra, Moscow (1969). (in Russian).
5. A.A. Ivanov and M.L. Voronova. *Halogen formations*. Nedra, Moscow(1972). (in Russian).
6. V.S. Konishchev. *Tectonics of halokinesis regions of the East Europeanand Siberian Platform*. Nauka, Minsk (1982). (in Russian).
7. S.M. Korenevskiy. *Complex of mineral resources of halogen formation*. Nedra, Moscow (1973). (in Russian).
8. V.I. Sedletskiy, N.I. Boiko and V.S. Derevyagin. About interrelation of halogen and bioherm sedimentation, *Sov. geol.* **12**, 8-21 (1977).
9. V.L. Shteingolts, V.G. Chaikin, Yu.V. Batalin and E.F. Stankevich. About paragenesis of domanicoid and evaporitic deposits, *Sov. geol.* **8**, 72-78 (1986).
10. N.M. Strakhov. *The principles of theory of lithogenesis*. Publishing House of Ac. Sci. of USSR, Moscow (1962). (in Russian).
11. R.N. Valyaev. *Riphean and Phanerozoic tectonics and minerageny of the East European Platform*. Nedra, Moscow (1981). (in Russian).
12. J. Wilson. *Carbonate facies in geological history*. Nedra, Moscow (1980). (in Russian).
13. M.A. Zharkov. *Paleozoic salt-bearing formations of the world*. Nedra, Moscow (1974). (in Russian).

An occurrence of primary inyoite at Lagunita Playa, Northern Argentina

C. HELVACI[1] and R. N. ALONSO[2]

[1] Dokuz Eylül Üniversitesi, Mühendislik-Mimarlik Fakültesi, Jeoloji Mühendisligi Bölümü, 35100 Bornova, İzmir, Turkey.
[2] Universidad Nacional de Salta, Buenos Aires 177, 4400-Salta, Republica Argentina.

ABSTRACT - The Lagunita Playa located within the borate district of the Argentinean Puna high plateau region is a Recent playa deposit characterized by primary inyoite [$CaB_3O_3(OH)_5 \cdot 4H_2O$]; the second reported occurrence in a playa. Mineralogical observations show that various dilution and concentration stages took place during borate mineral precipitation in the small-sized basin of Lagunita playa. Inyoite occurs as euhedral crystals and crystal aggregates in a bed up to 15 cm thick; it occurs beneath a ulexite-bearing bed and an efflorescent crust. Inyoite and associated ulexite crystals appear to have formed directly from brines penecontemporaneously with unconsolidated sediments enclosing the borates. Thermal springs and hydrothermal solutions associated with local volcanic activity are thought to be the source of the borates.

INTRODUCTION

The origin of borates is a subject of much discussion [1-7]. The main known reserves in the world are located in Turkey, the United States and Argentina in volcano-sedimentary Tertiary rocks. Borates in these deposits have undergone diagenesis and recrystallization. For these reasons the original minerals and primary features are usually altered. In the Central Andes of South America borates are presently forming, and in many cases they are found in playa-lake environments or in direct association with hot springs [7,8].
Muessig [9] described the first reported primary inyoite in Laguna Salinas in Peru. The new occurrence described in the present paper is the second occurrence of primary inyoite in a recent playa environment in South America. However, large amounts of inyoite occur in Miocene and Pleistocene rocks (around Pastos Grandes Salar) in the Puna region, Argentina [7,10,11].

LOCATION

Lagunita playa is located on the eastern side of the Coyahuaima volcanic complex, near Coyambuyo Volcano, in the Coranzuli region of Jujuy Province, approximately 65 km southeast of the point where the borders of Argentina, Bolivia, and Chile meet (Fig. 1). The Lagunita playa lies at 23°S. lat. and 66°30' W. long. at an elevation of approximately 4150 m and within the 1500 km long Altiplano-Puna.

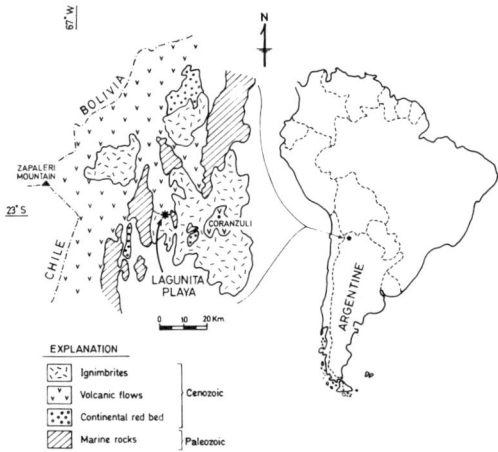

Fig.1. Location and geologic sketch map of Lagunita Playa.

The primary inyoite was found in a playa known as Lagunita (little lake) in Northern Argentina. It is a small, dry playa, with dimensions of 100 m in length and 50 m in width. The basin has an alluvial border and a saline deposit in the center. The saline deposit has dimensions 60 m in length and 30 m in width.

In this same area is an important region of boratiferous springs which are depositing thick ulexite deposits [8,12].

Presently there is no easy access to the Lagunita area. The closest roads lead to Volcancito "geysers" or the Alejandra mine, approximately one km to the east, where ulexite was formerly mined (Fig. 2).

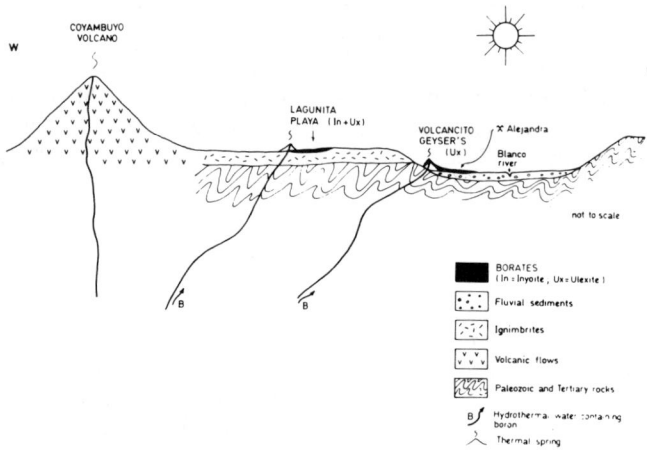

Fig 2. Geological schematic section and model of borate formation in Lagunita playa region.

GEOLOGICAL SETTING

Ordovician rocks, covered unconformably by volcanic and sedimentary Tertiary rocks, are the principal geological features of the region (Fig. 1). The Ordovician and Tertiary sedimentary rocks strike generally north-south while the volcanic rocks are emplaced along northwest-southeast lineaments. They constitute one of the effusive "fingers" that occur with the continental volcanic arc of this segment of the Central Andes. The alignment of stratovolcanic centers is called the Coyahuaima transversal volcanic chain [13].
The volcanism in the region is principally intermediate to acidic. The strato-volcanoes reach altitudes approximately 6 km above sea level and consist of extensive ignimbritic plateaus that cover the Ordovician and Tertiary rocks.

OCCURRENCE

Lagunita Playa is developed in a small flat basin in an ignimbritic flow of dacitic composition, that is partially covered by alluvial materials. The surface of the playa is dry, but water is encountered at a depth of 0.5 meter. The climate in the region is arid and evaporation is intense. Surficial efflorescence is indicated by intense white coloring. A vertical profile in the central part of the playa displays the following units from top to bottom: **a)** saline crust composed of sodium chloride, ulexite, and minor sulfates and carbonates, 5 cm thick; **b)** impure ulexite in a matrix of sand and mud, 10 cm thick; **c)** euhedral inyoite with minor ulexite, sand and mud, 10-15 cm thick; **d)** travertine, 5 cm thick; **e)** gravel and sand that extend down to the water table at a depth of 0.50 to 1 m (Fig. 3).

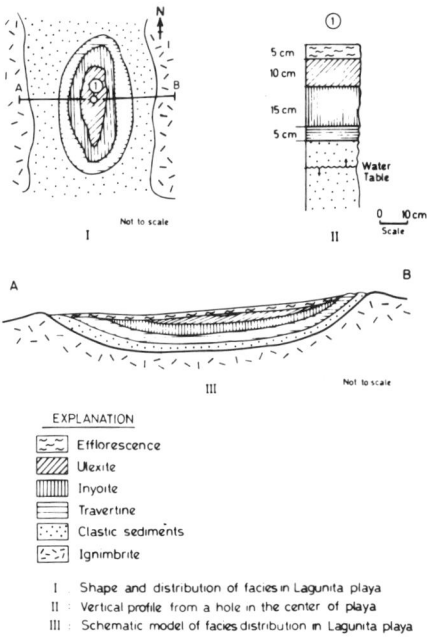

Fig. 3. Facies distribution, vertical profile and schematic section through Lagunita playa.

SAMPLE PREPARATION AND ANALYTICAL METHODS

All collected borate minerals have been found in bulk samples. Detrital material and mud fractions were eliminated from the borate minerals as much as possible. Borate identities were determined by Phillips directrecording X-ray diffractometer, using standard powder and oriented-sample techniques.
Fe, Al, Ca, Mg and Mn from the borate samples were determined by a 1272 Beckman atomic absorption spectrophotometer. Si, S and Ti were determined by a 100-60 Hitachi Spectrophotometer. Na and K were determined by the flame photometry, whereas B, H_2O and CO_2 were determined by wet chemical analytical methods. All samples were run in duplicate.

MINERALOGY

X-ray diffraction analyses of specimens collected from the Lagunita Playa bulk samples confirm that inyoite and ulexite are the borate minerals present. Associated minerals include calcite, sanidine, albite, montmorillonite and illite. Gypsum and halite occur within the efflorescent crust which forms the surface of the salar. The Lagunita salar occupies a closed basin underlain by ignimbrites and volcanic flows that originated from an eruption of the Coyambuyo Volcano (Fig. 2). The thickness of the inyoite bed is about 15 cm at the bulk-sampling site, whereas the ulexite bed reaches 10 cm (Fig. 3). Inyoite in the Lagunita salar occurs as: **a)** euhedral disseminated crystals in the mudstone and tuffite matrix, and **b)** crystal aggregates in a discontinous bed, whereas ulexite occurs as **c)** cauliflower - like nodules, and "cotton ball" textures in a clay and tuff matrix.

Inyoite $[CaB_3O_3(OH)_5 \cdot 4H_2O]$ is the only mineral of the calcium borate series found in this salar. It is the second reported primary occurrence of a calcium borate in a recent playa deposit, after the occurrence in the Laguna Salinas in Peru [9]. At Lagunita, it usually forms crystals aggregates, integrown crystal masses, and euhedral crystals reaching up to 1 cm in length but generally, 0.5-1 cm, with 110, 010, 111 forms; crystals are colorless, grey and light brown (Fig. 4a and 4b). It also occurs as clear, coarse-grained aggregates with perfect crystals (Fig. 4c and 4d) and it is clearly in contact with ulexite and calcite in the playa. X-ray diffraction patterns of three different impure inyoite samples are given on Fig. 5.
The occurrence of inyoite in the Lagunita salar is significant, because it shows that as found at Laguna Salinas, Peru, a calcium borate can form in playas, under surface or near surface temperature and pressure conditions, [1,4,14,15]. Inyoite displays its own crystal habit rather than pseudomorph of other borates; no relicts of other borates have been recognized in the inyoite crystals in this salar.
Table 1 lists analyses of three samples taken from the inyoite-bearing horizon, together with one ulexite sample. Analyses of inyoite samples, especially sample 1, display high content of CO_2 due to calcite impurities in the sample materials.

Ulexite $[NaCa_5O_6(OH)_6 \cdot 5H_2O]$ is the only mineral of the Na-Ca borate series that has been found in the Lagunita playa. It occurs in a 10 cm thick bed and always as cauliflower-like nodules and "cotton ball" textures in the mud and tuffitic matrix that overly the inyoite bearing bed in central part of the salar (Fig. 3). It is usually very soft, and pure ulexite concentrations

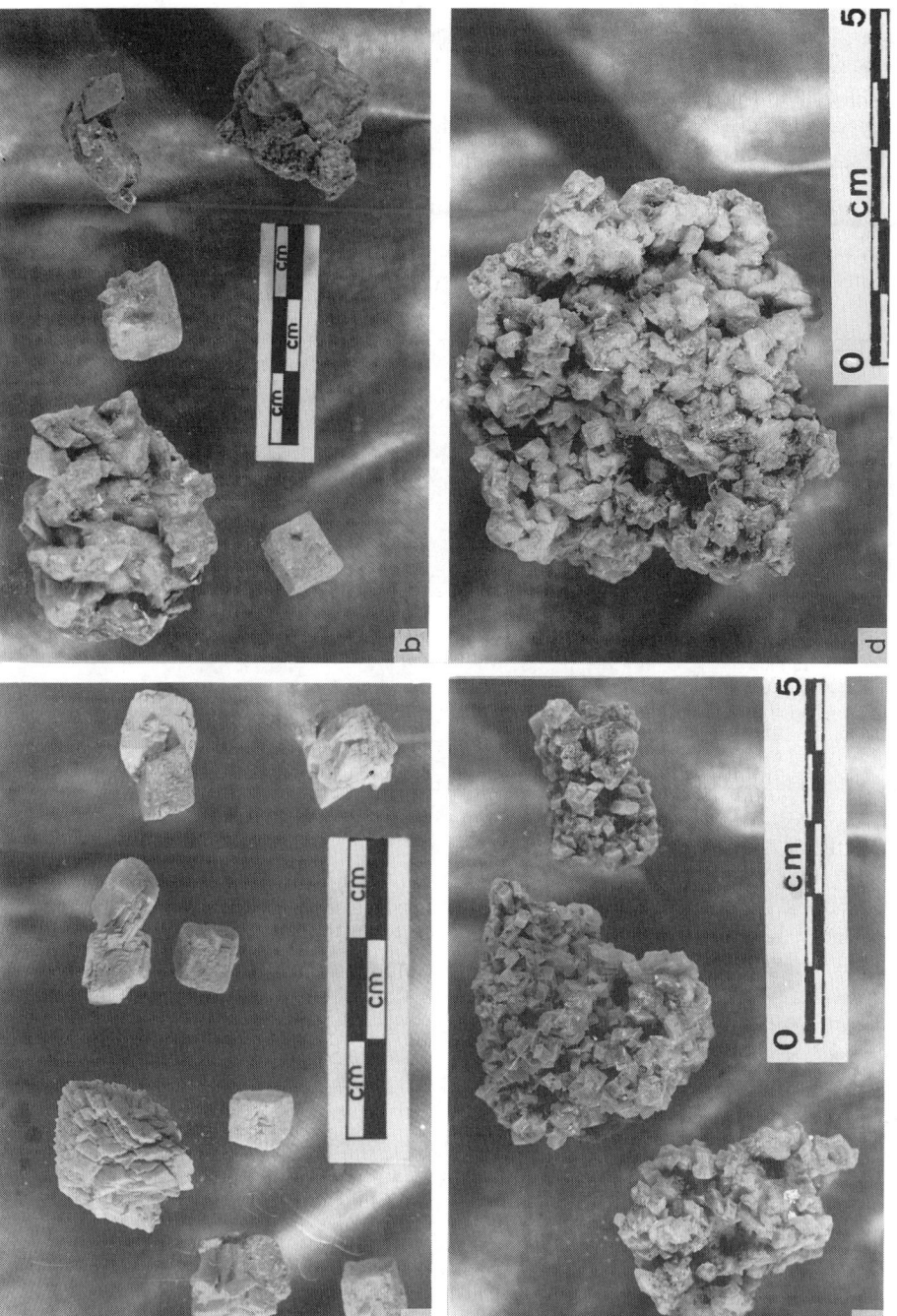

Fig. 4. a) Euhedral inyoite crystals from Lagunita Playa. b) Euhedral grey and brown inyoite crystals from Lagunita Playa. c) Aggregates of euhedral inyoite crystals. d) Aggregates of coarse-grained inyoite crystals from Lagunita Playa.

are white, but many are impure white and grey due to foreign material incorporated in the nodules that are growing in the mud.
Ulexite is commonly associated with inyoite and calcite. X-ray diffraction pattern of ulexite sample indicates that the ulexite crystals have 010, 110 and 101 forms (Fig. 6). An analysis of one ulexite sample is also shown in Table 1.
Calcite appears to have formed at all stages throughout the salar, but it occurs dominantly in travertine underlying the borate beds.

Table 1. Chemical analyses of inyoite and ulexite from the Lagunita Playa. 3a and 4a are calculated compositions.

Oxide Wt. %	Inyoite				Ulexite	
	1	2	3	3a	4	4a
B_2O_3	22.17.	32.06	35.69	37.6	44.41	43.0
CaO	30.60	22.71	19.58	20.2	11.59	13.7
Na_2O	0.36	0.27	0.50		6.44	7.7
MgO	0.27	0.06	0.01		0.00	
SiO_2	1.08	0.34	1.08		0.68	
Al_2O_3	1.42	1.11	0.25		0.25	
Fe_2O_3	0.00	0.39	0.00		0.00	
TiO_2	0.15	0.15	0.28		0.46	
MnO	0.00	0.02	0.00		0.12	
K_2O	0.17	0.17	0.25		0.14	
SO_3	0.24	0.27	0.24		0.25	
H_2O	25.12	36.29	40.15	42.2	35.69	35.6
CO_2	18.08	5.30	1.36		0.00	
Total	99.76	99.25	99.39	100.0	100.03	100.0
Cl(ppm)	9	14	11		7	

ORIGIN OF THE DEPOSIT

Travertine around the playa indicates the presence of springs, from which waters ascended to the surface and then flowed downhill to form a small lake.
Initially, the waters contained calcium and borate in composition, from which primary inyoite was precipitated. Afterwards, the composition changed to sodium and calcium borate and ulexite was formed. Such a change in the vertical zonation from calcium borate to calcium-sodium borate is very common in many Tertiary borate deposits, such as those of Kirka, Turkey [16,17]; Emet, Turkey [4,18]; Bigadiç, Turkey [19]; Boron, U.S.A.[5,20]; and Loma Blanca, Argentina [7,21].
Laterally, the evaporitic facies shows a crude zonation in which the inyoite occupies a central position, between ulexite and travertine at the center of the playa and the edge of the deposit respectively (Fig. 3)
There are no older, bedded-borate deposits in the Lagunita playa area, in contrast to some of the other Argentinean borate salars which have bedded deposits near by. Thus, there is no evidence that the source waters leached older buried borate deposits. Springs and hydrothermal solutions connected to the volcanic activity occurring at the edges of the Lagunita salar would seem to be the most probable source for the boron found in the deposit.

Volcanic materials, derived dominantly ignimbrites, deposited directly into the playa basin are probably also an additional source of boron.

Inyoite is the highest hydrate of the calcium borate series inyoite-meyerhofferite-colemanite. Inyoite, therefore, should not be rare in modern playas [14]. However, in similar geological environments of the Puna region, there are important concentrations of inyoite in Tertiary rocks, but not in the recent salars that contain principally ulexite and borax.

Meyerhofferite is not, and colemanite is very rarely, recorded from Recent salars [22,23,24,25]. A secondary origin for meyerhofferite and colemanite has been ascribed by many authors [8,14,17,26,27,28,29,30]. However detailed field investigations during recent years support the concept that some colemanite deposits may be primary in origin [1,2,16,18].

Inyoite is also considered as a secondary borate. Foshag concluded that the formation of calcium borates in playas is not possible because of chemical and physical conditions. Evidently, there are mineral transformations between inyoite-ulexite, inyoite-colemanite, inyoite-hydroboracite etc., but they are products of postdiagenetic transformations [4,7,15,31].

The distribution of inyoite in the boratiferous Province of the Central Andes [32] shows that the main concentrations are associated with Tertiary rocks (at Sijes, Loma Blanca, Tincalayu), some are associated with leistocene rocks (Pastos Grandes salar), and two are in playas (Lagunas Salinas, Peru and Lagunita, Argentina). In most, inyoite is of primary origin except in some deposits of Sijes district that contain secondary inyoite in veins that crosscut the hydroboracite beds [11].

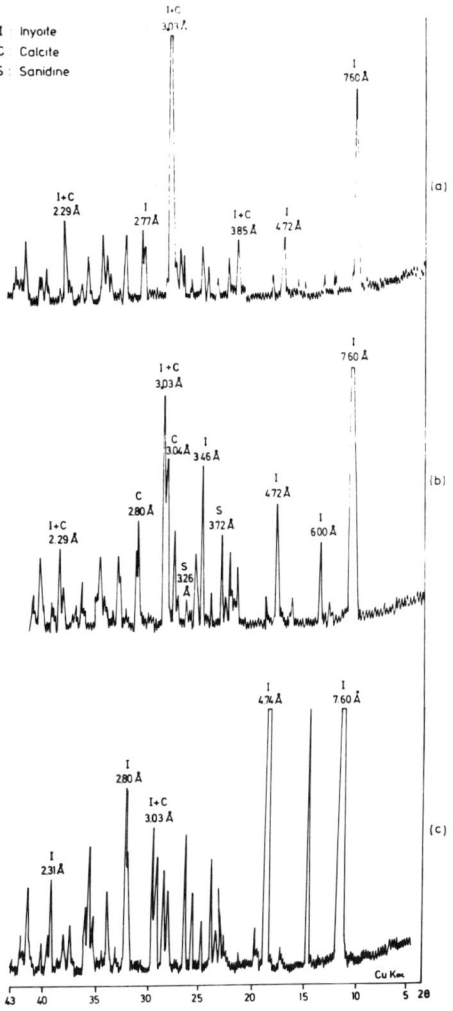

Fig.5. X-ray diffraction patterns of inyoite samples from Lagunite Playa. The letters a,b and c in the figure correspond to numbers 1, 2 and 3 in Table 1.

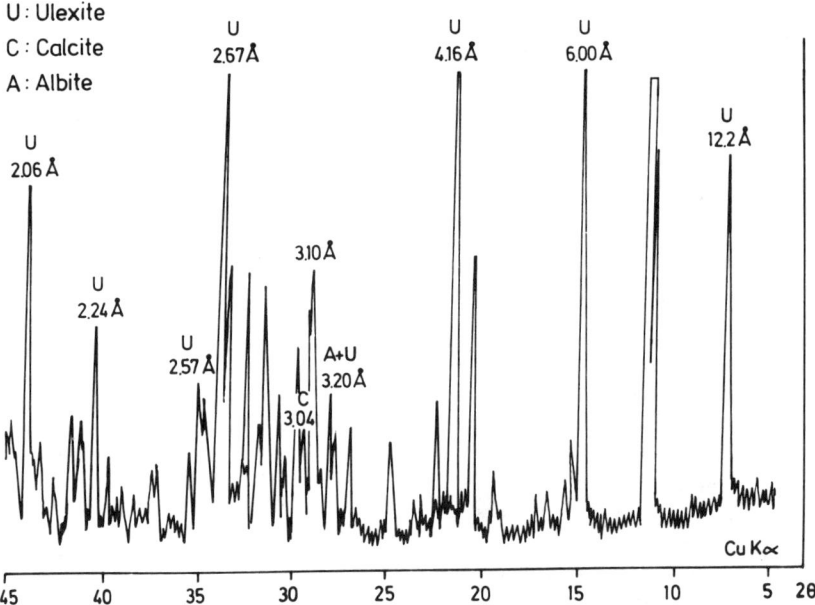

Fig 6. X-ray diffraction pattern of ulexite sample from Lagunita Playa.

CONCLUSIONS

The occurrence of inyoite in a Recent playa is recognized for the first time from the Puna region of northwestern Argentina, a region characterized by the most important borate deposits in the South American Andes. This occurrence of inyoite in a Recent playa deposit is the second known, after Laguna Salinas in Peru [7]. A comparison between the inyoite at Laguna Salinas and Lagunita shows similarities, e.g., in both cases the beds are covered by ulexite, at the center of the basin, and nearby hot springs appear to be indicated as sources of the borates. The presence of inyoite, and the lateral and vertical zonation with other borates and non-borates, permit a better understanding of the origin of borate deposits interlayered in Tertiary rocks that are effected by tectonism and diagenesis.

Acknowledgments

The authors wish to thank Industrias Quimicas Baradero S. A., principally Luis De Rito and Carlos Elias for their support of the field investigation, George I. Smith and George E. Ericksen of the U. S. Geological Survey, and Siegfried Muessig for their constructive criticism and review of the original manuscript. We also thank Kerime Nacakli and Erol Sanli of the Dokuz Eylül University for drafting and photographing assistance, respectively Author C. H. has been supported by a fellowship from the Argentinean Council for Scientific and Industrial Research.

REFERENCES

1. Barker, C. E. and Barker, J. M. A re-evalution of the origin and diagenesis of borate deposits, Death Valley region, California. *In* C.J. Baker and S.J. Lefond, (Ed.), *Borates: Economic Geology and Production*, 101-135. Society of Mining Engineer of the American Institute of Mining. Metalurgical and Petroleum Engineers, Inc. (1985)
2. Helvaci, C. Geology, mineralogy and geochemistry of the borate deposites and associated rocks at the Emet Valley, Turkey. Ph.D. Thesis, University of Nottingham, England, 338p. (1977).
3. Helvaci, C. A review of the mineralogy of the Turkish borate deposits. *The Mercian Geologist*, **6**, 257-270 (1978).
4. Helvaci, C. Occurrence of rare borate minerals: veatchite-A, tunellite, teruggite and cahnite in the Emet borate deposits, Turkey. *Mineral Deposita*, **19**, 217-226 (1984).
5. Bowser, C.J. Geochemistry and petrology of the sodium borates in the non-marine evaporite environment. Ph.D. Thesis, University of California, Los Angeles, 282p. (1965).
6. Kistler, R. B. and Smith, W. C. Boron and borates *In* S J. Lefond, (Ed.), *Industrial Mineral and Rocks, 4th Ed*, p. 533-560. Society of Mining Engineers of American Institute of Mining, Metallurgical and Petroleum Engineers, Inc., New York (1983).
7. Alonso, R. N. Occurrencia, posician estratigrafica y genesis de la los depositos de boratos de la Puna Argentina. Ph. D. Thesis (Spanish text). Universided Nacional de Salta, Argentina (1986).
8. Muessig, S. Recent South American borate deposits. *In Second Symposium on Salt. Cleveland, Ohio*: Northern Ohio Geological Society, 1, 151-159 (1966).
9. Muessig, S. First known occurrence of inyoite in a playa at Laguna Salinas, Peru. *American Mineralogists*, **43**, 1144-1147 (1958)
10. Aristarain, L. F. and Erd, R. C. Inyoita, $2CaO \cdot 3B_2O_3 \cdot 13H_2O$, de la Puna Argentina. *Anales de la Sociedad Cientifica Argentina*, 5-6, 191-211 (1971).
11. Aristarain, L. F., Rusansky, J. E., and Schoo-Lustra-de-Walker, M. F. Ulexite de Sijes: Provincia de Salta (Argentina) y caracteristicas generales de la especia. *Univ. Nac. La Plata, Fac. Cien. Nat. y Mus., La Plata, Geol.* **4**, 23-48 (1977).
12. Alonso, R. N. and Viramonte, J. G. Geyseres boratiferos de la Puna Argentina. Cuarto Congreso Geologica Chileno. *Antofagasta, Actas II*, 23-44 (1985).
13. Viramonte, J. G., Galliski, M. A., Arana, V., Aparicio, A., Garcia, L., and Martin, C. A. El finivulcanismo de lo depresion de Arizona, Salta. *IX Congreso Geologica Argentino*, 11, 234-253 (1984).
14. Muessig, S. Primary borates in playa deposits: minerals of high hydration. *Economic Geology*, **54**, 495-501 (1959).
15. Helvaci, C. Geochemistry and origin of the Emet borate deposits, Western Turkey. *Bulletin of the Faculty of Engineering, Cumhuriyet Univ. Serie A-Earth Sciences*, **3**, 49-73 (1986).
16. İnan, K. The mineralogy and geochemistry of the Kırka borate deposit, Turkey. Ph. D. Thesis, University of Manchester, England, 147p. (1973)
17. İnan, K., Dunham, A. C., and Esson, J. Mineralogy, chemistry and origin of Kırka borate deposit, Eskisehir Province, Turkey. *Trans. Inst. Mining Metall. (Section B., Appl. Earth Sci.)*, **82**, B 114-123 (1973).
18. Helvaci, C. and Firman, R. J. Geological setting and mineralogy of Emet borate deposits, Turkey. *Trans.Inst. Mining Metall. (Section B, Appl. Earth Sci.)*, **85**, B 142-152 (1976).
19. Helvaci, C. Mineralogy and genesis of Bigadiç borate deposits, Western Turkey (abst.) 28th International Geological Congress, Washington, D.C., USA, 2, 49-50 (1989).
20. Bowser, C.J. and F.M. Dickson, Chemical zonation of the borates of Kramer, California. *Second Symposium on Salt*, Northern Ohio Geological Society, Cleveland, Ohio, 1, 122-132 (1966).
21. Alonso, R. N., Helvaci, C., Sureda, R. J. and Viramonte, J. G. A new Tertiary borax deposit in the Andes. *Mineral Deposita*, **23**, 299-305 (1978).
22. Ahlfeld, F. and Angelelli, V. Las especies minerales de la Republica Argentina. *Univ Nac. de Tucuman, Instituto de Geologia y Mineria (Jujuy)*, Pub 548, 1-304 (1984).
23. Buttgenbach, H. Gisements de borate des Salinas Grandes de la Republique Argentine. *Anales Societe Geologique*, **28**, 99-116 (1901).
24. Catalona, L. Geologia economica de los yacimentos de boratos y materiales de las cuencas- Salar de Cauchari, Puna de Atacama. Direccion General de Minas, Geologia e Hidrologia, Publicacion no. 23, 1-100, Buenos Aires (1926).

25. Ericksen, G.E. Geology of the salt deposits and the salt industry of Northern Chile. U.S. Geological Survey, Open File Report, No. 698 (1963).
26. Rogers, A. F. Colemanite pseudomorphous after inyoite from Death Valley, California. *American Mineralogists*, **4**, 135-139 (1919).
27. Müehle, G. Colemanite pseudomorphs from the Corkscrew mine, Death Valley, California. *Mineralogical Record*, **5**, 174-177 (1974).
28. Gale, H.S. The origin of colemanite deposits. *U.S. Geological Survey Prof. Paper*, **85**, 3-9 (1913).
29. Foshag, W.W. The origin of the colemanite deposits of California. *Econ. Geol.*, **16**, 199-214 (1921).
30. Gulensoy, H. and Kocakerim, M.M. Solubility of colemanite mineral in CO_2-containing water and geological formation of this mineral. *Bull. Mineral Research and Exploration Inst. of Turkey*, No.90, 1-19 (1978).
31. Aristarain, L. F. and Hurlbut, C. Teruggite, $4CaO \cdot MgO \cdot 6B_2O_3 \cdot As_8O_5 \cdot 18H_2O$ a new mineral from Jujuy, Argentina, *American Mineralogists*, **53**, 1815-1827 (1968).
32. Alonso, R. N. and Viramonte, J.G. Provincia Boratifera Centroandina. Cuarto Congreso Geologica Chileno. *Antofagasta, Actas* II, 45-63 (1985).